THE GAP

THE GAP

THE SCIENCE OF WHAT
SEPARATES US FROM
OTHER ANIMALS

THOMAS SUDDENDORF

BASIC BOOKS

A Member of the Perseus Books Group
New York

Published by Basic Books,

A Member of the Perseus Books Group

Books published by Basic Books are available at special discounts for
bulk purchases in the United States by corporations, institutions, and
other organizations. For more information, please contact the Special
Markets Department at the Perseus Books Group, 2300 Chestnut Street,
Suite 200, Philadelphia, PA 19103, or call (800) 810-4145, ext. 5000, or
e-mail special.markets@perseusbooks.com.

Designed by Pauline Brown

Library of Congress Cataloging-in-Publication Data

Suddendorf, Thomas.

The gap : the science of what separates us from other animals /
Thomas Suddendorf.

pages cm

Includes bibliographical references and index.

ISBN 978-0-465-03014-9 (hardcover)—ISBN 978-0-465-06984-2
(e-book) 1. Psychology, Comparative. 2. Psychology. I. Title.

BF671.S86 2013

156—dc23

2013017538

10 9 8 7 6 5 4 3 2 1

For Nina, Timo, and Chris

C O N T E N T S

The Last Humans

THIS BOOK IS ABOUT YOU, what you are, and how you got here.

Biology puts beyond doubt that you are an *organism*. Like all living organisms, humans metabolize and reproduce. Your genome uses the same dictionary as a tulip and overlaps considerably with the genetic makeup of yeast, bananas, and mice. You are an *animal*. Like all animals, you have to eat other organisms—whether plant, fungus, or animal—for sustenance. You tend to approach things you want to eat while avoiding things that want to eat you, just as spiders do. You are a *vertebrate*. Like all vertebrates, your body has a spinal cord that leads up to the brain. Your skeleton is based on the same blueprint—four limbs and five digits—as that of a crocodile. You are a *mammal*. Like all placental mammals, you grew inside your mother and after birth received her milk (or someone else's). Your body features the same terminal hair as a poodle. You are a *primate*. Like other primates, you have an immensely useful opposable thumb. Your view of the world is based on the same color vision as that of a baboon. You are a *hominid*. Like all hominids, you have shoulders that allow your arms to fully rotate. Your closest living animal relative is a chimpanzee. Yet it would be prudent of me to call you an ape only from a safe distance.

Humans tend to think of themselves as better than, or at least separate from, all other species on this planet. But every species is unique, and in that sense humans are no different. In the tree of life each species is a distinct branch with characteristics that set it apart from others. Humans differ from chimpanzees and other primates in some notable respects. We can lock our knees straight, have longer legs than arms, and habitually walk upright, freeing our hands to do things other than carry our weight. We have a chin. Our body surface is covered in sweat glands that provide a more effective cooling system than those of other primates. We have lost our canines and much of our protective fur, leaving males with the apparently pointless, but persistent, growth of beards. The iris of our eyes is relatively small and surrounded by white rather than dark sclera, making it easy for us to identify the direction of another's gaze. Human females show no outward markers of their fertile phase, and human males lack a penis bone.

These are not exactly groundbreaking traits, compared to, say, the emergence of wings in birds, which predictably catapulted their bearers into a new sphere of possibility. Yet despite the paltry list of distinct physical attributes, we have managed to seize control of much of the planet. That is because our extraordinary powers do not derive from our muscles and bones but from our minds.

It is our mental capacities that have allowed us to tame fire and invent the wheel. They enable us to construct tools that make us stronger, fiercer, faster, and more precise, resilient, and versatile than any beast. We build machines that speed us from one place to the other, even to outer space. We investigate nature and rapidly accumulate and share knowledge. We create complex artificial worlds in which we wield unheralded power—power to shape the future and power to destroy and annihilate. We reflect on and argue about our present situation, our history, and our destiny. We envision wonderful, harmonious worlds as easily as we do dreadful tyrannies. Our powers are used for good as they are for bad, and we incessantly debate which is which. Our minds have spawned civilizations and technologies that have changed the face of the Earth, while our closest living animal relatives sit unobtrusively in their remaining forests. There appears to be a tremendous gap between human and animal minds, the nature and origin of which is the topic of this book.

WE HAVE BECOME SO SUCCESSFUL that many of us think a god singled our species out to run the world. Jewish, Christian, and Islamic traditions, for example, all share the fundamental belief that a universal god created humanity in his image, that only we are imbued with a soul, and that a glorious afterlife awaits those who follow a set of divine prescriptions. Nonhuman animals in these plots are cast as extras, and humans are given express rights to exploit them.

However, a couple of hundred years ago a range of inconvenient facts emerged to paint a very different picture of our place in nature. None were probably more profound than the extraterrestrial observations of Wilhelm Herschel. After moving from Germany to England, Herschel started to construct telescopes and study the night sky. His first breakthrough was the discovery of a new planet in our solar system, Uranus, in 1781. With the help of his sister Caroline and the royal support of King George III (before the madness), Herschel changed our view of the centrality of our Earth well beyond what Copernicus had done, cataloguing thousands of new star clusters and nebulae, and discovering the dynamic nature of the universe. He recognized that our solar system is traveling through space and that astronomical objects are born, change, and eventually die—a fate also in store for our own sun. He realized that starlight travels such enormous distances that some stars we see today have in fact already long died. The world turned out to be bigger, older, and more dynamic than anyone had anticipated.

Astronomy has demonstrated that we sit on a tiny speck in one of the billions of solar systems of the Milky Way, itself a galaxy among billions of others. This puts humanity, and all our troubles, in a radically new perspective—as Monty Python's "Galaxy Song" urges us to recognize, while memorably summarizing some of the key discoveries about our place in the cosmos:

> *Just remember that you're standing on a planet that's evolving and revolving at 900 miles an hour*
> *That's orbiting at 19 miles a second, so it's reckoned, a sun that is the source of all our power*

The sun and you and me, and all the stars that we can see are moving at a million miles a day

In an outer spiral arm, at 40,000 miles an hour, of the galaxy we call the Milky Way.[1]

Herschel's work gave humans a first glimpse at the really big picture. The realization that our planet and even our solar system are far removed from the center of anything cast serious doubt on previous theories that had put our species at the heart of a divine design. Indeed, with these discoveries, more secular views began to emerge. Pierre Laplace, for instance, proposed in 1799 that the sun, just as in other solar systems, originally condensed out of a nebulous cloud and then spun off the planets. When Napoleon asked him why there was no reference to God in his work, Laplace is said to have replied, "I had no need of that hypothesis."

Scientific approaches also began to threaten long-held beliefs about our special position on Earth. Again, the Herschel family played a pivotal role. Wilhelm's son John Herschel, who, like his father, served as president of the Royal Astronomical Society, wrote an influential book in which he promoted the new scientific approach, which allowed scholars to more effectively establish and accumulate knowledge. His inductive method of science had three parts: first, the gathering of data through observation and experimentation; second, the generation of hypotheses from these data; third, the testing of these hypotheses to see if they could be disproved. This systematic approach led to rapid progress across the disciplines, from astronomy to botany and from chemistry to geology.

Herschel's book, together with that of Alexander von Humboldt, another romantic founder of modern science, was a key influence on Charles

---

1 *Our galaxy itself contains a 100 billion stars, it's a 100,000 light-years side-to-side*
   It bulges in the middle, 16,000 light-years thick, but out by us it's just 3,000 light-years wide
   We're 30,000 light-years from galactic central point, we go round every 200 million years
   And our galaxy is only one of millions of billions in this amazing and expanding universe.

   The universe itself keeps on expanding and expanding, in all of the directions it can whiz
   As fast as it can go, at the speed of light you know, 12 million miles a minute, and that's the fastest speed there is
   So remember, when you're feeling very small and insecure, how amazingly unlikely is your birth
   And pray that there's intelligent life somewhere up in space, because there's bugger all down here on Earth.

Darwin, who was inspired to make his own contribution to our understanding of our place in the world. Our relationship to animals would
never be quite the same.

> Descended from the apes? My dear, we will hope it
> is not true. But if it is, let us pray that it may not
> become generally known.
>
> —REPUTED REMARK BY THE WIFE OF A CANON OF WORCESTER CATHEDRAL

DARWIN APPLIED HERSCHEL'S INDUCTIVE APPROACH in exemplary fashion. When he sailed around the world, he gathered enormous amounts of
data on plants and animals. These led him to a new hypothesis explaining
how species originated. *On the Origin of Species* was eventually published
in 1859, after years of subsequent observation and experimentation had
failed to disprove his theory of evolution by natural selection.

The theory is simple, elegant, and immensely powerful.[2] Most importantly, 150 years of subsequent attempts have still failed to disprove it. In
fact, science has unearthed a wealth of additional supporting evidence as
well as further aspects of evolution, such as a detailed fossil record and
the genetic foundation of life, which were unknown to Darwin. The implications of his work for humans' view of themselves did not escape him.
However, he only dared to make a brief reference to the human species in
his seminal work. The notion that we have evolved like all other animals,
that we share common ancestors with animals, that the same rules apply
to us and them, that we *are* them, was unthinkable for many at the time—
heresy even.

Nonetheless, twelve years later Darwin tackled the difficult but inevi-
table task of applying evolutionary theory to our own species head-on. In

---

2 From his travels Darwin noted that characteristics of animals and plants seem to fit their
function, and that populations vary in relation to their geographic isolation from each other.
He also noted that no two organisms are exactly identical—be that two dogs or two spiders.
Given finite resources and competition, some variants inevitably leave more successful descendants in the next generation than others. In other words, some inherited variations have
an adaptive advantage over others. If these are continually passed down the generations, the
number with this advantage increases and eventually replaces lineages without them. Over
large time spans organisms come to function better in their environment, and eventually,
especially in geographic isolation, descent with modification results in different species. This
is Darwin's theory in a nutshell.

The Descent of Man he argued that humans, like all other animals, are the product of evolution; he went so far as to propose that humans' closest living relatives are African apes. Today, various lines of evidence substantiate that this is indeed the case. Modern genetic comparisons have helped identify our animal family tree. Of all creatures compared to human DNA the two species of chimpanzee (common chimpanzees and bonobos) are clearly the closest match.[3]

In fact, the DNA of chimpanzees matches ours more closely than it does that of the African apes that look more like them: gorillas. In other words, from the perspective of chimpanzees, humans are their closest living relatives. Thus, by studying them we may perhaps learn more about the human condition than about the "animal condition."

Though it has become widely known that we are descended from apes, it is still often misunderstood to mean we have evolved from chimpanzees. We did not. Chimpanzees could equally be argued to have evolved from humans. Common descent means humans and chimpanzees share a common ancestor, just as you and your cousin do on a much shorter time scale. Both chimpanzee and human lineages have had equal time to evolve since their lines of descent split. Recent genetic analyses and fossil finds suggest that the split occurred some six million years ago.

IN THE ABSENCE OF MICROBIOLOGICAL and fossil evidence, Darwin's initial case for human evolution rested on signs of continuity. Descent with modification implies gradual change and therefore links between species. One can often find species with traits that are somewhat in between those of two other groups of species. Darwin, for instance, was most impressed

---

3 Note that figures such as a 99.4 percent match between chimpanzee and human DNA, though frequently cited, can be misleading. Common sense, for example, suggests that a zero percent match would indicate that two species are unrelated. However, given that genetic code of all creatures on Earth consists of only four molecules (adenine, thymine, guanine, and cytosine), the baseline is 25 percent and not zero percent. That is, if you compare a string of only one molecule, say, adenine, with the DNA of any species, be that rhubarb, porcupine, or human, you will find that on 25 percent of all DNA locations there is a match. About 25 percent of DNA is adenine. Furthermore, in addition to single base-pair substitutions, structural differences such as insertions, deletions, and duplications need to be considered when comparing genomes. For these, and some other reasons, one should take such figures like the 99.4 percent match with a grain of salt. Nonetheless, the relative match between different species can be interpreted in a straightforward manner.

with the Australian platypus, a so-called monotreme creature that appears to combine characteristics of mammals and reptiles (e.g., it has hair and lays eggs).[4] The importance of signs of continuity to Darwin's theory drove the search for so-called missing links, such as fish with rudimentary legs. To this day, almost every major fossil find is greeted in the press as a, or even *the*, Missing Link. (I will discuss the *found* links of human evolution in Chapter 11.) But even without fossils a strong case can be made for continuity between humans and animals.

The similarities in anatomy and bodily functions between humans and other primates are quite plain. We are made of the same flesh and blood; we go through the same basic life stages. Many reminders of our shared inheritance with other animals have become the subject of cultural taboos: sex, menstruation, pregnancy, birth, feeding, defecation, urination, bleeding, illness, and dying. Messy stuff. But even if we try to throw a veil over it, the evidence for continuity between human and animal bodies is overwhelming. After all, we can use mammalian organs and tissues, such as a pig's heart valve, to replace our own malfunctioning body parts. A vast industry conducts research on animals to test drugs and procedures intended for humans because human and animal bodies are so profoundly alike. The physical continuity of humans and animals is incontestable. But the mind is another matter.

How can anyone prove a gradual descent (or ascent if you prefer) from animal to human mind? This was arguably Darwin's greatest challenge. The seemingly vast gap between animal and human mind reeked of discontinuity. Even the co-discoverer of the principle of natural selection, Alfred Russel Wallace, and close scientific allies such as Charles Lyell were not convinced that natural selection could account for it.

Followers of René Descartes, who in the seventeenth century argued that animals are mere automata (machines governed by definable rules), thought that animals had no mental experience at all. Our own bodies may also be conceived of as mere machines; they are only the containers and vehicles of our exalted minds. In many cultures it is thought the mind governs and restrains the body. With the generous help of sanctions

---

4   This is especially interesting given that there are no truly native placental mammals in isolated Australia. Today the platypus is thought to be a surviving member of an early branch of mammalia, and a later branch is thought to have evolved into modern placental mammals and marsupials.

and taboos, the mind reins in the beast within. Now stop farting.[5] This dualism between mind and body still penetrates much of Western science and society.

Yet modern science has established that the mind is inextricably intertwined with the body. Lesions to your brain, say, from a tumor or a stroke, have predictable effects on your mind. For example, a lesion in the temporal lobes just behind your ear can destroy your ability to comprehend language. A subdiscipline of modern psychological science called "embodied cognition" examines more subtle links, showing that people's mental experiences and judgments change when their bodies are slightly manipulated. For instance, one finds the same situation more or less funny depending on whether one has a pen in the mouth or not. Try it while watching your favorite comedy. The pen prevents overt smiling and laughing, and thus reduces the subjective experience. People seem to judge a hill to be steeper when they are wearing a heavy backpack than when they are not. There are many ways to demonstrate that states of the body influence the mind. Ultimately, when our brain dies, all evidence points to the conclusion that our mind does, too.

What, then, about the brains of our primate cousins? Around the time *On the Origin of Species* appeared, Richard Owen, the founder of the British Museum of Natural History, argued that the human brain had unique structures, such as a hippocampus minor. However, Thomas Henry Huxley, who came to be known as Darwin's bulldog, won the subsequent scientific debate. He demonstrated that on close examination mammalian brains differ in size but share all major structures with humans. That conclusion has been influential to this day, although there have been recent challenges. For the time being, the case for Darwinian continuity between human and animal brains had been won.

The extreme position that animals have no minds at all became hardly tenable, given the continuity evident in brains, coupled with the evidence linking mind and brain. Animals' neurochemical and behavioral reactions to physical insult, for example, closely resemble ours. They apparently mind being injured. And like us, they do not seem to mind injury when under anesthetic.

---

5 Do you find such an example offensive or misplaced for a serious discussion of human nature? Well, that is part of the point. We often like to think of ourselves as better than that.

It is reasonable to assume, then, that many animals have the basic foundation of conscious experience. Yet people often reserve the word "consciousness" for higher functions of thinking. After all, Descartes was convinced of his own existence only on the basis of reflection: "I think therefore I am." But consider the Czech novelist Milan Kundera's astute reply: "'I think therefore I am' is the statement of an intellectual who underrates toothache." When you have a toothache, no further thinking is required to be sure of your existence and the fact you mentally experience things. The next time you doubt your own existence, go to the dentist (and decline anesthetic). The psychologist William James argued at the end of the nineteenth century that consciousness gives animals "interests." Because animals can feel, survival is made an imperative rather than a chance rule. They actively seek the experience of pleasure and relief from pain. Rats with inflamed joints, for example, will choose painkillers over usually preferred tastes when given the opportunity.

Even if we grant animals some mental experience, human minds appear to be vastly different from animal minds. In *The Descent of Man* Darwin tackled the problem of the apparent mental gap by comparing psychological characteristics such as emotion, attention, memory, and abstraction in animals and humans. Citing various anecdotal accounts he argued that animals have more sophisticated minds than is often assumed, and concluded that human minds differ only in degree and not in kind. The mental difference between ape and fish was greater, he argued, than that between ape and human. These conclusions remained controversial, even though Darwin predicted in *On the Origin of Species* that the study of the mind would be revolutionized by evidence for continuity: "In the distant future I see open fields for far more important researches. Psychology will be based on a new foundation, that of the necessary acquirement of each mental power and capacity by gradation. Light will be thrown on the origin of man and his history."

He must have peeked into quite a distant future, as over 150 years later psychology has not yet been placed on this foundation. Theories and scientific traditions from behaviorism to cognitive psychology, from Freudian psychoanalysis to ethology, have attempted to unravel various intertwined secrets of behavior, evolution, and the mind, yet there is still no consensus on what mental powers humans share with which other animals, nor have such questions been central to psychological inquiry. Even evolutionary psychology, which studies the nature of the human mind as a product of a long evolutionary history—"Our modern skulls house a stone-age mind," as two of its founders, Leda Cosmides and her husband, John Tooby, claim—has not yet taken seriously the challenge of investigating the apparent gap. Whole textbooks on evolutionary psychology barely mention our closest animal relatives, the great apes, or even our ancestral species.

Nevertheless, throughout the last century the work of some researchers, starting with pioneers such as Wolfgang Köhler examining the minds of chimpanzees, has been directly relevant to understanding the gap. In recent years, studies in comparative animal psychology have increased dramatically, and a clearer picture of the competences and limits of various nonhuman minds is finally emerging. Together with a more sophisticated understanding of the human mind and its development, I think we are finally in a much better position to tackle the question of what separates us from other animals.

Although signs of continuity between the minds of humans and animals were critical to Darwin's original case for human evolution, today we know that whatever the size and nature of the gap turns out to be, the evolutionary account is compellingly supported by genetic and fossil evidence. Even vast gaps need not be incompatible with evolution through descent with modification. Evolutionary biology can accommodate the possibility of profound changes, as illustrated, for instance, by Stephen Jay Gould and Niles Eldredge's arguments for rapid transitions followed by periods of relative stability. Most important, questions of continuity or discontinuity are, of course, about the evolutionary past and not about the present state of affairs. Current gaps are a function of what forms happen to have survived to this day. There is no need to assume that intermediate links must have survived (or that fossils of such links must be found). Indeed, most species that ever existed on Earth are extinct.

GREAT APES HAVE NOT ALWAYS been our closest living relatives. Only two thousand generations ago humans still shared this planet with several upright-walking, fire-controlling, tool-manufacturing cousins, including big Neanderthals (*Homo neanderthalensis*) and small "Hobbits" (*Homo floresiensis*). With its various bipeds it was a world reminiscent of Tolkien's Middle Earth. Our ancestors forty thousand years ago would have had much less reason to believe they were far removed from the rest of the Earth's creatures. We were but one of a group of similar species.

Perhaps because of the search for continuity and links, a picture persists of our ancestors evolving in a straightforward, single, and direct trajectory, up a stairway to *Homo sapiens*. This was not the case. For millions of years, many species of humans, technically called "hominins," wandered the planet and sometimes shared the same valleys. For example, between 1.6 and 1.8 million years ago, there were probably six or seven species in the human family,[6] ranging from the slender *Homo habilis*, makers of stone tools, to the stockier *Paranthropus robustus* with their massive, powerful jaws. There were also other types of apes such as the spectacular *Gigantopithecus*—an ape three meters tall that may have resembled Chewbacca from *Star Wars*. Our direct line of descent is only one branch on a flourishing, bushy tree of closely related species.

Some of these species were immensely successful. *Paranthropus boisei*, a heavily built hominin with a wide face, and the tall and large-brained *Homo erectus* each graced the planet for well over a million years. Modern humans have been here a mere fifth of that time. While there are clear signs of gradual change in, for instance, increasing cranial capacity and tool sophistication, diversity is also in evidence. Several new species have been described in the last decade. If the frequency of recent discoveries is anything to go by, archeologists will find many more fossils of hitherto unknown relations. We can expect an ever more complex family tree.

Yet today *Homo sapiens* is the only member of the human family left on this planet, and it so happens that a few species of great apes are our closest remaining relatives. A gap is defined by both of its sides. In an important sense, then, the answer to the question of why we appear so different from other animals is that all closely related species have become extinct. We are the last humans.

WHY IS OURS THE ONLY surviving lineage in this multitude of human forms? Why did the others die out? Radical environmental changes, such as ice ages and volcanic eruptions, are often responsible for extinctions. Such challenges have no doubt played a significant role in our relatives' past as well. The various extinctions were probably complex processes

---

6   Traditionally these are *Homo habilis, Homo erectus, Homo ergaster, Homo rudolfensis, Paranthropus robustus,* and *Paranthropus boisei.* In 2010, a seventh species, *Australopithecus sediba,* was described.

involving a multitude of factors, and the constellation of these factors likely differed in the demise of different hominin species. But for the disappearance of our close relatives we should consider another potential culprit: our ancestors.

Humans have been responsible for the demise of many species in recent times and, although we have no direct evidence, may well have had a hand in the extinctions of Neanderthals and other close relatives. Once our ancestors had managed to control much of the classic ecological challenges of survival, including predation by big cats and bears, members of the human family probably became their own primary adverse force of nature. We are more likely to be threatened, coerced, or killed by another human than by any animal. Aggression and conflict may have greatly affected hominin evolution.

A technologically advanced population can have devastating effects on other groups. People may be exterminated not only through killing but also indirectly through competition, habitat destruction, or even the introduction of novel germs. In his book *Guns, Germs, and Steel* the evolutionary biologist and geographer Jared Diamond vividly recounts the extraordinary case of a mere 168 conquistadors ransacking the Inca Empire in 1532. Most Incas were, in fact, killed by smallpox. The invaders brought the deadly disease, which swept ahead of their advance. The wholesale loss of life was an advantageous side effect for the Spanish—a result of hundreds of years of suffering from the disease in Europe. But some conquerors may have been aware of such causal chains and actively facilitated the process, guaranteeing widespread deaths as a result. British colonists, for example, have been accused of intentionally giving blankets infested with smallpox to native North Americans. It is unclear how common such callous acts were. What is evident is that humans are capable of them.

Yet we are also capable of extraordinary cooperation, empathy, and kindness. We can, and I would hasten to say should, make ethical choices that avoid exterminations of other people or species. As Steven Pinker recently documented in *The Better Angels of Our Nature*, violence has gradually declined over the course of history. In other words, war, blood feuds, murder, rape, slavery, and torture have been more commonplace in our past. Evidence of violent conflict goes back to prehistoric hunter-gatherers, but it is unknown when this dark side first emerged. Common chimpanzees are the only other primate species known to cooperate to directly kill members of their own species. So cooperative aggression may have ancient roots indeed.

No doubt our forebears at times would have also attempted to interbreed with close relatives, and they may have assimilated those with whom they could successfully reproduce. There is some anatomical evidence that humans and Neanderthals interbred, and in 2010 the first genetic evidence demonstrated that Europeans and Asians, in contrast to Africans, still carry some Neanderthal inheritance estimated at between 1 percent and 4 percent.[7] I am part Neanderthal. In December 2010, a thirty-thousand-year-old finger and tooth of another previously unknown human family member were described. Genetic analysis revealed that these so-called Denisovans were distinct from modern humans and from Neanderthals. They contributed about 5 percent to the genome of modern-day Melanesians.

Although sometimes represented as alternatives, making love and making war are not mutually exclusive possibilities. There is love as well as rape in times of war, and romance can result in conflicts. One way or another, it seems likely our forebears played important roles in the disappearance of at least some of our closest relatives. The reason the current gap between animal and human minds seems so large and so baffling, then, may be because we have destroyed the missing links. By displacing and absorbing our hominin cousins, we might have burned the bridges across the gap, only to find ourselves on the other side of the divide, wondering how we got here. In this sense, our exceedingly mysterious and unique status on Earth may be largely our own, rather than God's, creation.

What follows is the story of the chasm that currently separates human and animal minds. Chapters 2 and 3 first take a closer look at what is known about our closest remaining relatives and how we can establish animal mental capacities. In Chapters 4–9 I examine the major claims about what makes our minds unique: language, foresight, mind reading,

---

7   The logic is simple. When Neanderthal DNA is compared to different African groups of humans, they are equally different from each group. If, however, the DNA is compared to a European and an African, it more often matches the European than the African DNA. The same is true for tests with Chinese DNA. This suggests that when modern humans migrated out of Africa, they mixed with Neanderthals, presumably in the Middle East, where fossil evidence demonstrates prolonged cohabitation (see Figure 11.10), and then carried the Neanderthal component of their genome to the world outside of Africa.

intelligence, culture, and morality. I explain what we know about the nature and development of the human faculties and assess what is known about parallels in animals. Although some species have communication systems, can predict what happens next, can solve certain social and physical problems, have traditions and perhaps even empathy, we will discover that human minds are distinct for a small number of recurring reasons. Chapter 10 distills what is common about the gap in all of these domains and why. Our prehistoric forebears and clues about the evolution of our minds are the focus of Chapter 11. Finally, in Chapter 12 I consider the future of the science of what separates us from other animals, as well as the future of the gap itself.

T W O

Remaining Relatives

WE ARE PRIMATES. PRIMATES ARE generally adapted to life in trees and, because of the fatal risk of falling out of them, have evolved novel and sophisticated capacities for accurate seeing and grasping. Primates typically have forward-facing eyes and stereoscopic color vision, on which they rely more than on their noses. They have lost the whiskers characteristic of other mammals and also have a relatively reduced olfactory (smell-processing) brain.[1] Primates have evolved grasping hands with five separate digits, and many feature a versatile opposable thumb and fingernails rather than claws. We would not perceive and interact with the world the way we do were it not for our primate heritage.

Compared to those of most other animals, primate brains are large and primate minds are smart. When observing the lives of, say, gorillas,

---

1 Anatomically, primates are recognizable by a peculiar small bone covering part of the inner ear (the petrosal bulla).

one may well wonder why they would need smarts. Gorillas often do little more, it seems, than sit in the giant salad bowls of the forest, munching away. This observation prompted the philosopher Nick Humphrey to propose that it is social problems, rather than physical challenges, that might have driven the evolution of primate intelligence. This idea has increasingly attracted followers because most primates are indeed deeply social, and our closest relatives are especially so.

> It is hardly an exaggeration to say that a chimpanzee kept in solitude is not a real chimpanzee at all.
>
> —WOLFGANG KÖHLER

The intricacies of primate social lives have been extensively documented by field observations, which reveal that group structures are held together through relationships maintained by individual attention to others within the group. Primates are fond of grooming; it is relaxing and leads to release of endorphins and oxytocin, and sometimes to the groomed individual falling asleep. As tension and parasites are removed, social bonds are formed. Alliances are forged and repaired. The larger the group, the more time its members tend to spend grooming each other. While other animals aggregate in much larger numbers than primates do—wildebeest or sardines, for example, live in groups numbering many thousands— the individuals may be entirely anonymous to one another. Primates, on the other hand, know each individual group member.

In addition, primates appear to have some understanding of the relationships other group members have in terms of dominance, kinship, and friendship. Upon hearing the call of her infant, a vervet monkey mother will look toward the origin of the call, whereas other group members look to the mother—in apparent recognition that *her* infant is calling. Close observation of primate groups reveals that such knowledge is essential to their social lives. Therefore a fight between two individuals may affect the relationship between other members of the group. I once observed a young chimpanzee sneak up to an older female, holding a branch behind his back. When the female tried to groom the juvenile, he suddenly hit her with the branch and ran away, chased by the enraged female. This event had ripple effects throughout the community, as different chimpanzees appeared to take sides. Retaliations target not only perpetrators but sometimes also their kin or associates.

Primate social lives can be multilayered affairs. Achieving high rank is a function not always merely of brute strength but also of shrewdness. By grooming the right individuals one can gain support in power struggles. Social problems therefore require significant attention and consideration on the part of individual primates, especially as group size increases. In fact, evolutionary psychologist Robin Dunbar established that the greater the typical group size of a primate species, the bigger their brain—or, more precisely, the ratio between the neocortex and the rest of the brain. The larger the group the more cognitive power an astute social player may require to keep track of the increasingly complex web of information.

Primate foraging is also more complicated than is sometimes assumed. They feed not only on bananas but on a great variety of things including leaves, roots, sap, and meat in the form of insects and small mammals. To obtain such foods some primates employ tools. Capuchin monkeys, for example, open nuts with stones. Others have developed sophisticated processing techniques. Gorillas, for instance, carefully fold nettles to avoid being stung. Yet others cooperate, as chimpanzees do to hunt monkeys. In these ways primate species exploit a diverse range of niches.

Taxonomists subdivide primates into groups based on a variety of traits. Not all primates are monkeys. Much of this has to do with noses—there are two suborders, one that comprises prosimians such as lemurs and lorises, which have wet noses (called *Strepsirrhini*), and the other primates that have dry noses (*Haplorrhini*). The latter includes the two groups known as monkeys: the new-world monkeys, whose nostrils point away from each other and the old-world monkeys, whose nostrils are in parallel. New-world monkeys, as implied by their name, are only found in the Americas and include tamarins, marmosets, howler monkeys, spider monkeys, squirrel monkeys, and capuchins. Some of these species have evolved a prehensile tail allowing them to grasp and hang from branches while feeding using their hands. They sometimes use it like an arm: when I presented a choice task to a spider monkey, she made her selection with her tail about as often as she did with her hand. Old-world monkeys live in Africa and Asia and do not have a prehensile tail. They tend to walk on top of branches on all fours, rather than hang below them, and their tail only helps with balance. They can sleep sitting upright and therefore often have thick red calluses on their behinds (so-called *ischial callosities*). Well-known old-world monkeys include macaques, baboons, mangabeys, colobus monkeys, and langurs. Apes and humans belong to a group of dry-nosed, old-world primates that have lost their tails altogether.[2] Let's have a look at our closest remaining animal relatives.

APES ARE GENERALLY LARGER THAN other primates. They have relatively long arms, a broad chest, and no snout. Apes usually depend on trees for a living, but because of their weight they typically hang below branches rather than balancing on top of them. A rotational ability in the shoulders serves this mode of locomotion and is essential not only for our capacity to hang off branches or high bars but for throwing spears or balls with precision. Ancient apes were also relatively large and, up in the trees, probably quite safe from predation. Such security allows species to live longer and slower lives. Indeed, living apes grow up slowly, with long periods of gestation and of parental care. They reach sexual maturity late and have an overall long life span of up to about fifty years. This extended life history is a fundamental ape adaptation that provided the opportunity to grow bigger brains.

Apes were once diverse and widespread, but both their numbers and distribution have radically been reduced. The most likely culprit is the climatic changes that destroyed the rainforest habitat to which they were adapted. Humans, of course, became ground-dwellers and are responsible for much of the deforestation in recent times. We are the most widespread and abundant primate—over seven thousand million of our one species compared to a few hundred thousand apes of all other species combined.

When apes were first brought to Europe in the seventeenth and eighteenth centuries, people were immediately struck by their resemblance to humans, even if they were perceived as grotesque distortions. They were often considered half human and half beast. In German, monkeys are referred to as *Affen*, and apes are known as *Menschenaffen* ("humanapes"). In Indonesian and Malay, *orang utan* means "man of the forest," and an early European anatomist adopted a Latin version with the same meaning, *Homo sylvestris*, to describe an initial specimen from the African apes. Ever since Carl Linnaeus systematically classified organisms and placed humans among the primates, there have been debates about the appropriate groupings of apes and humans on the tree of life.

---

2 Barbary macaques also do not have a tail and are therefore sometimes called apes—but they are monkeys.

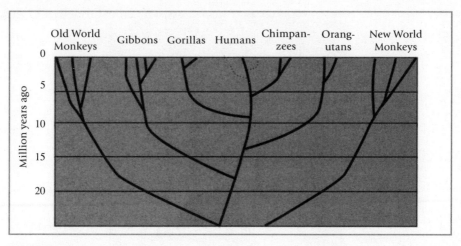

FIGURE 2.1.
Phylogenetic tree of humans and their closest animal relatives.

Although the relationships between primates are well established, the groupings and labels have been revised repeatedly in light of new data, most recently from genetics. In the latest and most widely used classification, and the one I will follow in this book (though only sparingly), humans and all apes are a group called hominoids. Humans and great apes, but not small apes (i.e., gibbons), are classified as a family group called hominids. Finally, the term "hominin" is used to refer only to humans and our extinct relations since the split from the last common ancestor with chimpanzees.

THE SMALL APES OR GIBBONS are, indeed, the smallest apes.[3] The first one I ever encountered used my arm like a tree branch as it swung past me to land on the path ahead. I stood flabbergasted somewhere in a Sumatran rainforest. Small apes have extremely long arms and are famous for their acrobatics. They are true *brachiators*, swinging from branch to branch (or arm to branch as the case may be). No other mammal comes close to their ability to rapidly move through the canopy. But when faced with one on the ground, you will see a much less elegant form of locomotion. The

---

3 Small apes are also often called "lesser apes," but I shall refer to small apes to avoid unfortunate connotations of inferiority.

FIGURE 2.2.
Siamang at Adelaide Zoo *(photo Andrew Hill)*.

gibbon was walking toward me on two legs with its arms swinging awk-
wardly in the air, in a style reminiscent of John Cleese's funny walks. I was
very much endeared by the cute little ape waddling up to me. I had been
told that there were some semihabituated white-handed gibbons around,
and so I was happy to make its acquaintance. Later on it even jumped on
my lap. I stroked it and promptly got bitten.

The line that led to the small apes split off from the ancestors of great
apes and humans some eighteen million years ago. Today small apes live
exclusively in Southeast Asia. They comprise four distinct genera: siamangs
(*Symphalangus*), hoolocks (*Bunopithecus*), crested gibbons (*Nomascus*), and
lar gibbons (*Hylobates*). These genera differ in various ways including the
form of their skulls and even in the number of their chromosomes (ranging from thirty-eight in hoolocks to fifty-two in crested gibbons). There
are several species of crested gibbons and lar gibbons, with little over-
lap between their habitats—only siamangs (the largest small apes) and
white-handed-gibbons (a lar gibbon species) share territory in the forests
of Sumatra and peninsular Malaysia.

Unlike great apes, but like many humans, gibbons live in small, monogamous families composed of a mated pair and several dependent offspring. Pairs stay together for years, and males invest time and effort in the rearing of the young. Siamang males even become the primary carrier of their offspring when the infant has reached the age of one year. When gibbons become sexually mature, they leave their "nuclear" family to start their own.

Each family typically has a territory of several hectares, which it vehemently defends. Gibbons mark their territory through their characteristically clamorous songs, which tend to last several minutes and are typically produced at specific times of the day. Different species of small apes can be identified through their songs, which range from booming and piercing cacophonies to haunting wails. Pairs often sing different parts in duet. The vocalizations of the small apes are far more diverse in range than those of the great apes. It has been argued that their ability to voluntarily produce such a diversity of sounds may be a better model of what our hominin ancestors turned into speech than the vocalizations of great apes. However, the functions of these small ape vocalizations seem to be restricted to territory marking, mate attraction, and, in case of the duets, pair bonding.

Surprisingly little is known about gibbons' cognitive abilities. What work has been done is largely limited to studies on lar gibbons. Even these studies have been few relative to the myriad of psychological tests conducted on great apes. I found this a curious omission, given their close relationship to us, but soon learned the reason. It turns out that psychological testing of gibbons is rather difficult. Unlike great apes, captive gibbons do not tend to sit still opposite a human experimenter and then readily follow an experimental procedure. Perhaps because they are more vulnerable to predation, they are generally more fearful and flighty. In the studies my colleagues and I conducted, they frequently interrupted testing by indulging their penchant for energetic, high-speed acrobatics.

Small apes used to be widespread in Asia. Early Chinese texts report their presence as far north as the Yellow River. Today, however, they are confined to ever smaller areas of primary forest in Southeast Asia. Habitat destruction, hunting, and illegal trade have made them critically endangered. In fact, several gibbon species are alarmingly close to extinction in the wild.[4] The silvery gibbon (*Hylobates moloch*) is down to about four

4 To learn about gibbon conservation, see http://www.gibbonconservation.org

thousand individuals and the western black crested gibbon (*Nomascus concolor*) to fewer than two thousand. The worst situation, however, is that of the Hainan black crested gibbon (*Nomascus hainanus*). This species, at last count, consisted of twenty-two individuals.

THE GREAT APES COMPRISE THREE genera: orangutans, gorillas, and chimpanzees. As the name suggests, they are larger than the small apes, with some male gorillas reaching over 200 kilograms.[5] Great apes differ widely in the way they make a living, where they spend their time, and their social structure. However, unlike the small apes, they build nests in trees or on the ground to sleep in. They also make vocalizations akin to laughter when they engage in chasing games, wrestling, and tickling. As in humans, the armpits and the belly are particularly ticklish. The apparent joy of these activities does not seem to disappear with age—nor does their immense curiosity.

In the 1960s the curiosity of Louis Leakey, already famous for his discoveries in paleoanthropology, spawned long-term field studies of our closest living relatives. He sent out three female researchers to study each of the great ape genera in the wild. They are sometimes affectionately called "Leakey's Angels." Jane Goodall went to Tanzania to study chimpanzees, Dian Fossey to Congo and Rwanda to research gorillas, and Birute Galdikas to Borneo to investigate orangutans. Their approaches were initially considered unorthodox, but their persistence in recording behavior over long periods had a lasting influence. While Fossey's efforts ended in tragedy in 1985 (as dramatized in the film *Gorillas in the Mist*), Goodall's and Galdikas's projects are ongoing. These long-term studies, together with a few other committed undertakings such as Toshisada Nishida's work in Tanzania and Christophe and Hedwig Boesch's work in the Ivory Coast, have yielded an extraordinary amount of new knowledge about our closest relatives.

ORANGUTANS ARE RED APES THAT live in the remaining jungles on the islands of Sumatra and Borneo. They are classified as two species (*Pongo abelii* and *Pongo pygmaeus*), but they can interbreed in captivity. Orangutans hardly ever descend from the trees. They prefer the lofty heights and

---

5 Humans can be even greater apes: the heaviest recorded specimen reached more than an astonishing 500 kilograms.

FIGURE 2.3.
A male orangutan in Ketambe, Sumatra *(photo Emma Collier-Baker)*.

have a corresponding air of aloofness about them when they gaze down at us ground dwellers. At least that is what it felt like to me, as my neck started to ache from prolonged upward staring. Male orangutans can weigh over 80 kilograms, making their climbing relatively slow and considered. I sympathized with their climbing efforts, as the branches move or break in roughly the way they would if I were trying to do the same thing. They have one distinct advantage over us, however: they have four hands, each with an opposable thumb at their disposal.

Orangutans spend a good deal of the day in search of fruit. They supplement their diet with juicy stems and insects. Occasional meat eating has also been observed. For example, Sumatran orangutans sometimes kill slow lorises—a slow-moving, wet-nosed primate. Perhaps because fruiting trees

FIGURE 2.4.
Utama, a female Sumatran orangutan at Perth Zoo *(photo Andrew Hill)*.

can only feed so many orangutans at a time, they do not live in social groups like other great apes. Instead, adult males usually live a largely solitary life.[6] Females travel with their offspring, as males do not partake in the rearing. Adult males grow to twice the size of females, and some develop large cheek flanges and throat sacs that they use to emit long calls. They range over a particular territory, and receptive females join them occasionally for a period of up to three weeks. They tend to repeatedly copulate during this time, including face-to-face intercourse. It can look very intimate.

For some time researchers thought there was another smaller species of orangutan. Today, it is clear that this is a second form of mature male, sometimes called the Peter Pan morph. These mature males are transient and stay in the subadult (unflanged) stage for many years. In spite of the usually gentle nature of orangutans, these males may physically force females to copulate with them. As they look like adolescents, they may escape the ire of resident adult males. Upon some cues, possibly when an

---

6 Orangutans are an exception to the social brain hypothesis that links large brains to large group sizes. One reason for this may be that the ecological niche of modern orangutans is quite different from the context in which they evolved. In captivity, orangutans show more social affiliation, so it is also possible that the complexity of their social lives high up in the trees has been underestimated.

old flanged male disappears, a Peter Pan morph can grow into a flanged full-grown male.

The tendency to attempt forceful intercourse in some circumstances appears not to be restricted to Peter Pan morphs, or even to males. One researcher who had regular contact with a female orangutan once told me that he had to turn away a direct advance, upon which the snubbed orangutan would no longer cooperate in his research efforts.

Apart from holding large leaves as umbrellas when it rains, tool use among orangutans was long thought unusual. In recent years, however, primatologist Carel van Schaik and colleagues have documented a variety of cases. For example, some orangutan groups in Sumatra use stick tools to get to the seeds of Neesia fruit and to obtain insects from holes. This appears to be socially maintained behavior. In Borneo, Anne Russon and Birute Galdikas documented orangutans' capacities as imitators. At the reintroduction center of Tanjung Puting, orangutans sometimes even copy peculiar human behavior. One orangutan, called Supinah, has been particularly curious about humans' ability to light and control fire—not unlike King Louie in the Disney film adaptation of *The Jungle Book*. The ensuing experiments with kerosene and other materials understandably raised some concern but were ultimately unsuccessful.

Orangutans are able to solve a range of other problems. I once observed a subadult male reaching over from a somewhat isolated tree to the outer branches of its neighbor. He then stayed in a rather awkward

FIGURE 2.5.
Orangutan bridge in Ketambe, Sumatra.

horizontal position between the trees for what seemed to be far too long for comfort. He hung on until a juvenile, maybe three years old, made its way down from the top of the trunk and used him as a living bridge across the gap. In this way, the subadult offered himself as a tool to the juvenile.

In spite of all the attention the enigmatic red ape receives from the public and governments, the remaining populations are rapidly declining. The latest population estimate for Sumatran orangutans is a mere 7,300 individuals. The figures for the orangutan of Borneo are slightly better, with estimates of between 45,000 and 69,000 individuals, but their numbers are also declining. Continuing habitat destruction (especially for palm oil plantations), bush fires, hunting, and the pet trade are reducing the population. According to the International Union for the Conservation of Nature (IUCN), Bornean orangutans are endangered, and Sumatran orangutans are critically endangered; that is, they are likely to be extinct in the near future.[7]

GORILLAS ARE THE LARGEST OF the great apes, so it is not surprising that King Kong was cast as a gorilla. In spite of the size of male silverbacks and their impressive chest-beating displays, gorillas are largely placid vegetarians. I once had the pleasure of visiting a habituated group of mountain gorillas in Uganda (something I highly recommend both as an experience and for the sake of conservation), as they were relaxing in the forest. The silverback was lying on his side studying his fingernails. He then casually grabbed his butt cheek, lifted it a little, and let one rip—a parallel to human behavior seldom discussed. The only chest beating I saw came from a one-year-old infant.

There are two generally recognized species of gorilla: the Western gorilla (*Gorilla gorilla*) and the Eastern gorilla (*Gorilla beringei*). In 2012 the first draft of the gorilla genome was published and suggested that these two species diverged some 1.75 million years ago, albeit with some subsequent gene flow. The majority of both species live in the lowland, but the Eastern species include the mountain gorillas. There are only a few

---

7 As I write this, fires apparently set by a palm oil company are raging through one of the remaining few Sumatran orangutan habitats, the Tripa Swamp forest. For orangutan conservation, see the Borneo Orangutan Survival Foundation (http://orangutans.or.id) or the Orangutan Project (http://www.orangutan.org.au).

FIGURE 2.6.
A mountain gorilla spying on us from
behind a bush in Bwindi Impenetrable
National Park, Uganda.

hundred mountain gorillas left, and their habitat differs starkly from the typical rainforest most Eastern and Western lowland gorillas call home. Much of what is known about natural gorilla behavior comes from detailed studies of mountain gorillas first initiated by Dian Fossey.

Mountain gorillas live in small family groups comprising an adult male, a few females, and their offspring. As they reach maturity, females leave to join another group. A full adult male is larger than the females in his harem and develops the distinctive silver-gray hair patch on his back. "Bachelor" adult males tend to live solitary lives until they can take over such a group.

In spite of their colossal weight, gorillas are generally good climbers. On the ground, gorillas usually move on all fours using their hand knuckles. Mountain gorillas mainly eat ground-level roots, shoots, and leaves. Lowland gorillas eat somewhat more fruit, which they gather in trees. Recent fecal analyses suggest they occasionally also eat mammalian meat. To maintain their size on a predominantly plant-based diet, gorillas, unsurprisingly, spend a lot of their time feeding. Some plants have serious defenses, such as thorns; others only have tiny cores that are edible. The psychologist Dick Byrne has documented gorillas' painstaking procedures to get to these juicy bits without getting hurt. These techniques can be quite complex, involving multiple steps, and young gorillas acquire them by observing their elders.

FIGURE 2.7.
Kigale, a female Western Lowland
gorilla, at the National Zoo,
Smithsonian Institute, Washington, DC
(photo Emma Collier-Baker).

In captivity, gorillas have long been known to be adept at using tools, much like orangutans and chimpanzees. In the wild, such behavior had not been reported until quite recently. In 2005, a gorilla was observed using a stick to dig up tubers and to check water depths while wading through a swamp. One reason for the paucity of tool use in the wild may be that the gorillas' immense strength provides them with alternate avenues to solve problems. My occasional collaborator Andrew Whiten told me that he once gave a puzzle box to a gorilla, a box that had previously been used to investigate chimpanzees' imitative capacity at manipulating cogs and levers, and quickly learned that the gorilla had an easier way of getting at the treat inside: he simply smashed the entire thing open.

Mountain gorillas are critically endangered, with an estimated 680 individuals surviving in the wild.[8] The Eastern lowland gorilla population comprises only a few thousand animals. The IUCN lists Eastern gorillas as endangered and Western gorillas as critically endangered. The Western subspecies known as Cross River gorillas are down to only about 250 individuals. The total number of Western lowland gorillas used to be estimated at more than 90,000 individuals, but the population appears to be rapidly declining. In 2006, an Ebola epidemic killed some 5,000 gorillas. Yet for once there has been welcome news about numbers in the wild. In 2008 a previously unknown large population, possibly up to 100,000 individuals, was found in the Congo—a fantastic discovery. Alas, there is little hope that any other uncharted habitats will be found in the future.

---

8 For protection of the gorillas, see the International Gorilla Conservation Program: http://www.igcp.org.

CHIMPANZEES ARE SUBDIVIDED INTO TWO species: the common chimpanzee (*Pan troglodytes*) and the pygmy chimpanzee or bonobo (*Pan paniscus*). Because chimpanzees are our closest living relatives and bonobos are in some aspects quite distinct from common chimpanzees, I shall introduce them separately.

Common chimpanzees live in large areas across the sub-Saharan forests of central Africa and are often subdivided into four subspecies (Western, Eastern, Central, and Nigerian chimpanzee). Young chimpanzees regularly feature in movies (from Tarzan's Cheetah to Ronald Reagan's Bonzo), and this has created a public impression that chimpanzees are small and cute. Adult males can in fact be large and ferocious. They are a lot stronger than most humans, and it is dangerous to get in their way when they are excited. They can also be temperamental. We sometimes work with two male chimpanzees at Rockhampton Zoo in Australia, and they are exceptionally friendly, sitting on the other side of the mesh attending our psychological tests. However, a loud lawn mower, a bus, or some other trigger can set them off into a frantic rage. They scream, beat the mesh, jump and charge hysterically through the enclosure, and spit. In other words, they go ape-shit. You certainly do not want to be carrying one of them on your shoulders like Tarzan.

In the wild, common chimpanzees are by far the best studied of all the apes. There are now data from several study sites that have observed specific groups for dozens of years. Chimpanzees live in large groups with typically forty to sixty members (some with over a hundred), but they usually travel and feed in smaller subgroups. Membership of these subgroups changes, and the whole group comes together only from time to time. This flexible system is known as fission-fusion—a lot of coming and going. It provides opportunity to spend time with selected individuals but poses special social cognitive problems of keeping track of changes in relationships and in the social hierarchy.

Females migrate to join another group when they become sexually mature, whereas males tend to stay in their group of birth. Achieving high rank is important to males, as it secures greater sexual access to females. Female chimpanzees advertise their sexual readiness through distinctive pink genital swelling and during that time may mate between five and an astounding fifty times a day. Mind you, intercourse usually only lasts

FIGURE 2.8.
The common chimpanzee Ockie at
Rockhampton Zoo.

seven seconds. Copulating with many males, even when already pregnant, may be a strategy to reduce the likelihood of males in the group killing the baby. Unlike gorilla silverbacks, which monopolize access to females, chimpanzee males cannot be certain about their paternity.

As Jane Goodall famously documented, the social lives of chimpanzees are complex and intriguing. For instance, gaining the support of the alpha female may elevate a male's position. Coalitions may allow lower-ranking males to topple the reigning alpha male. In turn, other alliances may retaliate against usurping individuals and their associates. As the primatologist Frans de Waal aptly observed, there is good cause to refer to "chimpanzee politics"—which is what he called his seminal book on the subject. The notion that humans are the only political species, as Aristotle proposed, can only be upheld with appropriate restriction of the term's definition.

War is often said to be an extension of politics. The territory of each chimpanzee community has distinct boundaries that male groups patrol. Goodall observed how chimpanzees on these patrols killed chimpanzees from a neighboring group. At that time it was widely believed that only humans had that deplorable trait to cooperate in the murder its own kind. The ferocity of the killing shocked a lot of people. The outnumbered victims were held down while assailants beat and mauled them, dragged them back and forth along the ground, and attacked them long after they had stopped defending themselves. Many examples of chimpanzee raids and violence against neighbors have been recorded since. Given that humans and chimpanzees share this cruel potential, it may be an ancient trait indeed, which, as already noted, may have had an important role in the very creation of the gap.

Chimpanzees also share the human penchant for hunting. Some groups supplement their diet with small animals and even baboons. Hunting primates, such as swift colobus monkeys, seems to involve sophisticated cooperation. For example, one chimpanzee may drive the prey toward others that appear to hide in ambush. What cognitive abilities are involved, however, is subject to considerable debate. (Various pack animals, such as lions and wolves, also engage in cooperative hunting.) In 2007 chimpanzees in the savannah of southeastern Senegal were reported to sharpen sticks and spear bushbabies—small nocturnal primates—hiding in tree holes.

Although significant, meat is by no means a major food source for most chimpanzees. About half of what they typically consume is fruit, and other vegetable matter such as leaves and bark is also commonly eaten. Some groups have been observed digging for tubers. There have even been reports that chimpanzees seek out medicinal plants when sick. Many chimpanzees obtain protein by eating ants and termites and have developed ingenious ways of fishing them out of their holes. Small sticks are stripped of their leaves, inserted, and retracted once insects have gathered on the tool.[9] Protein requirements are sometimes further supplemented by consumption of nuts. In the Tai forest, for instance, chimpanzees spend considerable time cracking open nuts with stone hammers and anvils, and appear to have done so for a long time in this region. At one site an archeological study indicates that chimpanzees used such stone tools 4,300 years ago—the chimpanzee stone age predates human farming in the area. Chimpanzee foraging is much more complex and diverse than once thought. They have developed ingenious ways of obtaining varied food sources, and some of these methods are passed on, apparently over thousands of years, through social learning.

---

9   When Jane Goodall first documented this tool manufacture in 1964, it caused quite a stir. Curiously, this observation was not as novel as had long been believed. It turns out that a Liberian stamp from 1906 already depicts a chimpanzee using a stick to forage for termites long before any scientist documented such behavior.

Groups differ in what tools they use and how they use them. Leaves are used for cleaning, branches are shaken and stones are hurled in aggressive displays, rocks and logs are used as hammers and anvils, and sticks are employed to obtain objects otherwise out of reach. Some tools are made to suit the task (e.g., stripping and sharpening a stick and adjusting its length), and the same kind of objects may be employed for a range of different purposes (e.g., leaves may be used as toilet paper, sponges, and umbrellas). In short, as we will see throughout the book, our closest animal relatives have rather clever minds.

Common chimpanzees used to be, well, common in Equatorial Africa, living in over twenty countries. Current estimates put the total chimpanzee population size to above 170,000 but less than 300,000. This comprises over 20,000 Western, 90,000 Eastern, and 70,000 Central chimpanzees. There are fewer than 6,500 Nigerian chimpanzees remaining. Though these numbers are higher than for most of the other apes, it may be instructive to compare them to the populations of a few towns and to imagine those settlements were all that is left of humanity on the planet. Numbers are declining primarily as a result of habitat destruction and degradation, as well as hunting for the bush meat trade and the pet trade. The common chimpanzee is thus classified as endangered by the IUCN.[10]

BONOBOS, OR PYGMY CHIMPANZEES AS they used to be known, were only described in 1929. They are more petite than their better-known cousins; have a black, relatively flat face, pink lips, and a higher forehead; and look well groomed, with their hair often parted neatly down the middle. Together with a relatively straight posture when standing upright, which they do about a quarter of the time when on the ground and especially when carrying things, they look eerily like one might imagine an early human ancestor.

Bonobos live in a limited region south of the river Congo. It may have been river barriers that led to their separation from common chimpanzees between one and two million years ago. They eat mainly fruit, complemented with some leaves and, on occasion, a small amount of animal protein. Only in 2008 were bonobos first described to collaboratively hunt

---

10 For chimpanzee conservation, see, for example, the Jane Goodall Institute: www.jane-goodall.org.au.

FIGURE 2.9.
Young adult male bonobo Kevin *(photo by and courtesy of Frans de Waal).*

monkeys and share the spoils. There has been no evidence yet of bonobos using tools in the wild, but that may only be a matter of time and patient observation. They certainly do use them effectively in captivity. Little is known about wild bonobos, with only two permanent study sites currently in operation.

Like common chimpanzees, bonobos live in fission-fusion societies and can typically be found in subgroups of up to twenty-five individuals. The entire group may be as large as two hundred individuals. What little study there has been of bonobos in the wild suggests distinct traits that separate them from common chimpanzees. They show far less aggression, are less male-dominated, and have a lot more sex. There is sex between all ages, genders, and ranks. Bonobos seem to enjoy sex and indulge in a variety of positions, including face-to-face intercourse, tongue kissing, and even oral sex. Bonobo sex, like human sex, is not just used for procreation. In certain situations sex seems to serve a tension-reducing function. After conflicts, for example, sex is often used as a means to establish reconciliation. Perhaps there is a lesson here.

As enthusiastically documented by Frans de Waal, bonobos have a peaceful—some might call it "utopian"—society. De Waal argues that bonobos have compassion, empathy, and kindness. The relationship between the sexes is quite egalitarian. There is little violence compared to the frequent outbursts observed in common chimpanzees. Recall, though, that we know a lot less about bonobos than we do about common chimpanzees. More extensive research is highly desirable and may well reduce some of the apparent differences between these species of chimpanzee. It took years of observation before Jane Goodall found that common chimpanzees engage in collective killing.

Estimates of bonobo population size range between 30,000 and 50,000 individuals. Bonobos are classified as endangered by the IUCN and face similar pressures from human activity as the other great apes. Since they only live in the Democratic Republic of Congo, the local political situation is crucially important in the continuing survival of this fascinating hominid.[11]

AS OUR CLOSEST REMAINING RELATIVES, these species of ape provide context for our discussion about the nature and origin of the apparent gap that separates animal and human minds. Given that minds are generated by brains, I'll close this chapter with a comparison of human brains and those of our animal relatives.

Thomas Huxley found that mammalian brains are broadly equivalent in structure and differ primarily in size. Size matters. Even within our own species, there is some evidence that people with larger brains are more intelligent than those with smaller brains, at least as measured by IQ tests. Do humans, then, simply have the largest brains?

The brains of small apes weigh about 80 grams and those of great apes between about 300 and 450 grams. Humans have by far the largest brains of all primates, typically weighing between about 1.25 and 1.45 kilograms, and containing some 170 billion cells, about half of which are neurons. Metabolically we heavily invest in brain activity. Our brain comprises about 2 percent of our body mass but consumes some 25 percent of our energy. (Thinking is exercise and costs you 20 to 25 watts to run. Yes, you are exercising right now.) However, brain size alone cannot explain the gap—alas, we do not have the biggest. Elephant brains can weigh

---

11   For bonobo conservation visit, for example, the Bonobo Initiative: http://www.bonobo.org.

over 4 kilograms, and whales have much larger brains still, weighing up to 9 kilograms.

However, if you take overall body size into account, humans have much larger relative brain sizes than these giant creatures, whose brains comprise less than 1 percent of their bodies. Yet, in spite of its initial intuitive appeal, it is not entirely clear why relative brain size should matter. Perhaps larger bodies require larger brains in terms of innervations and neural management, but shouldn't cognitive processing be independent of body size? After all, we don't become more or less smart when we change relative brain size through losing or gaining weight, do we? Furthermore, some large animals, such as crocodiles, do fine with walnut-sized brains. Why should we adjust brain measures according to body size when large bodies can be run by small brains?

In any case, the outcomes of relative size comparisons have also not been supportive of humans' sense of superiority. Some shrews and mice, it turns out, have brains that are up to five times larger than ours relative to body size. They can have an extraordinary 10 percent of their body be brain, compared to our 2 percent. This may please fans of Douglas Adams, whose fictional laboratory mice were smarter than us and conducted experiments on human scientists who thought they were experimenting on them, but there are no signs of extraordinary mouse intelligence I am aware of.

Since we get beaten by large mammals in the first scheme and small mammals in the second, a third scheme has been devised that takes into account that as mammals get larger, brains get *absolutely* larger but *relatively* smaller. The psychologist Harry Jerison calculated so-called encephalization quotients, or EQs, which compare the actual size of a species' brain to the size one would expect for an average animal of its size from the same taxon. Among mammals the average animal is calculated to be a cat. Table 2.1 lists some examples of mammalian EQs. In this influential scheme, humans emerge on top with a brain over seven times larger than that predicted for the average mammal of our size. Many other animals seem to be ranked in line with common assumptions. Some findings, however, are unexpected. Capuchin monkeys, for instance, have surprisingly large EQs that put them well ahead of chimpanzees. One may also worry about the influence of the reference group. If instead of comparing us to the average mammal, we narrow this down to a comparison to the average primate, or expand it to a comparison to the average vertebrate, the results

TABLE 2.1. Some sample encephalization quotients

Species	EQ
Human	7.4–7.8
Dolphin	5.3
White-fronted capuchin monkey	4.8
Chimpanzee	2.2–2.5
Gibbon	1.9–2.7
Old-world monkey	1.7–2.7
Whale	1.8
Gorilla	1.5–1.8
Fox	1.6
Elephant	1.3
Dog	1.2
Cat	1
Horse	0.9
Sheep	0.8
Lion	0.6
Ox	0.5
Mouse	0.5
Rabbit	0.4
Rat	0.4
Hedgehog	0.3

change. Not surprisingly, then, debate continues over which measure is most informative.[12]

Given the limitations of both absolute and relative measures, Andrew Whiten and I combined the two. We took Jerison's EQs for primates and computed the absolute brain mass that is in excess of that predicted for an average mammal of the same body size (see Figure 2.10). Lo and behold,

---

12 Other schemes have been proposed, such as ratios between different parts of the brain. For example, we have already encountered the ratio between the neocortex and the rest of the brain; Robin Dunbar used it when relating group sizes in primates to cognitive powers. There is some difficulty in demarcating different sections of the brain in different species, so this also has not been universally accepted. Concern has been raised about ignoring absolute brain matter altogether in favor of ratios or quotients. Surely there must be something about absolute neuronal resources that enable or limit cognitive capacities. If you have only one thousand neurons, there is only so much computation that can be done regardless of EQ or relative brain size. Indeed, it has been argued that absolute size is the better predictor of capacities in primates. But given that body size appears to have some effect on brain size, we cannot entirely ignore this factor either.

FIGURE 2.10.

Average excess brain in grams over and above that predicted by body size.

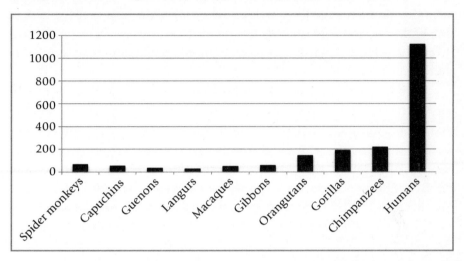

humans came out on top, with our closest relatives lined up in an order that appears to make intuitive sense. The great apes enjoy absolute neural resources well beyond that typically evident in a mammal of their size, and substantially more than monkeys. Humans have disproportionally more computing power still.

This makes us feel good, perhaps, if a little uneasy. Are we just massaging data to get the results we want? A lot of ink has been spilled on the pros and cons of various measures of brain size comparison, but it remains uncertain whether any of them have unearthed some hidden truth or whether we are simply using statistics to confirm our preconceptions.

The obsession with size may be misleading. When the neurologist Korbinian Brodmann produced his seminal comparative brain maps at the beginning of the twentieth century, Huxley's claim about structural similarity between mammalian species was confirmed. However, Brodmann appreciated that more sophisticated methods might eventually find differences in internal organization.

New methods are indeed beginning to reveal some subtle differences. For example, new techniques for estimating the number of brain cells in mammals suggest that different scaling rules link brain size and number of neurons for rodents, insectivores, and primates. It turns out that ten grams of monkey brain contain a lot more neurons than ten grams of rat brain. However, these data also suggest that human brains are simply linearly scaled-up primate brains in terms of cell numbers.

Brain characteristics other than cell numbers and overall size may be responsible for our peculiar minds. The human olfactory bulb (the brain area that processes smell) as well as the primary visual area at the back of the brain are relatively smaller in humans than they are in our close relatives. Such changes may reflect some brain reorganization that occurred in human evolution.[13]

Modern neuroscience is beginning to identify finer differences. The first documented microscopic distinction between ape and human brains has been a unique cell organization in a layer of humans' primary visual cortex—not an area typically associated with higher cognitive functions. There is also some suggestion that humans differ in the neural connections in the prefrontal cortex, an area very much associated with higher cognitive functions. While neurons in the back of the brain have relatively few connections, and human brains differ little from other primates in this respect, there are many connections in the prefrontal cortex. Their density is much higher in humans than in other primates that have been examined thus far. Much more fine-grained counting will be necessary to determine exact quantitative differences.

Future research may well identify characteristics that distinguish various mammalian brains. Some evidence suggests that certain species even differ in the types of cells that constitute the brain. The brains of great apes and humans, for instance, contain a distinct type of large cell that is unusual in other species.[14] We are only beginning to unravel the mysteries of brains, so we may hold out for as yet undiscovered features that set human brains critically apart from the rest.

As it stands, however, it is not clear what it is about our brains that causes our minds to be special. The study of the brains of different primates has not clarified what their minds are capable of and what their limits are. To find out about the nature of the gap we need to return to behavior as an indicator of mind. Long-term field studies have increasingly provided details about the ecology and natural behavior of our closest animal relatives. Controlled experimental studies aim to infer their mental capacities. So next we turn to comparative psychology and the tools of studying minds.

---

13 One idea is that the primary direction of information flow has reversed in humans from the typical flow from the back to the front of the brain, to one biased toward flow from the front to the back.

14 For a while these so-called Von Economo neurons were thought to be unique to great apes and humans, but recent work has documented them in the brains of elephants, whales, and macaques.

THREE

Minds
Comparing Minds

MIND IS A TRICKY CONCEPT. I think I know what a mind is because I have one—or because I am one. You might feel the same. But the minds of others are not directly observable. We assume that others have minds somewhat like ours—filled with beliefs and desires—but we can only infer these mental states. We cannot see, feel, or touch them. We largely rely on language to inform each other about what is on our minds. But even when someone says what is on his mind—that he is, say, sorry or happy—you might question whether or not he is telling the truth. Still, when verbal and behavioral indicators point in the same direction, we can generally be confident about another's mind.

Similarly, we can use behavior to infer the minds of animals. In the absence of verbal self-reports, however, we may be less certain about what goes on in their minds, as they lack the ability to confirm our conclusions or set us straight. So people sometimes maintain opinions about animal minds that are in stark contrast to each other. At one extreme, humans imbue their pets with all manner of mental characteristics, treating them as if they were little people in furry suits. At the other, humans regard animals as mindless bio-machines—consider the ways animals are sometimes treated in the food industry. Many people vacillate between these interpretations from one context to another.

Scientists are supposed to guard against preconceived ideas that bias their research. Nonetheless, the philosopher Daniel Dennett notes that comparative psychologists also gravitate toward opposing positions: the "romantics," who ascribe complex, humanlike abilities to animals, and the "killjoys," who are reluctant to do so.[1] In other words, some researchers favor what I like to call "rich" interpretations, while others prefer "lean" accounts. One might suspect, as Dennett does, that the truth typically can be found somewhere in the middle. But when scientists themselves have such biases, we have a problem—and a clue why progress on establishing what we share with other animals and what sets us apart has been somewhat slow.

To get to the truth, we need to go beyond preconceptions and apply methods and criteria that can compellingly establish animal mental capacities. In this chapter I will illustrate the approaches of modern comparative psychology by reviewing evidence for some mental faculties apes seem to share with us. Only a prudent and cautious analysis will allow us to make systematic progress on identifying the gap. It is about time that we do. If we can establish scientific consensus about what we share with which other animals, it will have important implications, for instance for establishing the genetic and neurological bases of mental capacities and, arguably, for animal welfare. For us humans, nothing less is at stake than our place in nature.

DARWIN TRIED TO SUPPORT HIS case for continuity with anecdotes of behavior purportedly showing that animals possess the rudiments of many facets of the human mind. He wrote that "the senses and the intuitions, the various emotions and faculties, such as love, memory, attention, cu-

---

1 There are many reasons why researchers may have such leanings. For instance, if you are dedicated enough to work, and perhaps live, with an animal for many years, it should not be surprising that you would become attached and prefer to see successes rather than failures. I have seen a prolific researcher cry over a sick animal and proclaim its life worth more than that of a human. Others may be attracted to the role of the detached, hard-nosed defender of parsimony in a world too easily swayed by romantic, wishful thinking.

riosity, imitation, reason, &c, of which man boasts, may be found in an incip-
ient, or even sometimes in a well-developed condition, in the lower animals."

However, anecdotes are difficult to verify and may be biased by preconceived ideas of the person reporting them. For instance, one nineteenth-century
scholar cited an anecdote of a dog burying the remains of a duck to argue
that the dog was aware that murdering the duck was a crime and that it
was trying to conceal the evidence. Maybe so. Yet we can explain why dogs
bury bones without having to suppose that they scheme about thwarting
law enforcement. Conway Lloyd Morgan, one of T. H. Huxley's students,
argued that we should not interpret animal behavior as the outcome of a
higher "psychical" faculty when it can be fairly explained as the result of
a lower one. This principle came to be known as Lloyd Morgan's canon
and is a favorite of killjoys.

Even impressive and unusual animal behaviors often have simple explanations. A classic cautionary event occurred at the beginning of the
twentieth century. A German schoolteacher named Wilhelm von Osten
had trained his horse to give apparently intelligent answers to a great variety of questions. The horse became a sensation and was widely known as
Clever Hans (Der Kluge Hans). He could answer questions about simple
addition (e.g., What is 5 plus 7?) by stomping his foot the right number of times. Hans could also correctly reply to questions about calendar
days (If Monday is the eighth, what is Friday's date?). And it did not stop
there. Herr von Osten developed a table that would translate stomps into
letters, allowing Clever Hans to answer questions in words. The idea that
this horse understood the German language and how calendars work may
seem a bit rich. But no one could explain how else the animal might do
what it did. A scientific committee, headed by the leading psychologist
Carl Stumpf, examined the case and, failing to come up with a leaner alternative explanation, endorsed the equestrian genius.

One of Stumpf's students, Oskar Pfungst, pursued the issue further.
He eventually discovered that Hans could only answer the questions correctly if the person asking the question knew the answer. He also noted
that when he made von Osten ask questions from a place where the horse
could not see him, performance broke down. Pfungst concluded that the
horse stopped stomping in response to subtle cues, such as a jerk of the head.
The horse was clever in picking up on these cues and learning the association with reward, but did not understand German or mathematics. There is
a long history of people, such as those working in a circus, training animals
to respond in particular ways to subtle movements. What is remarkable in

this case is that Herr von Osten was apparently unaware that he had cued the animal into acting in the desired way.

Such unintentional cue-giving is a serious problem for those studying animal mental capacities, and this case had a dramatic influence on comparative psychology. It highlighted the need for researchers to actively guard against such cuing, as merely trying not to give cues may not be good enough. One way to achieve an unbiased result is to have the experimenter unaware ("blind") of what the desired response of the animal is. In addition to enhancing methodological rigor, the Clever Hans effect, as it came to be known, drove home the fact that behavior that looks intelligent can be the result of simpler learning processes. It powerfully illustrated Lloyd Morgan's canon. When confronted with a rich interpretation, one that supposes complex mental capacities, we should consider carefully whether leaner accounts could do the trick.

Anecdotal approaches and rich interpretations subsequently went out of fashion. Behaviorism became the dominant force in psychology for much of the twentieth century and focused on explaining behavior in terms of general associative learning rules, rather than species-specific mental capacities. For example, Edward Thorndike's studies on the time it took animals to escape from a puzzle box led to the "Law of Effect": consequences of an act determine its future probability in similar situations. Lean interpretations were favored, and any attribution of mentality to animals was greeted with suspicion, if not ridicule. The pendulum had swung from the extremes of romantic anthropomorphism to a "mindless" zeitgeist. Despite Morgan's openness to evidence for it, many behaviorists threw the baby out with the bathwater and seemed to deny mental capacities to animals *a priori*.

Today, more romantic positions have regained popularity. Comparative psychologists try to establish the mental capacities of animals, and reports of amazing feats once thought to be uniquely human appear frequently. However, academic debates between romantics and killjoys are commonplace. Both can make important contributions to a more precise understanding of animal minds. Romantics challenge lean views by searching for evidence demonstrating rich competences in animals, and killjoys provide lean alternative explanations for claims about rich animal capacities. The hope is that with each of these battles we get a little bit closer to the truth.

Indeed, moderate explanations are beginning to take much clearer shape, as we will see throughout this book. Old dichotomies and extreme

f bolting into the distance is pretending to chase a rab-
nning. Indeed, it is exceedingly difficult to distinguish
m other types of play in the absence of props such as
that human children use (which in itself may hint at
is one published report of purported pretend play with
ing a toy from a chimpanzee the researchers called Ka-
ing with his pregnant mother, Kakama carried a log for
he seemed, according to the observing researchers, to
ike a baby. This behavior included making a nest and
side it. When two field assistants, who did not know
served Kakama some months later engaging in similar
lected the (new) log and labeled it as his "toy baby."
ggestive, there may be reasons other than pretend play
for the observed behavior in this anecdote.
 many convergent anecdotes, researchers may become
 a richer interpretation. Andrew Whiten and Dick Byrne
cted and categorized many reports of what they called
in primates. These are cases of apparent social ma-
suggest that primates at times pretend in order to get
oid punishment. For example, a lower-ranking male
almost caught by the alpha male in a sexual act with a
ered his erect penis with his hands, as if to hide the ev-
ts suggested not just concealing but active misleading.
oon being chased suddenly stopped, jumped up onto
eered intently into the distance, as if he had spotted a
ant baboon then also halted and stared in the same di-
e chase. Was the pursued animal pretending that there
g the chase to a halt? Such examples certainly make it
 animal acted as if something is the case when it knows
hiten and Byrne's survey, complex tactical deception
quent in great apes. Only gorillas and chimpanzees,
ed cases in which an individual spotted a half-hidden
ut then ignored it until a competitor disappeared from
s safe to secure the treat. There were even accounts
, where a competitor would not completely leave the
ickly to snatch the food from the would-be deceiver.
tes' maneuvering in their social worlds requires some
oting to conclude that great ape minds are capable of

positions are sometimes not as far apart as they at first appear. It is quite possible, for instance, that there is a vast gap separating animals and humans and that animals have richer and more diverse mental lives than is often assumed.[2] Even the apparently clear-cut distinction between qualitative and quantitative differences, between the belief in a fundamental difference in kind that sets us apart from animals and Darwin's belief that human and animal minds only differ in degree, can be a bit of a red herring. We know that changes in degree frequently produce distinct attributes that may well be regarded as a difference in kind. When temperature gradually increases, the properties of H_2O change radically as ice turns to water and then to gas. By the same token, gradual changes in, say, processing capacity could lead to radically different possibilities of thinking.

Apparently stark disagreements about the gap at times merely reflect differences in opinion about the relevant criteria rather than about matters of fact. For example, to demonstrate language, is it sufficient to show that other animals communicate, or must they be able to tell us a story? As long as we can establish what exactly animals can and cannot do, the rhetoric does not really matter. Throughout this book I shall focus on finding scientific answers rather than on securing human superiority or dispelling human arrogance.[3]

The question to answer is this: Which characteristics of the human mind, be that distinct traits or gradual differences, enable and motivate us to do the great diversity of things that other animals do not? What makes us human? On the face of it, human behavior differs from animal behavior in countless ways. For instance, we play football, sell insurance, go to school, build bicycles, and have BBQs. What are the fundamental bases for our many peculiar behaviors? The answer should shed light on related

---

2 The opposite is also possible. That is, there may be a smaller gap than many people think, and yet animal minds can be explained in lean ways. This case has been argued on the basis that we may overestimate our own mental faculties and that much of human behavior may be based on very lean associative mechanisms that we share with other animals.

3 The title of this book may raise suspicion that I might be biased toward putting the bar high to exaggerate differences between animals and humans. I am known in the discipline (at least to some) for proposing that a key advantage of humans can be found in our capacity to mentally travel in time (see Chapter 5). However, I have no particular interest in establishing or defending an overly wide—or overly narrow—gap. I have advanced some killjoy explanations as alternatives to romantic claims in the literature, but I have also made cases for richer abilities in primates than were previously known. For example, my colleagues and I reported the first evidence to suggest that chimpanzees can notice when a human copies them—something to keep in mind the next time you ape an ape at the zoo.

questions, such as: what allowed us to dominate this planet, and why do we ponder such questions?

In the chapters to follow I will discuss the domains that are most commonly claimed to be, or to contain, key uniquely human attributes that may have led to a profound shift in human behavior: language, mental time travel, mind reading, intelligence, culture, and morality. Each topic is multifaceted and requires some careful scrutiny. Therefore, about half of each of these chapters deals with what the science of the mind has taught us about the human faculties. In particular, I discuss what mental traits are fundamental in these domains and important to our success. To identify the basic building blocks of our minds, it is useful to consider carefully how infants and children first acquire them.[4] I will therefore reflect considerably on development along the way to identifying the gap.

In the other half of each chapter, I examine what is known about animal capacities in these domains that challenge claims of human uniqueness. We will discover many sophisticated parallels to human behavior as well as some strange capacities that are unique to certain species. We are only beginning to scratch the surface of the plethora of colorful mechanisms various animals deploy. The repeated comparison of human abilities with those of other animals will bring into sharper view what it is about the human mind that sets us apart. Before we get into these analyses, however, I will first address how we can infer the presence or absence of mental capacities in animals.

CONSIDER ONE OF THE MOST fundamental aspects of our human mind: we can imagine things other than what is available to the senses. We can picture past, future, and entirely fictional worlds and think about them. William James noted that it is the capacity to conceive of alternatives that allows us to question why things are the way they are. Humans ask big questions like: what are we, where do we come from, and where are we going? Most cultures have elaborate creation myths that children are told when they start raising these questions. Some of the oldest are those of

---

4 The minds of preverbal infants are also difficult to establish, and here researchers too sometimes fall into opposing camps of romantics, who lean toward accepting indications of sophisticated mental abilities in babies, and killjoys, who lean toward dismissing such claims. For instance, evidence that babies assess individuals on the basis of seeing them help or hinder others has been challenged by simpler explanations based on associative learning.

indigenous Australians, w
was molded, for instance,
"dreamings," belong to sp
with meaning. Questionin
man mind. (Why else are
they ponder past, future, o
ing of life? Can they conju
and now? Do they have e
find out?

Darwin argued that va
Aboriginal sense of the v
Perhaps Darwin was right:
(REM) sleep, we typically
show REM. However, a ki
tail dreaming in other ani
a capacity to imagine whe
not told us their dreams (

In human developm
expression of an imagina
dlers typically start to dis
many subsequently spen
sies. When children prete
worlds: in addition to per
imaginary scene. A block
pen a comb. The hand n
impenetrable wall. Child
of an object with its real
tempted to eat their mud
a representation of realit
situation. Therefore, if w
gage in pretend play, we
have a basic capacity for

Alas, animal field st
tend play in the wild.

---

5 It is still not known why
gests that dreaming helps me
events.

whether, say, a wo
bit or is simply ru
imaginary play fr
the dolls and toy
differences). Ther
something resemb
kama. While trave
several hours that
treat suspiciously
placing the log in
of these events, ob
behaviors, they co
Although this is s
that could accoun

When there ar
more confident in
systematically coll
tactical deception
nipulation, which
an advantage or a
chimpanzee, wher
female, quickly cov
idence. Other repo
For example, a bab
his hind legs, and
predator. The purs
rection, giving up
was danger, to brin
seem as though the
that it is not. In W
was particularly fr
for instance, produ
favorite food item
the scene and it w
of counterdeceptio
scene but return q
It is clear that prim
sinarts, so it is tem
pretending.

A killjoy, however, would remind us that these are all merely anecdotes. We simply cannot be sure what is driving any of these overt behaviors. Perhaps the apparent counterdeception was mere coincidence. The
ape came back and was lucky to stumble across the other ape as he was attempting to retrieve a choice food item. Similarly, the baboon that was
chased might have actually thought, mistakenly, that there was a predator
in the distance. Or it might have been in a similar situation in the past and
subsequently learned that stopping suddenly and staring in the distance
could lead to the end of an annoying chase. Neither of these options requires that the primate pretended that something she knows is not true is
actually the case. As with Clever Hans, behavior can look sophisticated and
yet be driven by simpler means.

There are some famous reports of pretense from great apes held in
human captivity. Researchers raising young great apes have offered rich
interpretations of their charges playing with dolls or toy animals as if
they were real things. For instance, Sue Savage-Rumbaugh has written that
chimpanzees bathe dolls and sometimes make a toy animal "bite" others.
The gorilla Koko has chased people around the room with a rubber snake.
To cite a personal anecdote, I once brought a small rubber crocodile to the
two chimpanzees, Ockie and Cassie, we sometimes test at Rockhampton
zoo. They were intrigued, and when I pretended that the crocodile was trying to bite my co-investigator Emma Collier-Baker, they quickly jumped to
her side and appeared to play along—even to the extent of reassuring her
when the "danger" had passed. At least, that is how I remember the event,
although frankly I cannot be sure whether any pretense took place from
the chimpanzees' perspective.

The most often-cited case comes from the chimpanzee Viki, who is
said to have repeatedly pulled an imaginary toy. Sometimes Viki acted as
if the pretend cord had tangled around an obstacle and then placed one
fist above the other, seemingly strained backwards repeatedly until, with
a little jerk, she walked on as if she had loosened the stuck imaginary toy.
She would then continue on her way, trailing the pretend pull toy behind.
At least this is what it looked like to the human observer. This anecdote
is perhaps stronger behavioral evidence because the ape seems to follow
through with the logical implications of the imaginary events. Children
do this a lot. For example, if an empty glass filled with imaginary juice is
spilled, they then proceed to pretend to clean up the resulting imaginary
mess on the floor. I know of only one other incident, reported at a conference, of a chimpanzee possibly following through with such pretend

implications, in a case involving play with what appeared to be imaginary building blocks.

Human children, of course, use words to elaborate and share their pretend play. The gorilla Koko, having been trained in human sign language (more on that in the next chapter), has been documented to use appropriate signs in her apparent pretend games. In response to being given an ape doll, for instance, she hugged it and then signed "drink." She then made the doll sign "drink" by taking its thumb to its mouth. Again, though, there remain obvious difficulties with interpretation. Did the gorilla intend to make the doll sign "drink," or did she merely put its thumb into its mouth?

What are we to conclude? Can captive great apes really be said to engage in pretend play, or are they merely mimicking human behavior they see around them? A romantic perspective might point out that we would probably be content to accept many of the reported anecdotes as evidence for pretend play had we seen them in a child. A killjoy perspective would question what part of the alleged pretense is in the mind of the ape and what is in the mind of the human observer. Clearly, then, evidence of pretense, whether in play or in deception, remains tenuous.

A neutral observer might highlight that neither extreme view should overlook inconvenient facts. On the one hand, although there are reports of pretend play in great apes, the list is relatively short, and even if the reports are accepted at face value, great apes do not show anywhere near the sophistication or amount of pretend play that human children do. Children often put their heads together in pretense, creating one scenario after another. On the other hand, while members of all great ape species have been repeatedly recorded performing apparent pretend activities, monkeys and other animals, even if reared in similarly intimate human contexts, have not. As it stands, the current data merely raise the possibility that our closest animal relatives have some ability to entertain alternative worlds in their minds. To unequivocally establish that great apes can think about things other than what they directly perceive, we need stronger evidence. We need carefully controlled experiments.

IN RECENT YEARS EXPERIMENTAL APPROACHES for animals have increasingly been adapted from human developmental psychology. Most children acquire their sophisticated mental capacities in a predictable manner, and developmental psychologists have created nonverbal tests to identify the steps involved. Performance on these tests often predicts later capacities and performance on other measures (including on verbal tasks that cannot be adopted to test animals). The emergence of pretend play tends to coincide with the development of various other capacities, such as reasoning about hidden objects and recognizing one's image in mirrors. Systematic experimentation has bolstered the case for imagination in great apes with research from these domains.

The Swiss child psychologist Jean Piaget, who produced what is still the most comprehensive and influential theory of how human minds develop, constructed a progressive series of simple search tasks to measure children's growing capacity to reason about objects that they no longer directly perceive.

For young infants, "out of sight" quickly turns to "out of mind." A brief distraction is typically enough to stop them from searching for something they had previously pursued. Children seem to only gradually acquire an understanding of the fact that things exist independent of whether they can still perceive them or not. Even adult philosophers wonder whether a tree falling in the woods really does make a sound when no one is there to perceive it.[6] Philosophers aside, adults generally trust that people and objects continue to exist regardless of whether someone senses them or not.

Piaget called this "object permanence" and proposed developmental stages through which infants progressively pass over the first two years of life.[7] At around twelve months of age infants can reliably find a target object if they witness it being hidden beneath one of several boxes. They remember where it is and so have achieved what Piaget called stage 5 object permanence. However, when the target is first put in a small container, and this container is then used to move it somewhere else (so-called invisible or hidden displacement), they are at a loss. Imagine putting a small toy in your hand, closing it, and then putting it in the pocket of your jacket. You then pull out your hand and show your toddler that the hand is now

---

6   This riddle is often used not just to raise the question of unobserved reality but the distinction between sound waves and sound perception.

7   Developmentalists have expended much research effort over the years trying to show where Piaget had been wrong. Though most of his object permanence stages have been confirmed, research suggests that infants can achieve some stages earlier than originally proposed. However, crucial to the present argument are only the final stages, and these are still only completed by children at about the end of the second year of life.

empty. It is obvious that, unless there is some trickery, the object must now be in the pocket. Yet young infants and many animal species do not know where to look for the object. As long as there are no other clues, such as the smell of the treat or someone pointing to the pocket, they are at a loss as to where the toy has gone. They either do not remember the past situation or cannot relate this memory to the current situation to infer the solution. Toddlers only pass such tasks from around age eighteen to twenty-four months. This is stage 6 object permanence and demonstrates reasoning about the movement of the target that could not be directly perceived.

Because Piaget's search tasks are nonverbal and relatively easy to adapt to the needs of different species, they have become among the most widely used comparative tests ever. All you need are some boxes and a cache of desirable treats. Many species have been studied over the years, and several species have demonstrated stage 5 object permanence. Species that have formally passed this test include cats, chimpanzees, dogs, dolphins, gorillas, magpies, orangutans, parrots, and various species of monkeys. However, the sixth stage, assessed with the invisible displacement task and careful controls, has only been passed by a select group. Monkeys typically fail it, whereas members of all the great ape genera have passed these tasks repeatedly.

You may want to try out the invisible displacement task on your pets. In fact, domestic dogs were one of the few species that had been reported

FIGURE 3.1.
Test apparatus for standard displacement task;
the chimpanzee Ockie selecting a box.

to have passed these tasks. However, be careful. When our research group had a closer look, we found that dogs had "cheated." They can find the target only under specific conditions. For example, they consistently perform well when the container (the displacement device) is put next to the hiding place, but fail when this cue is not available. In other words, they would only pass if you, after having put your hand in one of, say, four pockets to hide a treat, placed your empty hand next to the pocket where the food is. This suggests that the dogs simply learned this association and relied on this cue in their search. When we set up the experiment so that rules such as "search next to the displacement device" did not lead to success, dogs performed randomly. When we gave these carefully controlled versions of the task to chimpanzees and twenty-four-month-old children, on the other hand, they continued to pass consistently.[8]

Indeed, chimpanzees can find objects even if you invisibly displace them more than once. Imagine you put the treat in your hand and your closed hand then goes in and out of two of several available pockets before emerging empty. The target can thus logically be expected to be in either of the visited locations, but not in the pockets you did not put your hand in. In a formal test of this sort, subjects are given a second choice if they select one of the visited hiding places and it happens to be empty. When Emma Collier-Baker and I gave such a task to chimpanzees, they selected as their second choice the logical alternative hiding place rather than one of the other options, passing all of Piaget's object permanence tasks (this final one is called 6b).

To confirm that the task really measures what we intend it to measure, we also tested children with these same object permanence tasks, and, as expected, they solve it by age two. Emma also has tested orangutans and gorillas, and these data suggest that they too are able to pass all these tasks. Like human two-year-olds, but unlike other primates, great apes have demonstrated that they can think about things they did not perceive. These results are in line with several other problem-solving experiments that have been done by the comparative psychologist Josep Call and others in recent years. For instance, great apes can work out the location of hidden food from clues about where it is not.

---

8 This raises questions about other claims of competences in animals, such as marmosets and gibbons. They warrant further examination with careful control conditions.

EVEN WHEN WE HAVE ESTABLISHED that an animal can reliably solve certain problems, there can be disagreement about what that behavior entails in terms of mental capacities. Take research on mirror self-recognition. Adult humans commonly spend considerable time looking in mirrors and fussing with their appearance. Just think of the multibillion-dollar cosmetic industry. Although a lot of animal species adjust their posture and appearance to suit different situations, such as puffing up to increase their size to predators, it is not clear if they are actually aware of what they look like. When our pets see themselves in mirrors, be they cats, dogs, fish, lizards, or birds, they do not stop and use their reflections to groom themselves. Instead, even when they can learn to use a mirror to find or avoid things, they seem confused about their own reflection, often treating it as if it were another individual or ignoring it altogether. Cats sometimes check behind the mirror, just to be sure. Even monkeys do not seem to get it. Great apes, on the other hand, can show interest in their image and may use reflective surfaces to investigate parts of their body that are otherwise out of sight, such as facial skin aberrations or their nether regions.

Darwin briefly described the reactions of apes and children to their mirror image, but it was not until 1970, when Gordon Gallup developed an objective test for mirror self-recognition, that research moved from collections of anecdotes to systematic experimentation. Gallup anaesthetized chimpanzees and placed an odorless red mark on their face. Upon recovery he presented them with a mirror and found that the chimpanzees used it to inspect the unusual marks with their hands. He concluded that they must hence recognize their image. This experiment has been replicated many times since, though typically using surreptitious marking rather than anaesthetization. Orangutans and gorillas have also passed the task repeatedly.[9] Versions of this simple test have been widely used with a variety of other animals as well as with children.

This is a great little game to play with toddlers. You can wipe rouge on children's faces under the pretense of cleaning, or you can surreptitiously place a large sticker in their hair in the process of patting their head. Wait a little to ensure that the child has not noticed the mark directly and then present a mirror. Though they are typically interested in their

---

9 Initial research with gorillas failed to find successes, leading to speculation that they may be the only great ape to have lost an ancestral capacity. Subsequent studies, however, have found gorillas pass the test.

FIGURE 3.2.
The chimpanzee Cassie looking at his mirror
image.

reflections, you will find that toddlers younger than about fifteen months
do not touch the mark on their face. They will be surprised when you take
the sticker out of their hair even if they have studied the sticker's reflection
extensively in the mirror. Only from the middle of the second year onwards do toddlers investigate their own head upon seeing their reflection.
By twenty-four months close to all children retrieve the mark instantly.[10]
They know they are seeing themselves and may also start to talk about that
fact. A recent direct comparison of human and chimpanzee infants found
that they develop the capacity to pass Gallup's task at comparable ages.

If you try to do the same task with your pets, however, you will find
that they do not catch on. They do not use reflections to investigate
changes to their appearance; so, like many other species that have been
tested, they fail the mirror mark test. Monkeys such as baboons, capuchins,
and macaques all fail the task even after hours of exposure. Yet there have
been occasional headline-grabbing claims that animals other than the
great apes recognize their mirror image. In the first instance, the influential behaviorist B. F. Skinner (famous for his work on the effects of reward
and punishment he studied using the boxes that bear his name) and his
colleagues conditioned pigeons to peck on their body in response to mirrors. Pigeons displayed this self-pecking only after extensive reinforcement

10   There is some variation in terms of the onset between cultures, but even Bedouin children, said to have had no prior exposure to reflective surfaces, pass the task by the end of the
second year.

over hundreds of trials[11] and do not spontaneously use mirrors to explore their bodies.

Much media coverage followed the announcement that two dolphins demonstrated mirror self-recognition. Note, however, that because they lack hands with which to touch the mark, one cannot give dolphins the standard test. It's true that the dolphins in this study spent more time at the mirror when marked than when not visibly marked. Yet does this behavior demonstrate mirror-guided self-exploration? It would not be too surprising if dolphins were able to recognize themselves; after all, they have big brains and probably see their own reflection more often than any other mammals do, as they frequently jump out of the water. However, the current evidence for dolphin mirror self-recognition is not watertight and at minimum requires replication. Replication is also a problem for the final two claims. Two magpies and one elephant were recently reported to have passed the task. In another study, however, all elephants failed. So far we only have strong evidence, replicated in different laboratories, from experiments with members of the great ape genera. They can examine themselves in mirrors like humans do.[12]

In spite of its intuitive appeal and widespread use, the interpretation of this task has been controversial. After all, in some sense many animals recognize themselves. Most animals, except for some dogs, do not chase their own tails. Various animals adjust their appearance to camouflage themselves within the environment. Bobtail squid even generate light from their underside, effectively canceling out their own shadow. Many mammals, including dogs, use urine to mark their territory and must be able to distinguish their own smell from that of others. So why do most animals fail to pass the mirror test, and what does it mean that great apes and humans can pass it?

---

11 Extensive conditioning attempts have failed to make capuchin monkeys "pass" the task.

12 Even among the great apes tested, not all passed the task. In fact, only 43 percent of chimpanzees (42 of 97), 33 percent of gorillas (5 out of 15), and 50 percent of orangutans (3 of 6) did in published reports so far. What should we make of the great apes that have failed the task? It may mean that some do not get it. It is possible, for instance, that some of the tested animals were too young, just as humans under eighteen months tend to be too young to pass. But there are also many other reasons why an individual may have been classified as not passing the task, given that studies have employed different criteria (e.g. in terms of the required frequency and accuracy of reaching for the mark). Some subjects may simply not be motivated to retrieve the mark, may not have paid attention, and so forth. What is clear, though, is that mirror self-recognition is within the capacity of these species.

As usual, there are advocates of rich and of lean interpretations. On the richer side, Gallup argues that to pass the task one has to become the object of one's own attention: one has to be self-aware. The term "self-awareness" refers to much more than knowing what one looks like (even for people who are very concerned about their appearance). It implies awareness of where you come from and where you are going, what you are good and bad at, your personality, your likes and dislikes, your values, and so on and so forth. Can we infer this knowledge from the mirror test? Gallup suggests that we can and argues that passing the test indicates a capacity for inner reflection, thinking about self, and even awareness of one's own mortality. And in fact, children's passing of the task has been found to be associated with the emergence of self-conscious emotions, such as embarrassment, as well as with the use of personal pronouns. A killjoy skeptic, however, might wonder if any form of mental reflection, other than as a play on the word "reflection," is necessarily involved. Can you use a mirror to shave or clean yourself without engaging in deep inner reflection?

Then there are lean accounts. The comparative psychologist Celia Heyes is skeptical of rich interpretations and argues instead that all one needs to pass the task is an ability to distinguish feedback from other types of sensory input, and thus she questions if anything special is measured by the mirror self-recognition task. In her view, any animal that manages to avoid bumping into things, or that avoids biting itself in a fight, has demonstrated such ability. Heyes's perspective, however, fails to explain why only a few species can pass the task (or why young infants fail it) when in other contexts they evidently distinguish feedback from other input.

Between these very lean and very rich proposals lie interpretations that try to claim the middle ground. For instance, cognitive psychologist Ulric Neisser argued that only when infants notice that their facial appearance matters to other people do they become interested in learning about their own faces through mirrors. The mark test may therefore indicate a change in children's (and apes') attention to faces.

The developmental psychologist Josef Perner posits that the test measures a more general ability to mentally juggle two different ideas about the same thing. Just as in pretend play, where children simultaneously hold in mind, and can compare, two distinct notions of the same object (e.g., a banana as a fruit and as a telephone), here they need to consider and compare their reflection and their expectation of what they look like to discover that something is new—such as that novel mark.

How could empirical tests distinguish between these interpretations? My colleagues and I developed versions of the mark test with children that have provided some clues. First of all we surreptitiously marked our participants' legs, rather than their heads. The children sat in a highchair with a tray that prevented direct view of their legs. A mirror was placed in front such that they could see their own legs but not their upper body. Children performed exactly as they do in the classic task, reaching for the mark on their leg at the same age as they do for a mark on their face. This suggests that there is nothing specific to cognitions about faces, as Neisser suggested.

This set us up for the main manipulation. We sewed baggy tracksuit pants to the highchair and slipped a new group of participants in them without allowing them to see the pants directly, as the tray blocked their view. When children were then presented with the mirror and marked on the now tracksuited leg, most children failed. They did not seem to recognize that it was their own legs they saw in the mirror. However, when we allowed other children to directly see these baggy pants for just thirty seconds before affixing the tray to block their view, they performed very well—as well as children did in the first study or in the classic task. This suggests that young children have a mental expectation of what they look like and that this expectation is rapidly updatable. The thirty seconds of prior exposure was enough.

Thus it appears that the classic mirror mark test measures more than the lean accounts proposed,[13] though it need not necessarily demand the higher cognitive capacities that the rich accounts conjecture.[14] The results suggest that passing indicates subjects have a mental expectation of what they look like—they recognize themselves and examine violations of expectations, like that peculiar mark.

A moderate account, such as Perner's view, appears to be most strongly supported by the current evidence. His explanation proposes nothing spe-

---

13 The feedback in the last two conditions was the same, yet children retrieved the sticker in one but not the other condition. So there is more to passing the task than distinguishing feedback from other sensory input.

14 Passing the test may indicate self-awareness about one's appearance in a mirror, but there is little to suggest that it entails awareness about other aspects of self. Research demonstrates that even awareness about one's physical appearance depends on the context. Visual self-recognition in live video is more difficult for children than in mirrors. Self-recognition in delayed videos only emerges at around age three to four. In a recent study we found that in adults self-recognition in mirrors and photos involves different brain activity.

cific about the self but puts forth a more general capacity to mentally entertain and relate multiple models of the world, in this case the perception of one's image and the expectation of what one ought to look like. As noted, toddlers begin to recognize themselves in mirrors around the same time as they begin to pass stage 6 object permanence tasks and to pretend play. Though on the surface these skills are dissimilar, all three involve a capacity to consider more than what is directly available to the senses: an expectation of what one ought to look like, inference of the transfer of a hidden target object, and the imaginary pretend identities of objects or actions.[15] The comparative evidence suggests that great apes share this basic capacity to mentally go beyond the here and now, although the extent to which they can do this may well be limited in various ways.

AS THESE EXAMPLES ILLUSTRATE, SYSTEMATIC study can establish what animals can do. There will remain some uncertainty, and alternative explanations will be put forth, but with appropriate controls, replications, and careful comparisons we can become increasingly confident about the mental capacities of animals. Evidently, establishing what animals can do is difficult, but determining what they cannot do, what is uniquely human, can seem even more challenging. How can we be certain that a mental trait is not present in animals? When animal behaviors appear to display the trait in question, we face the difficulties associated with the aforementioned rich and lean interpretations. But even when there exist no obvious candidates for animal competence, we cannot simply conclude that the trait in question is uniquely human. *Absence of evidence* is not the same as *evidence of absence*. It remains possible, for instance, that we have not looked carefully enough.

When we say that no other animal can, for example, learn to play chess, we are making a claim known as a "universal negative." In principle, it should only take a single conclusive case to the contrary, such as one octopus that can give you a halfway decent game of chess, to reject the claim. To absolutely prove the truth of the claim, on the other hand, you would

---

15 The middle of the second year is a watershed in children's cognitive development. In fact, various other expressions of this ability appear around this time in other domains. In a review of the research literature Andrew Whiten and I found that great apes, unlike other animals, have provided at least some evidence for capacities in all of these domains.

in theory have to test every living animal to show that absolutely none of them can do it. Even then, you would have to presuppose that there exists a foolproof way of ascertaining that a particular animal does not have the capacity. After all, the octopus genius may, for instance, have been tired or otherwise not motivated when we tested it. Failure on a task is difficult to interpret in general. There are typically a number of reasons why the subject might not have performed well, aside from not having the capacity in question, and so negative results are seldom published. Should we give up and accept that we can never know for certain that some animal might not have the capacity after all?

The situation is not that dire. Though proving the absence of a characteristic is troublesome, we can, of course, draw conclusions after reasonable attempts. Once we have tested, say, thirty octopuses, and none of them can pass the task, it is reasonable to assume (for now) that other octopuses also cannot do it. In fact we agree on universal negative claims regularly. Until proven otherwise, we are happy to say the dodo is extinct. Until proven otherwise, we might be equally happy to accept that only humans can learn to play chess. Although we cannot directly prove the absence of mental characteristics, we can become increasingly confident about such claims when people have tried to prove the idea wrong but failed. The more people have explored the world without coming across a living dodo, the more likely it is that they are extinct.

To bolster claims of absence, then, we need to give animals the opportunity to falsify the claim and demonstrate competence. Consider the question of mirror self-recognition again. Given that humans and great apes show competence, small apes, the next closely related primates, are of special interest for our understanding of the origin of this trait. Three previous small-scale studies on gibbon self-recognition had produced equivocal results. There was no clear evidence that they could recognize themselves, but the researchers remained open to the possibility. So Emma Collier-Baker and I set out to test more individuals, as well as more diverse small apes than the other studies.

Over the course of a two-year project at zoos in Australia and the United States we tested seventeen gibbons (seven siamangs, three dwarf gibbons, and seven crested gibbons). Each animal was first exposed to mirrors for over five hours. Next we conducted motivation checks in which we offered each gibbon some cake icing of a color that we later used to mark their face. All of them eagerly consumed the icing. We then surrepti-

FIGURE 3.3.
White-cheeked crested gibbons inspecting a mirror
(photo Emma Collier-Baker).

tiously smeared some icing on one of their limbs. When the subjects later discovered the icing on their arm or leg, they all immediately examined and then consumed every last bit of it. Gibbons generally engage in less self-grooming than great apes, but this shows that they are clearly motivated to retrieve icing from their bodies. They should therefore also be keen to inspect a mark of the same color as the icing on their head, if they could discover it via the mirror. Yet none of our subjects passed the mark test. Instead, most of them looked or reached behind the mirror, in what looks to most human observers to be a search for that apparent other gibbon with a mark on its head.

Because failure may be due to a variety of reasons, we conducted a battery of subsequent tests to see if the gibbons would pass eventually. We repeated the test, this time putting actual icing on their heads at the considerable risk of producing false positives—that is, the apes might have smelled the icing on their head rather than inferred the location using the mirror. We jumped up and down behind them to emphasize the nature of the mirror, marked them with large stickers, and so forth—all without success. Perhaps most tellingly, when we smeared icing on the mirror surface itself, the apes would scrape or lick off every last bit of icing; yet they ignored the big blob of icing on their own head that was clearly

visible in the mirror. We thus concluded that we now had more than absence of evidence: these results amount to evidence of absence. Until there
are new data to suggest otherwise, it is appropriate to conclude that small
apes, like monkeys, do not recognize themselves in mirrors. Systematic
research can inform us not only about competences but also about the
limits of animal minds.

People are typically more excited about a finding that shows that an
animal has succeeded in doing something we did not think it could do
than about a report that notes that an animal could not do a task. It is also
generally more difficult to get negative findings published than positive
results (though it can be done). There are good reasons for this, given
that failures are often difficult to interpret, as they may be caused by any
number of factors. But we need to consider that a persistent flow of published positive findings and a lack of acknowledged negative results may
seriously skew our perception of animal capacities.

BY ESTABLISHING THE PRESENCE AND absence of mental traits in various
animals, we can create a better understanding of the evolution of mind.
The distribution of a trait across related species can shed light on when
and on what branch or branches of the family tree the trait is most likely
to have evolved.

In general, species may share similar traits for two distinct reasons:
convergent evolution (leading to analogous structures) and common descent (leading to homologous structures). The wings of birds and insects,
for example, both solve the problem of flight. A closer look at the structure of these types of wings, however, shows that they are quite different:
they are independent solutions to a similar adaptive problem. There is no
reason to believe that these two types of wings evolved from one original
body plan and that all the nonwinged animals between insects and birds
on the tree of life somehow lost that ancestral trait. Such convergent evolution has occurred many times. The wings of bats, for instance, are analogous to those of both insects and birds. Cases of convergent evolution
tell us about the selection pressures that may have driven the evolution of
certain characteristics. On the other hand, homologous traits, traits shared
because of common descent, tell us about their origins on the tree of life.
Bird species have similar wings because of common descent. Some birds,
such as penguins, have lost parts of the trait and cannot fly, and others,
such as kiwis, have barely any wings at all, but they are all descended

from a feathery winged ancestor. Biologists come to this conclusion by comparing the likelihood of potential explanations of the current distribution of the trait. Again, parsimony is important: the simplest explanation, the one that requires the least number of assumptions, is the most likely explanation. It is less parsimonious to propose that each bird species independently invented a feathery solution to flying than to assume they inherited the trait by common descent from a single ancestor that evolved this solution.

Given that chimpanzees, gorillas, and orangutans have repeatedly demonstrated mirror self-recognition, we can ask how such potential may have evolved on this part of the tree of life. If this trait evolved independently, it would entail that it emerged at least four times: once in the ancestors of each of the genera of great apes and humans. However, if today's great apes and humans share this capacity because of common descent, we would have to assume only one change event in the past: that over fourteen million years ago, before the line leading to orangutans split off, the common ancestor to great apes acquired the trait and passed it on to all its descendants. Homology, in this case, makes fewer assumptions and is therefore a more parsimonious explanation than convergent evolution.

Given that small apes and monkeys do not appear to have the capacity to recognize themselves in mirrors, we can narrow the emergence of this trait in primates down further. The great ape ancestor probably evolved the trait after the split from the line that led to modern gibbons some eighteen million years ago (see Figure 3.4). Based on current evidence,

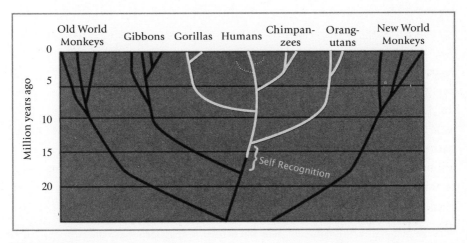

FIGURE 3.4.
The likely origin of the hominid capacity for visual self-recognition.

then, the potential for mirror self-recognition evolved between eighteen million and fourteen million years ago in the shared ancestor of hominids. We can make these assertions without ever having to lay eyes on a fossil of the ancestor that first evolved it. We do not know what this creature looked like, but it is likely to have known what it looked like.

Reasoning by homology is a powerful tool. It allows us to make inferences about the minds of long-extinct ancestral species.[16] Given what we have reviewed, the common great ape ancestor and its descendants probably could reason about things that were not directly available to the senses. As we establish the capacities and limits of more species, we can develop a clearer picture of how mind evolved on the various branches of the tree of life. This method, of course, only works for traits that are shared between species. To establish the evolution of traits that are uniquely human we need to examine the fossil record—I will defer discussion of that topic until Chapter 11.

We are now ready to examine what mental traits make up the gap. What are the essential mental capacities that set us apart and led to the explosion of new behaviors that characterize the human condition? When I ask my students this question, their most common answer is language. This is also the most prominent answer in the literature. So that is where we shall begin.

---

16 Note that it is possible that a trait inherited from a common ancestor is subsequently lost and that there are trait distributions where competing models make the same number of assumptions, so that parsimony cannot distinguish between them.

F O U R

Talking Apes

Thanks to words, we have been able to rise above
the brutes.[1]

—ALDOUS HUXLEY

UPON SEEING A LIVING CHIMPANZEE, the bishop of Polignac in Paris
in the early eighteenth century is reported to have proclaimed: "Speak and
I will baptize thee." The idea that language is distinctly human is common,
and versions of it can be found in the writings of many influential scientists, including Noam Chomsky, Michael Corballis, Terrance Deacon, and
Steven Pinker, to name a few. But is language really the essential quality
that elevates us to humanity (and, so qualifies us, at least according to the
bishop's view, for a soul worth saving)?

Language is a universal part of the human condition. Every human group has one or more languages with which it communicates,

---

1 Huxley goes on: "and thanks to words, we have often sunk to the level of the demons."

and language permeates most of what we think and do. You would be in deep trouble if you had lost your faculty for language; for starters, you would not be reading this. Nineteenth-century postmortem studies on the brains of people who had lost it established the left side of the cortex as crucial for language.[2] With dedicated areas for the production and comprehension of language, we seem to be hardwired for it—for connecting our brains so that we can exchange vast amounts of information between each other. This information flow is essential to human cooperation and culture, and it is not surprising, then, that many scholars argue that language is *the* defining characteristic of the human mind.

However, animals also have—sometimes quite sophisticated—means of communication. A bee signals the location and abundance of a food source to its hive, meerkats stand guard and alert their group about predators, birds dance with each other before selecting a mate, and in most mammals mother and offspring inform each other about their whereabouts. There are important reasons to have reliable transfer of information between members of a species, and animals have therefore evolved acoustic, haptic (touch-based), visual, and chemical ways of communicating. Are these then not languages? Are we merely biased because we have not fully deciphered what birds, monkeys, or whales are saying?

To find out what might be unique about the human communication system, we need to have a closer look at what characterizes human language. The first thing to note is that we speak not one but over 6,000 different languages. Furthermore, some people do not speak but instead use sign languages to express the same wealth of information with gestures. Others can use touch to read Braille with their fingertips. Our faculty of language transcends different modalities. To examine the essential characteristics of human language, we therefore need to look deeper than the ability to utter words—which, of course, parrots share with us.

THE MOST FUNDAMENTAL FEATURE OF language is that it allows us to exchange thought. In conversations we connect the private world of our

---

2 This is the case for virtually all right-handers and two-thirds of left-handers. In particular, Paul Broca discovered that people who had lost the capacity to produce speech had damage to the left frontal lobe, now known as Broca's area, and Carl Wernicke found that people who had difficulties understanding language had damage slightly further back on the left, in what came to be known as Wernicke's area.

minds to the minds of others as we share attitudes, beliefs, desires, knowledge, feelings, memories, and expectations. This book is designed to put my thinking about this subject into your head through the symbols of the written word. You have tens of thousands of words in your vocabulary. They are arbitrary conventions. The word "walking" has nothing about it that in itself refers to a type of locomotion. We could equally have called it *gehen*, as Germans do. Human languages differ in the symbols they employ for the same meaning; if you want to learn a foreign language, you need to study which symbols speakers of this language use to represent which concepts.

A symbol, be it a sound, a drawing, or a gesture, is a thing that is intended to stand for something else. Language works because individuals agree on the meaning of a set of symbols and how they should be used. Symbols are *about* something. That is the key to any representation: it is about something other than itself. Consider the following three paintings.

FIGURE 4.1.
A painting by the chimpanzee Ockie.

The first picture (Figure 4.1) is a painting made by the chimpanzee Ockie as I sat next to him and supplied him with materials. Ockie likes smearing paint almost as much as he likes eating it. I could easily have chosen from among the paintings my daughter, Nina, made when she was one year old to make the following point: these pictures are what they appear to be—pieces of paper with paint on them. They may be beautiful, but they are not *about* anything else. Ockie's painting may remind you of abstract art. Some of these pictures may, in fact, be indistinguishable from such art and are sometimes even sold as such. However, while the artist can tell you what the painting represents—even if it may sound rather absurd or far-fetched to you—as far as I know the chimpanzee paintings, like those of my daughter, represent nothing more than themselves.[3]

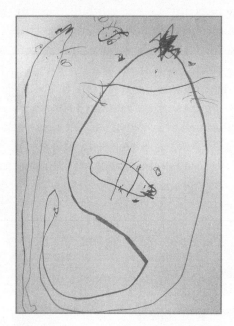

FIGURE 4.2.
A picture of four whales by my son, Timo (age thirty months).

The second picture (Figure 4.2) is one of the first representational drawings ever produced by my son, Timo, when he was two and a half years old. He explained to me that it was of a group of whales. You may be able to decipher where the Papa whale and the Mama whale are; I was told that there was also a Timo whale in front of Papa and a baby whale at the top. This, then, is representational; it is symbolic. Putting aside concerns about realism, in addition to the primary reality of the color on a piece of paper, the picture represents something else, namely a family of whales.

Language is built on such "representational insight." Very young infants are not bothered if you read them a picture book upside down. (Try it out if you have a child younger than about eighteen months.) They do not care, presumably because they do not interpret the pictures as we do. A simple way to test whether children understand its symbolic character is to give them the opportunity to learn from a picture about something else. For instance, one can point out the location of a hidden object in a picture of a room, and then observe whether children can use this cue to find the object in the actual room. For a long time results of such studies suggested that children develop this representational

---

3 Animals may be trained to paint in specific ways that resemble human art. I have heard that some elephants in Thailand, for instance, paint what looks like flowers for tourists. But these actions are shaped by the trainer, so this does not suggest that the animals intend the picture to represent something else.

insight surprisingly late, between age two and a half and three—about the time they start to draw symbolically. However, after some study I found even twenty-four-month-olds can, at least initially, use a picture to successfully guide their searches.[4] They can learn from what they see represented—so be mindful about what you let your toddler watch.

The third picture (Figure 4.3) is by Rory, the then four-year-old daughter of my colleague, developmental psychologist Virginia Slaughter. It is a drawing of herself drawing a picture of what looks to be herself. This picture within a picture demonstrates an important capacity: the ability to build models of models. In this case, the drawing shows that Rory could reflect on the relation between a symbol and what it stands for. In psychological jargon, this is known as forming a "meta-representation," and the ability to do so has many uses, as we will discover. It must have been essential in the evolution of any language that is comprised of arbitrary symbols. After all, to propose, understand, and agree that an arbitrary symbol has a

FIGURE 4.3.
A picture by Rory (age four) of Rory drawing a picture.

specific meaning, people must have had an ability to reflect on the relationship between symbols and referents in the first place.

So how did we initially come up with the words to symbolize concepts in our languages? In some cases, as in the ticktock of a clock, the meaning of a word is related to its sound. This is known as "onomatopoeia." Such words are like paintings, in that they somehow share salient characteristics

4 As it turned out, the children's problem in the previous studies was not with understanding the relationship between the picture and the room, but with persevering over repeated trials. They typically did well on their first trial, but on subsequent trials they had a tendency to return to a previously successful location. When I presented children not with the typical four trials in one room, but with one trial each in four different rooms, even twenty-four-month-olds could solve the task.

of the referent. They may hence be relatively easy to agree on,[5] and it has therefore been argued that spoken words originate from parallels between sound and meaning. However, why then are words for the same thing so different from each other in different languages? Most words do not sound anything like the thing they stand for. When it comes to abstract concepts such as "justice" or "evolution," it is not clear how they possibly could.

Agreeing on things we can all see may be relatively easy, since you can point to an object or action and make the proposed sound or other symbol. Even if the sound is nothing like the thing in question, one can imagine that repeated pointing and co-occurrence would establish the connection between symbol and referent in a group of people. But how did our ancestors manage to establish words for concepts that are not directly observable at all? Much of that, it turns out, has been accomplished by piggybacking on concepts that are observable. This is an important idea to *grasp*. We *milk* concrete contexts, such as what happens in a kitchen, as a source for a multitude of metaphors that allow us to talk about abstract concepts. Perhaps you need to let this idea *simmer*. Let it *ferment*. Do not *regurgitate half-baked* ideas. It is better to properly *digest* it. It is *food for thought*. You can *cook up* your own examples. But please do not *stew* over this for too long, even if you may have *whet your appetite* for more. These *samples* should be enough to get a *taste* of the nature of metaphors. They can be *delectable*. But they can also be less *palatable*, especially when they are *mixed*.

Metaphors allow us to use the concrete to express the abstract. In many cases we are so used to them that we may not even think of them as analogies. For instance, in English we often use words for spatial relations to refer to abstract relations in time. Here are a few examples starting with the first letter of the alphabet: *about, across, against, along, among, around*, and *at*. There are also several starting with *b*. Our spatial understanding *scaffolds* our communication about time.

We may also tell extended metaphors, such as allegories or parables, to make a point about a subject that is not directly observable. These may be individually concocted stories, but useful ones are repeated and become part of the cultural-linguistic heritage. Our metaphorical heritage includes many idioms—phrases that have particular nonliteral meanings. The

---

5   Using a task similar to the game Pictionary, researchers are beginning to examine how graphic pictures can become conventionalized into abstract symbols.

English language has thousands of idioms. We might say that someone is "like a bull in a china shop" to offer a concrete sense of what destruction clumsiness can bring in a fragile world, even if the context to which it is applied refers to destruction that is entirely social or emotional. To recognize the analogy, one must be able to map the relevant dimensions of the concrete situation (bull moving about, pushing over precious china) to the situation at hand (even if it does not involve large animals or porcelain). This ability to invent and understand novel metaphors again illustrates the importance of reflecting on representational relations.

Because speakers have to establish agreement on meaning, languages are regional. Languages were not established by expert committees or by decree, but gradually evolved out of people's interactions with each other and their desire to communicate. Separation, physical or social, breeds new dialects. Yet as long as there is interaction, language boundaries are more fluid than names for languages or national borders might suggest. For example, I grew up in Germany a stone's throw away from the Dutch border. My parents' version of German, their local dialect rather than what they were taught at school, is very similar to the local dialect on the Dutch side. The Germans might not understand high Dutch, and the Dutch may not understand high German,[6] but the farmers on either side of the border use pretty much the same language. They are neighbors, after all. Their language is part of the West Germanic dialect continuum.

The language of the Germanic tribes that settled the British isles more than 1,500 years ago has been separate from mainland German for long enough, and exposed to sufficient Celtic, Norman, and other influences, to be mutually unintelligible. Yet a good proportion of English words are still identical to German. I hear myself mutter the German/English words "arm," "hand," "finger," and "ring," as I look down in search of examples. Even many of the idioms used in both languages are similar, though at times strangely different. For instance, in German it is an elephant, not a

---

6 Though they usually do understand high German too. Being multilingual is another solution that allows people to interact across cultural divides. Traders who interact with many different peoples, such as the Dutch traditionally did, benefit from being able to speak a variety of languages. It is also not uncommon for people whose language has relatively few speakers, such as tribal groups of Australians and New Guineans, to speak numerous other languages spoken by surrounding tribes. Incidentally, people who speak more than one language, it has recently been documented, are less likely to suffer from Alzheimer's dementia in old age—so if you are worried about that, go on, learn another language.

bull, that is messing about in the china store. Curiously, the dialect of my parents sits in many ways between modern English and modern German (e.g., *Schwester* is German for "sister," but my parents would say *süster*)— so when they visited me in New Zealand, despite not speaking any English, they at times managed to make themselves understood speaking their low Saxon dialect (Plattdeutsch). This is because some of the tribes that invaded Britain, and so founded the English language, came from Old Saxony. Languages are the product of history, of past social interactions between individuals who needed to communicate. We inherit them socially from our forebears.

Though dictionaries and linguistic pedants may make it appear as if languages are fixed, they are alive and constantly changing. Old words, phrases, and pronunciations disappear, and new ones are added. Much has happened since English and German parted ways, and speakers of these languages have established new conventions. New words are introduced, become useful or fashionable, are abbreviated or merge with other words, and then are superseded by yet newer expressions. We can propose new words, drawing for instance on analogous relations, contrast them to similar symbols, and negotiate their precise definition. We could, for instance, agree that a new word, "gappist," shall from now on refer to someone who exaggerates the mental gap between animals and humans, whereas a "gapanier" is someone who understates the gap between animal and human minds. Naturally, this only works if the words catch on and people use them.

WE CAN UTTER AND UNDERSTAND novel expressions. While English is fairly restricted in admitting new words, other languages allow speakers to create new words on the fly. There cannot be a dictionary that contains all German words, for example, because separate elements seamlessly merge into larger words. I once saw a street sign with the word *Astabbruchgefahr*— which literally means "Branchbreakingofdanger." I had never seen this word before but understood enough not to linger next to the sign.[7] Instead

---

7   Some languages can make compounds virtually without limits. Consider this extraordinarily long Maori place name: "Taumatawhakatangihangakoauauotamateaturipukakapik-imaungahoronukupokaiwhenuakitanatahu"—which translates to "The hilltop where Tamatea with big knees, conqueror of mountains, eater of land, traveller over land and sea, played his flute to his beloved."

of combining syllables into words, most of the creative work in the English language is done by combining and recombining words into novel sentences. You can understand almost any coherent sentence, such as this one, even if you have never heard or read it in your entire life. And you can make up new sentences as you please. Human languages are *generative* in this way. Imagine a world in which we were limited to a fixed set of, say, ten words or sentences. Such a scenario might make a plot for a quirky movie, but it would quickly become constraining.

Fortunately, language is open-ended, even though the building blocks of language are limited. How can this be? Languages are based on finite sets of arbitrary units, symbols such as sounds and words. Grammar rules govern how these units are combined and recombined to generate innumerable expressions. A phoneme is the smallest unit of speech that can change meaning, and the English language has some forty-four of them. The difference between the word "car" and "bar" is one phoneme. Across all human languages there are only about 150 different sounds in use. Some languages, such as the click languages of the bushmen of the Kalahari, have over a hundred phonemes. Others, like the Maori in New Zealand, have little more than a dozen. In any language, phonemes can be combined through a branch of grammar rules, known as phonology, to express any meaning. Languages with fewer phonemes need to employ more repetitions to create new words (in Maori there are words like "whakawhanaungatanga"), but there is no limit to the meaning that can be expressed in any language.

The smallest units of meaning are called morphemes, and they are the building blocks of words. They include stems (e.g., joy, man), prefixes (e.g., after-, anti-), and suffixes (e.g., -able, -ful), which are combined into words (e.g., joyful). There are also functional morphemes, known as inflections, that have a grammatical function but little meaning in themselves. For instance, ending a noun with an *s* in English indicates plural, and ending a verb with an *ed* indicates past tense. English has few inflections, but other languages use them a lot. The grammar rules that govern all this are called morphology.

Finally, syntax rules govern how we combine these words into phrases and sentences. You may remember some of them from school (or you may remember that you have forgotten them). Even if you are not able to explain them, you still know when they violated being are—unless the order of the last three words seems correct to you. We use these rules to produce and decode novel sequences generated from the same limited set of units. For instance, instead of having millions of words for every possible concept, we have thousands of words that can be combined. Instead of having single words for "big table" and for "small table," we have words for concepts like "table" and attributes like "big" and "small" that we can then also use in conjunction with other words. Perhaps the best illustration of the generative power of human language is the quest for the longest ever sentence. Whatever monumental sentence you might produce, one can always add a relative clause and extend it further. You could add at the start: "You think the longest sentence is. . . ." Or you may go further and add: "I am not convinced by it, but you think the longest sentence is. . . ." Which may draw the retort: "But I insist it is true that, although you are not convinced by it, the longest sentence really is. . . ." There is no end to the possibilities; the generative nature of language allows us to continually expand.

The same is true of the search for the highest number. I remember as a child arguing about the concept of infinity with a playmate who, not easily swayed, plainly countered that he knew a higher number: infinity plus one. Though technically incorrect, the reply nicely illustrates the mechanism that gets us to produce ever-higher numbers. The trick is another variant of nested thinking called "recursion." In mathematics, a formula is said to be recursive when the next term in a sequence is based on a preceding term. So in counting, our ten Arabic numerals are combined and recombined on the basis of a simple set of recursive rules that allow us to continue building larger terms (0, 1, 2 . . . 9, 10, 11, 12 . . . 99, 100, 101 . . .). There is no natural end to counting. Nonetheless, we can reason about infinity (typically given this symbol: ∞). You can create endless other sequences with different kinds of recursive rules. The following series (1, 1, 2, 3, 5, 8, 13, 21 . . .), for instance, is determined by a recursive rule that states that each new number is the sum of the previous two [$F_n = F_{n-1} + F_{n-2}$].[8] Recursion is a procedure in which output and input are linked, creating open-ended loops. It enables us to generate novel combinations from finite resources.

Recursion is considered a key property of grammar in language. A relative clause can be defined as a relative clause plus an optional further

---

8   To get this going, one first needs to assume $F_1 = F_2 = 1$.

relative clause. Therefore, relative clauses can be strung together, or be embedded, such as this one, virtually indefinitely (though practically speaking, there are limits to what one can follow). Grammar rules allow us to point back to previous parts, and these can be merged into bigger structures. For instance, in the sentence "The monkey I watched fighting by the lake tried to steal my purse," we can relate the stealing back to the monkey after having conveying other information such as the primate's fight. Phrases and sentences can be embedded into larger narratives. Language is thus in principle open-ended, and we can construct communications of whatever complexity required. According to the most influential psycholinguist of the previous century, Noam Chomsky, this generative grammar is a human universal that underlies all languages.[9] Recursion, he and his colleagues argue, defines the language faculty in its narrowest sense.

Chomsky's original ideas contributed enormously to the start of the so-called cognitive revolution in psychology and the decline of radical behaviorism. Rather than being the result of general associative learning rules, as contended by behaviorists such as B. F. Skinner,[10] Chomsky argued that humans are innately predisposed to develop language. Numerous lines of evidence support this claim. Children acquire language rules effortlessly and without explicit instruction. They are not predisposed to learn a particular language—a Japanese infant brought up in an Italian household will become fluent in Italian and vice versa—but they are able to distill the rules that govern their linguistic environment. They can then apply these rules in entirely new contexts. For instance, my son, Timo, when two and a half years of age, spoke confidently of one shoe and two

---

9 Languages, Chomsky argues, differ in terms of their surface expressions rather than their deep structure. For instance, the simple sentence "The computer copied over my files" can be considered as a noun phrase ("The computer") and a verb phrase ("copied over my files"). The latter comprises a verb ("copied") and a prepositional phrase ("over") and another noun phrase ("my files"). While English tends to structure these elements such that subjects precede verbs, which in turn precede objects, Japanese has a subject-object-verb surface structure.

10 Skinner had been enticed to provide a behaviorist account of human language following an encounter with the philosopher Alfred North Whitehead. In 1934 Skinner found himself seated next to Whitehead at a dinner and boasted about the power of his learning theory. Whitehead challenged: "Let me see you account for my behavior as I sit here saying, 'No black scorpion is falling upon this table.'" It took Skinner twenty-three years before he published a book attempting to explain language in terms of associative learning. In an appendix he responded to the challenge, surprisingly citing Freud, by suggesting that Whitehead said what he did because he was afraid behaviorism, symbolized by the black scorpion, would take over. Yet something else was about to take over. In the same year, Noam Chomsky published his entirely different, and ultimately far more influential, theory of the nature of language.

shoes, and then equally confidently about one foot and two foots. Even though I am reasonably confident he never heard anyone say "foots" before, he generalized the rule that typically is employed when generating plural in English. Exceptions to rules, as we all have to find out the hard way, must be learned individually.

Most children acquire language in a similar way regardless of differences in intelligence, schooling, and culture. You probably started pronouncing, that is babbling, the phonemes of your environment by about eight months. Though young infants can distinguish phonetic contrasts of any language, they quickly zoom in on the sounds that make up their language. By the end of the first year, you produced your first words. My son Timo's first word was "cheers," and he uttered it with great enthusiasm as he demanded a clinking of cups. By the end of the second year, you accelerated your word acquisition dramatically, learning about one word every two hours. At around the same time, you began to string the first two-word expressions together. Recursive syntax, then, develops over the next two years. Of course, every child requires linguistic input—an environment of people who use a language—as the tragic case of the girl Genie illustrates. Having been neglected and not spoken to for most of her childhood, she failed to acquire normal fluent language. There thus appears to be a critical period of language acquisition during which children learn the language of their group.

Such a period becomes evident when we try to learn a second language. A young child can quite comfortably learn two or three languages. My brother spoke English to his children, and his wife spoke German, while everyone else was speaking Dutch, since they were living in Holland. Both children acquired all three languages effortlessly, until they moved away from the Netherlands and dropped back to two languages. Bizarrely, in most countries a second language becomes part of the school curriculum in fifth grade or higher—approximately the age at which effortless language acquisition stops and the hard work begins. Learning a second language after puberty usually results in an accent that is virtually impossible to overcome. Alas, I will retain my German accent no matter how hard I try and how long I live in English-speaking countries. If language was acquired simply through general associative learning, one would not expect to find such a critical period. Furthermore, rules would not be overgeneralized, and there would not be universals in grammar or developmental stages. If Skinner had been right, I should be able to lose that German accent, and we should be able to learn our languages like everything else. Indeed, we should be able to teach versions of our language to other animals who can learn associatively. But Chomsky argued that humans' language instinct was unique in the animal kingdom. He suggests that a mutation, perhaps a mere one hundred thousand years ago, resulted in the great leap forward, giving our ancestors the precious gift of open-ended language.

Although the Chomskian views on language reigned supreme for half a century, the last few years have increasingly seen new challenges to his linguistic gospel. For example, some researchers argue there is very little that can be said to be universally true across all human languages. The languages spoken on Earth today differ widely from each other. Some put verbs at the beginning, others at the end; some are built on short words, while others can create long composite words. There are languages that appear not to have basic forms such as prepositions, adjectives, articles, and adverbs. Even recursive syntax, according to Chomsky the core of language narrowly defined, is perhaps not present in all human languages. The Piraha of the Amazon and the Bininj Gun-Wok of Arnhem land in Australia are said to lack it. Thus a simple sentence like "They stood watching us fight" can only be expressed successively as "They stood; they were watching us; we were fighting." These reports require further systematic examination, but they are threatening received wisdom. Linguists are increasingly questioning whether there are any grammatical constructions or markers that are truly universal.

Part of the problem may be that in the past, language was studied primarily by examining the major written Indo-European languages. Yet most of the world's languages do not have a writing system. Australia, New Guinea, and Melanesia comprise well over a thousand oral languages. In Vanuatu alone there are over one hundred different languages with an average of two thousand speakers each. Many of these languages are dying out, but they probably provide us with a much better sense of the nature of human language and how it emerged than a small selection of European written languages could.

Any claim about universals should depend on a careful examination of the breadth of human languages. Recent studies on the diversification of languages have begun to apply computational models from evolutionary biology. For example, one study compared word order in hundreds of languages and concluded that the current existing rules of a language are shaped by its cultural history rather than any innate universal grammar.

People gradually developed grammatical rules that suited their needs, and over time the rules change, are modified, or replaced. As is the case with words and their meaning, grammar rules are the product of a history of social interaction. Different descendant groups create diversity by developing in their own idiosyncratic ways, especially if isolated. Much of this cultural evolution seems to follow the same logic of descent with modification as natural selection in biological evolution. Indeed there is much debate about the relationship between the two (see Chapter 8).

Even though the current critiques suggest Chomsky is wrong and humans do not have an innate universal grammar, this is not to say that we are not biologically prepared for language in ways that other animals are not. Only a mind capable of nested thinking, of meta-representation and recursion, should be able to establish arbitrary symbol meanings and grammar rules that enable efficient combining and recombining of these finite units into open-ended sentences. It requires a mind that wants to understand and be understood.

> Language is the source of misunderstandings.
>
> —ANTOINE DE SAINT-EXUPÉRY

LANGUAGE IS ABOUT COOPERATION. IN conversations we exchange information by taking turns being speaker and listener. To have an effective conversation, one needs to keep track of what is known, desired, and believed by the communication partner. There is little point just repeating what the other already knows—though that does not stop everyone. One has to quickly compute what is being said, given the current context, and how one might respond or add to it. Conversations are pragmatic encounters and typically follow some fundamental rules.

The philosopher Paul Grice identified four maxims we tend to adhere to in our conversations. The first is that we should say what we believe is true. If we all lied all the time, there would be no point in conversing with anyone else. That is not to say that deception and self-deception do not abound (more on that later). The second states that we should provide the appropriate level of information as required by the situation. When asked about the temperature, we usually are not expected to give the degrees to five decimals. The third maxim states that your contribution should be relevant to the goals of the conversation. Digressions to other topics should be avoided, which reminds me of a conversation I had last week where . . . Well, you get the point. The final maxim states that your contributions should be clear and avoid obfuscations. We should customize our talk to what the audience knows and avoid unnecessary jargon. I have thus used phrases such as "distinctly human traits" instead of using the technical expression "human autapomorphies."[11] Words should be chosen that are likely to be understood by one's audience. We have all encountered awkward conversations in which something was not quite right because one or the other of these maxims were violated. (Try counting the violations the next time you hear a politician being interviewed. It might make it a lot more interesting.) Still, we largely adhere to these maxims. To do this, we must take many things into account; especially important is the mind of your conversation partner.

Minds can be considered representational systems themselves—no obfuscation intended. Consider the book or screen you are looking at. Light hits your retina and triggers nerve cells to fire. This activation is passed on to the back of your brain, where various parallel processes establish the composition of the scene in terms of color, orientation, and so forth. These are then integrated into your visual experience of the writing in front of you. You are forming a mental representation that you can still access, to some extent, if you stop the input, for instance, by closing your eyes. We represent visuals but also sounds, concepts, and beliefs. People differ in their representations of the world, and we must take that into account in our conversations.

You may, for example, believe that a banana is on the counter in the kitchen. I may know that you represent the world in this way but may also know that you are mistaken (because I have eaten it) and may hence volunteer new information. This, again, requires nested thinking: I (meta) represented your (mis) representation and adjusted my communication accordingly. If someone else knows that I believe that you think your banana is located in the kitchen, yet another layer of complexity is added. This embedding can go on and on, but I will postpone further discussion of mind reading to Chapter 6. Suffice to say that human conversation

---

11 I did this even though I actually like the word "autapomorphy." The term refers to a derived trait that is unique to one member of a clade, and not found in any others, not even those most closely related or a common ancestor.

involves a lot of reasoning about what the other knows, desires, and believes to function as the efficient cooperative information exchange system
that it is.

The content of many of our conversations involves reflections on past
events and potential future events. Human language is exquisitely capable
of representing meaning that goes beyond the here and now. As we will see
in the next chapter, imagining future events can involve the construction of
novel scenarios by combining and recombining basic elements (not unlike
the combining of words into new sentences). For this and other reasons,
Michael Corballis and I have argued that language and our capacity to travel
mentally through time evolved hand in hand—although the emergence of
content likely preceded the means of communicating that content.

> There is between the whole animal kingdom on the
> one side, and man, even in his lowest state, on the
> other, a barrier which no animal has ever crossed,
> and that barrier is—Language.
>
> —FRIEDRICH MAX MÜLLER

IN 1873, TWO YEARS AFTER the publication of Darwin's *The Descent
of Man*, Friedrich Max Müller, chair of philology at Oxford, posed a
counterargument that no other animal had anything remotely like human language and hence there was no sign of gradual evolution, as Darwin's theory seemed to predict. He raised this issue in defiance of the 1866
ban on discussions of the evolution of language by the Linguistic Society
of Paris. In fact, Müller's argument was perceived to be a serious threat
to Darwin's theory of evolution by natural selection. Recall that in the
absence of genetics and a detailed fossil record, the debate centered on
evidence of continuity between living species. Thus Müller's claim about
the language barrier not only was relevant to humans' purported unique
position but turned into an early battleground about the very theory of
evolution. At the time little was known about primate communication,
and Darwin himself wrote: "I wish someone would keep a lot of the most
noisy monkeys, half free and study the means of communication."

Enter Richard Garner—a young man from Virginia who in the 1890s,
with the help of Edison's newly invented cylinder phonograph, went out

to decipher the vocalizations of primates through playback experiments. The idea was to record primate vocalizations in various circumstances and then play them back to other individuals to study their responses. Garner conducted his initial work in zoos and, to wide acclaim, reported early success in identifying the vocabulary of different primate species. He claimed, for instance, to have identified capuchin monkeys' "words" for things ranging from "food" to "sickness." He believed that the primate tongues he discovered were limited to names for concrete things but that they were the building blocks from which human abstract notions evolved. Not surprisingly, these conclusions attracted a lot of attention from both the public and the academic world.

When Garner boldly proposed to take a phonograph to Central Africa, sit in an electrified cage, and study simian tongues in the wild, there was hope the language evolution debate could be settled once and for all. Alas, the expedition did not turn out as planned. Garner's quest to find convincing evidence in the African jungle was a failure even before it had begun. In spite of links with Edison, he did not manage to obtain a phonograph for the mission. When he returned, he faced an avalanche of accusations in the popular press about inconsistencies in his reports about what had occurred on the expedition. Rumors spread that he did not spend months in the deepest jungle but instead lived in or near the comforts of a mission; the stories led to widespread suspicion and ridicule. In any case, Garner did not return with new evidence but with a chimpanzee that could not speak or be deciphered—and that quickly perished. There was no support for his claims about chimpanzee and gorilla language. Instead of making a mark in the annals of evolutionary theory, Garner's dream to learn the languages of animals became firmly entrenched in the realm of fiction, most notably in the character of Hugh Lofting's Dr. Dolittle. Serious scientific attempts to discover simian languages, or to teach them ours, were put on hold.

In a fascinating account of Garner's story, the science historian Gregory Radick suggests that one reason for this change in zeitgeist might have been the rise of ideas about evolution proceeding in leaps, as Stephen Jay Gould maintained, rather than in small, gradual steps. This reduces the need to find precursors of human language in apes. As I clarified earlier, Darwin's theory does not have a real problem with the existence of radical discontinuities in the current record. Creatures that may have had precursors of human language, such as *Homo erectus*, are now extinct (leaving an apparent discontinuity between extant species). Furthermore, precursors of language need not even have been in the domain of vocalizations. It is possible that language first evolved in gestural form. Indeed, this idea of a gestural origin of human language, moving subsequently from hand to mouth as it were, is increasingly gathering momentum. The search for language precursors in primate vocalization may hence be a case of barking up the wrong tree.

Garner's abandoned playback approach had its revival in the 1980s, when ethologists Dorothy Cheney and Robert Seyfarth conducted seminal studies on vervet monkey alarm calls at the foot of Mount Kilimanjaro. By recording alarm calls and then playing them back to the unsuspecting group at a later stage, these researchers established, for the first time, that nonhuman animal vocalizations may have meanings not entirely unlike human words, even if there exist far fewer of them. The animals make different alarm calls when they see a snake, an eagle, a leopard, or a human. When played back the monkeys tend to react differentially and appropriately to such calls. That is, they hide under a tree if the call is the eagle alarm, but they run up the tree if it is the leopard alarm.

These calls are gradually learned but limited to alarms only. There is no evidence for vervets stringing them together, let alone generating open-ended sentences. They do not show any evidence for recursion in their communication. A monkey may on occasion falsely utter the alarm call for "leopard," making the rest of the troop run up the tree while he stays behind and eats the food the others discarded. This seems like a pretty clever form of tactical deception, but it also illustrates the lack of reasoning about what the others know. The monkeys up in the tree do not seem troubled by the fact the individual that cried wolf did not flee himself and instead took their food. They do not seem to reflect on (that is, meta-represent) the discrepancy between what the individual's alarm represents and what his lack of running away represents. There is no evidence of reflective embedding in other monkey communication either. Although perhaps a building block on which human language was constructed, the calls of monkeys are limited in terms of flexibility, meaning, and use.

Indeed most animal vocalizations seem to be under emotional rather than cognitive control. When researchers stimulate a subcortical part of the brain known as the periaqueductal gray, it causes meowing and growling in cats, shrieking and barking in rhesus monkeys, echolocating in bats, and laughing in chimpanzees and humans. Destruction of this area causes muteness. The area is indispensable for animal vocalization and nonverbal human vocalization. As we have seen, human speech, on the other hand, is primarily driven by cortical areas of the left hemisphere that allow for voluntary control and extreme flexibility. Animal vocalization may hence not be closely related to human speech.[12]

Nonetheless, some animal communication systems are quite sophisticated. Bee dance, for instance, communicates the size of a food source, the distance, and the horizontal direction. However, close examinations of communication systems of animals have found them to be restricted to a few types of information exchanges, typically to do with reproduction, territory, food, and alarm. There appears to be little content transferred beyond these realms. There is as yet no sign that animal communication features the open-ended flexibility that typifies human language.

What about whales? you might ask. Humpback whales sing in the most curious manner, they have large brains, and there is even evidence that they learn the songs from each other. Are they talking about us behind our backs? Disappointingly, the answer is "probably not." The possible information content of a humpback whale song is estimated to be low indeed—just enough to say, "Hello, baby. Check me out." Researchers now think it likely the songs are serving a simple mate-attraction function.

Thus far, all attempts at deciphering animal communication systems point to narrow repertoires, devoid of the recursive characteristics that make human language so flexible. Yet our knowledge of animal communication systems is still quite limited. Perhaps there is sophisticated communication going on that we have not figured out. For instance, only recently has work on the alarm calls of prairie dogs suggested that they may signify more details than do vervet monkeys. Cephalopods, such as squids, octopus, and cuttlefish, change their color and patterns not only for camouflage but perhaps also to communicate surreptitiously. There is evidence to suggest that they use reflections of polarized light from their

---

12 Speech involves complicated deliberate motor acts. Great apes have much more voluntary control of their hands, which they use, for example, to extract foods, than over their articulations. If our common ancestor had such control, it would appear easier for natural selection to tinker with such control than for it to invent vocal control de novo. Gestural communication also has the advantage of being more iconic, where words are almost entirely arbitrary. Indeed, some things are much more intuitively communicated by gestures: try to explain to someone what a spiral is. So perhaps language evolved first in the gestural domain, as Michael Corballis argues, and only later was complemented and largely displaced by articulations.

skin to signal others without predators being able to perceive it. Given that more might be going on in animal communication than meets the eye, it is prudent to remain cautious. Recall that absence of evidence is not equivalent to evidence of absence.

Another approach is to tackle the problem from the other end. Can we teach human language to animals? Linguists are increasingly questioning whether humans are innately wired for a universal grammar; they highlight cultural learning instead. Perhaps, then, animals can acquire human language in the right cultural circumstances. From fairytales to serious literature, our folklore is replete with stories in which such attempts succeeded. Is there any truth to these imaginings?

> I cried out a short and good "Hello!" breaking
> out into human sounds. And with this cry I sprang
> into the community of human beings, and I felt its
> echo—"Just listen. He's talking!"—like a kiss on
> my entire sweat-soaked body.
>
> —FRANZ KAFKA

IN FRANZ KAFKA'S FAMOUS SHORT story "A Report to an Academy," a chimpanzee explains in eloquent prose how he learned the human ways in captivity. In the real world nothing like this has ever happened. In spite of attempts by Garner and others, no ape has learned to speak. Great apes do not seem to have the voluntary fine-motor control of the face and voice that we use in speech. Their vocal apparatus does not enable articulation of the required vowels.

Parrots, however, can mimic human words. Indeed, the perhaps most famous speaking parrot, the African grey parrot Alex, was said to have been able to produce some 150 English words. His trainer, the comparative psychologist Irene Pepperberg, worked with him for thirty years and reported that Alex could name some fifty objects, count to six, and form contrasting concepts such as over/under, bigger/smaller, and same/different. Although there was no evidence of recursive grammar, some of his communications included turn taking and appropriate use of phrases such as "I am sorry."

Other species have been examined with mixed success. The comparative psychologist Louis Herman trained bottle-nosed dolphins to follow complex gestural commands. The dolphins evidently understood a range of instructions and even responded appropriately to changes in symbol order. Seals have also been trained similarly. A border collie, Rico, was shown to understand verbal instructions of its owner beyond what was previously imagined. From ten months onwards, the owner regularly placed three items around the house and asked for Rico to fetch them. At age ten Rico was able to discriminate about two hundred object words. Rico would even learn new names for objects on the basis of exclusion. That is, when asked to fetch something with a new name, the dog would select the only object which name it did not know. This ability is called fast mapping and is sometimes supposed to be a key to children's rapid word acquisition. Impressive as these cases are, none of these efforts include any symbol production on the animals' part. There is hence as yet no conversation with these animals. Great apes, on the other hand, have demonstrated capacities in both comprehension and production.

Although vocal language is impossible for great apes, researchers have had some success in teaching them sign languages. Great apes have been taught to produce and understand several hundred signs. Famous examples include the chimpanzee Washoe, the gorilla Koko, and the orangutan Chantek. An alternative approach is to teach the apes arbitrary visual symbols mounted on a board that can be touched to communicate. Again, up to a few hundred such symbols have been learned, for example, by the chimpanzee Sarah and the bonobo Kanzi.

The initial enthusiasm that great apes could learn language was challenged by the psychologist Herbert Terrace. His work with a chimpanzee named Nim Chimpsky—a pun on Noam Chomsky—suggested that the apes simply and slowly learn to use the symbols through associative learning rather than through a real understanding of the representational function of words.[13] He concluded that Nim had nothing like human language and cast doubt on previous studies. A heated dispute ensued between rich and lean interpretations of the ape language projects.

One new line of evidence on this debate comes from research on great apes' understanding of pictures and models. Even if they cannot draw

---

13   The revealing 2011 documentary *Project Nim* tells the sad story of Nim and his fate following Terrace's research effort.

symbolically themselves, one can examine whether they have a capacity for representational insight. As we have seen above, in developmental research this faculty has been investigated by asking a child to find an object in a room after being shown the hiding place in a picture of that room. There is now evidence that chimpanzees can pass such tasks. The psychologists Valerie Kuhlmeier and Sally Boysen showed that their chimpanzees were able to find a hidden object in a room after being shown on a photograph or scale model of the room where the object was going to be hidden. The apes appear to be able to interpret these images and models as a source of information about the real thing.

Such results may offer some consolation to the advocates of rich interpretations of the "ape language" projects, as these studies are sometimes known. It is generally accepted now that great apes have demonstrated some significant understanding of the symbols they use but also have limits.[14] They can pick up symbols without explicit training, name objects appropriately, and comprehend novel requests. Alas, the apes themselves have contributed somewhat less to the debate than one might have hoped. They have not told us much about their worldview. Their talk is nothing like the reflections Kafka imagined. Rather, the apes tend to use a single word, or a string of two or more words, to deal with their present situation: "give apple," "tickle chase," and the like.

Ape signs are primarily requests, and Terrace has argued that apes only utter imperatives and no declaratives. However, a recent analysis of decades of data (some hundred thousand utterances) of three language-trained apes found that 5.4 percent of signs may be classified as comments or statements (i.e., declaratives). The same analysis found a mere eleven incidences in which the apes named something to show, offer, or give it, and none in which they simply tried to get attention. Though the apes make some reference to actions that have just happened or are about to happen, they have not acquired tense or symbols to refer to temporal displacement—we cannot enjoy a mutual trip down memory lane or debate the remote future with them. However much we may want them to, they

---

14 Chimpanzees tend to have severe problems with some basic elements of symbol-referent links. For example, after intensive training they may learn that the color red is to be matched to a symbol for red. However, as soon as the relation is reversed, they struggle. That is, they find it difficult to select between red and blue colors when presented with the symbol for red. The bidirectional correspondence seems to look obvious only to us.

do not ask us the big questions about where they come from, what they are, and where they are going.

This is not for want of trying. At least a few accounts of curious and elaborate responses to philosophical questions have been reported. The gorilla Koko, for example, is said to have responded as follows to five instances of being asked, "Who are you?"

1. Me gorilla nipples tickle.
2. Polite-Koko Koko nut nut polite.
3. Koko polite me thirsty.
4. Polite me thirsty feel Koko love.
5. Koko polite sorry good frown.

These responses all share at least one reference to the animal's proper name, Koko, or even to "me" and "gorilla." This is interesting, given great apes' capacity to recognize themselves in mirrors. Yet this example also illustrates the utter lack of syntax that is emphasized by scholars to support a lean interpretation. There is no structure or embedding of phrases in their production—no evidence for the open-ended generativity that is so typical of human language. In spite of the trainers using syntax, none of the animals seems to have picked this up.[15] So perhaps Chomsky and colleagues are right in arguing that recursive syntax is distinctly and uniquely human.

There have been challenges to this conclusion. The bonobo Kanzi can generate two- and three-word combinations that appear to follow simple rules, such as putting the verb before the noun. However, his most frequent three-word sentences are all similar requests (the top five are: chase person person, person pat person, person person pat, person chase person, person grab person). He apparently can understand quite complex verbal English commands. For instance, he will perform unusual requests such as "put the keys in the refrigerator." His comprehension is better than his production and has been likened by comparative psychologist and trainer Sue Savage-Rumbaugh to the level of competence of a two-and-a-half-year-old

---

15 Tamarins have been found to be able to learn a nonrecursive acoustic sequence but not a recursive one. Subsequently research has used this test to suggest that starlings can learn recursive rules. The test itself, however, has been criticized by Michael Corballis because it can actually be solved in nonrecursive ways. In this case the killjoys appear to have the upper hand over romantic explanations.

child—an impressive performance. Kanzi is probably the linguistically most competent nonhuman alive.

Yet linguists continue to argue that not even these apes have the hall-marks of language. Steven Pinker, for example, insists that they "just don't get it." None of them, not even Kanzi, shows convincing evidence of recursion and the generativity that it provides. There is no use of grammatically important characteristics like inflections and tense, for example, and there is no distinction between statements and questions. Grammar aside, they show no sense of the real logic of having a symbolic communication system. They do not regularly teach each other, point out things for others' benefit, or ask for the names of things. If they had understood the principle, one would expect them to want to learn more useful words. This limit to their competence is also evident in the otherwise impressive attempts at teaching apes to count. Sally Boysen, for example, taught her chimpanzees to count using Arabic numerals up to nine. Each number took equally long to acquire. What's more, they did not get the recursive rule that allows us to use these numbers and reassemble them to represent virtually any quantity. There also appear to be limits to the pragmatic side of their communication, as I will discuss in the next chapters. We still wait for an animal to tell us about what it is like to be what they are; to tell us their views on life, politics, philosophy; or to even tell a simple story (let alone to report to an academy).

In sum, there are indeed some characteristics of language that appear to be uniquely human. On current evidence, it is fair to say neither in their natural communications, nor in our attempts at teaching them human linguistic systems, have animals provided evidence for a full-fledged language. Animals do have communication systems, and they do form concepts. They can learn humans' arbitrary symbols, and some appreciate the fundamental attribute that a symbol can inform about another object or event. Some species, like parrots, can produce speech sounds, though many lack adequate multitasking capacities and voluntary control of face and vocal tract to establish vocal conversations. However, sensory-motor skills may not constitute an absolute barrier, given that a language faculty could be expressed in other ways.

What appears to be lacking, even in great apes, is a motivation to find means to exchange what is on each other's minds. Animals have not demonstrated the capacities required to invent and agree on arbitrary symbols for concepts or on grammar rules that allow for efficient combination of such symbols. They have not developed an open-ended, generative communication system comparable to a human language, nor have they been able to learn one of ours. It looks increasingly likely that humans do not have an innate universal grammar that enables language; instead, we culturally inherit a specific language from people who, based on more general capacities for embedded thinking, manage to establish such symbols and rules for the practical purpose of exchanging the matters on their minds—minds filled with thoughts about past and future, about others' minds, about problems and opportunities, about cooperation and morality. Without such complex mental content there would be little use for an open-ended communication system like ours. To these mental contents, therefore, we turn next.

FIVE

Time Travelers

Forethought is the most important of all the causes
that make human life different from that of animals.

—BERTRAND RUSSELL

IMAGINE SOMEONE HAS FINALLY INVENTED a time machine. Where in time would you go? Would you want to travel to the distant future, or witness a special event in the past? Perhaps you are the type of person who is content to stay where you are. Regardless of your personal preferences, the idea of a time machine has long had a curious appeal—its endless possibilities tickle our fancy. Alas, modern physics suggests time travel will never become a reality. We must make do with time traveling in our minds. We can recall past episodes and imagine future events, including entirely fictional scenarios (such as the invention of an actual time machine). Much of my research has centered on this fundamental human capacity. It therefore seems fitting to begin with some reminiscing of my own.

As a child, I struggled to come to terms with the most unwelcome of all realizations: the fact that, one day, I would die. I lay in bed, staring at the ceiling, and attempted to imagine "not being." I thought I might be able to do this, as the state seemed little different from dreamless sleep. However, I couldn't get my head around the idea of never existing again—never to wake up; to be gone forever. Even now the thought makes me feel queasy, and I appreciate why people seek comfort in the idea of an afterlife. I think I realized even then that this dilemma was entirely due to our ability to project ourselves forward in time. I am not sure if I considered whether this was a uniquely human problem, but this was to become an important question in my studies.

In my first year of undergraduate psychology at a German university, we read a book by Norbert Bischof, a student of Konrad Lorenz, on the nature of incest avoidance. Buried a few pages from the end was his claim that only humans have "time representation." This notion resonated with me, and I began to investigate the questions it raised. What is the nature of the human ability to think beyond the here and now? How does it develop in children? What are the temporal capacities of animals? Don't they ponder the good old days? Don't they imagine what the future might hold? In a radical shift in lifestyle, I ended up moving to New Zealand to research these questions in my master's thesis.

I eventually found an idyllic houseboat in the mangrove swamps of an offshore island and a wonderful mentor, Michael Corballis, at the University of Auckland. He gave me an early laptop that I powered with a couple of truck batteries that in turn were charged through a solar panel. I had time to focus—there were few distractions other than the mosquitos. One day, after I had almost completed my thesis, WordPerfect was in the process of saving my files when the power supply seized up. I lost much of my work. Michael consoled me with the wise counsel that rewriting would probably improve the whole flow immensely. I dutifully rewrote what I had done—only to lose it again, this time on a computer at the university.

I learned a thing or two from this experience about backing up my work, something we do thanks to our capacity to think about the past and the future (the backup software on my current computer is in fact called Time Machine). Finally we turned my thesis into a monograph in which we proposed that mental time travel into the past and mental time travel into the future are two aspects of the same faculty. We could not find compelling evidence for anything like this faculty in other animals and argued that its emergence must have been a prime mover in human evolution. Mental time travel, as it turns out, explains many of the most peculiar human traits, from celibacy to suicide and from diverse expertise to greed.

"It's a poor sort of memory that only works
backwards," the Queen remarked.

—LEWIS CARROLL

WE HAVE SEVERAL DISTINCT MEMORY systems. We can experience problems with one system without impairment of the others. The English musician Clive Wearing, for example, developed amnesia after an infection of the herpes simplex virus destroyed a part of his brain known as the hippocampus, which, as you may recall, plays a role in mental maps. He has retained many skills (such as how to play the piano) and knows facts about the world (such as what a piano is), yet he cannot remember a single event that has ever happened to him (such as his having given a concert). He knows he is married but cannot remember getting married. Because of such dissociations, researchers commonly distinguish the following memory systems: memory for how to do things (procedural memory), memory for facts (semantic memory), and memory for events (episodic memory).

Episodic memory is probably closest to what we typically mean when we use the word "remember" rather than "know." The psychologist Endel Tulving first proposed this concept to refer to memory of one's own past experiences. When you retrieve episodic memories, you travel mentally back in time and reexperience perceptions, actions, emotions, or thoughts of a past episode of your life. You may revel in reliving your past successes or mourn your failures. Clive Wearing, on the other hand, continually reports that he has just woken up from unconsciousness and is seeing and experiencing things for the first time—only to forget about them and then experience yet another resurrection.[1] Without episodic memory, he is stuck in a time bubble.

---

1 In a dramatic documentary Clive is presented with a video recording from earlier in the day in which he can be seen conducting a choir. He acknowledges that it is him in the video but refuses to identify with his image. Instead, and as always, he insists that he has just woken up from unconsciousness. The fact that people can show him his past on tape makes no difference to him. Tulving argued that retrieval of episodic memories is characterized by a particular "self-knowing" consciousness and contrasted it with the "knowing" consciousness involved in semantic memory of facts, and the "not-knowing" consciousness associated with procedural memory.

Although you may treasure many of your episodic memories, it is not entirely clear what the ultimate function of this capacity is. At first glance, you may think that its job must be to provide a faithful record of our past. Indeed, memory research in the past 150 years has focused almost exclusively on factors affecting memory accuracy. What this research has found, however, is that our episodic memory system is not particularly comprehensive or reliable. Perhaps this comes as a bit of a relief—it certainly makes me feel better about my own apparent memory shortcomings. We tend to struggle with simple requests. Try to recall what happened on your birthday three years ago. Now try to recall what happened at lunch two days before or after that birthday. Our memory for events is nothing like a recording; we cannot simply rewind and press play. In fact, it seems we forget most events. Think of all the hours at school you sat on a chair in class. You may remember your classmates, the teachers, and some special interactions, but for the most part your memories blend together. I suspect you would struggle to recount the exact events of any single day or even the details of a specific hour.

We do, of course, remember some events in great detail. Perhaps you recall the moment when you found out about the 9/11 attacks on New York. Yet even here there is reason for pause. Research shows that we can be wrong about events we think we recall clearly. Naturally this poses a serious problem for all those situations in which someone's memory is all we have to go by—as often happens in the courtroom. Jennifer Thompson thought that Ronald Cotton had raped her at knifepoint in 1984. Her identification led to his conviction in 1986, and he served eleven years in jail before his innocence was revealed. After a DNA test, another man, Bobby Poole, whom the victim did not identify when he originally appeared in court, admitted to the rape. Thompson's memory had been wrong. There has been extensive work on the reliability of eyewitness testimony, and the results are not exactly reassuring. Untrue information about crucial details, be it the presence of a stop sign or a mustache, is sometimes falsely recalled. People's accounts can be influenced by information introduced after the event. Suggestions by other witnesses, police, or lawyers, for instance, may be incorporated into eyewitness reports. What is more, confidence in the accuracy of one's memory is a surprisingly poor predictor of which detail in the eyewitness report is false and which is correct. Even when you are positive you remember something correctly, you might be wrong.

Remembering episodes is a reconstructive process that draws on some stored gist that is then actively expanded as we rebuild the scenario of the past.[2] We may embellish to create a better story or adjust our reconstructions to make them consistent with our prevailing attitudes. Repeated acts of reminiscing and retelling can lead to increasing distortions. So we bolster our faulty memories with external storage systems, such as diaries, photos, drawings, books, and data sticks—in acknowledgment that our memory of past events is not a terribly reliable record keeper.

How did such a flawed system evolve? Unlike a memory researcher using a video recording, natural selection cannot go back in time and check your memory matches accurately with the original event. If a false memory or a memory bias improves fitness (the capacity to survive and to propagate one's genes), the memory system that produced it has a selective advantage no matter how inaccurate. For example, you typically recall your own good behavior better than your bad behavior, but show no such bias when recalling the behavior of others. This partiality leads to a flawed reflection of the past, but if it improves, say, the chances of impressing potential mates, then those with the distorted view may on average have more children in the next generation than those with an accurate view. And so the bias spreads. Evolution works only on how memory influences fitness, not for how accurately memory reflects the past per se.

In this sense, accurate prediction is more important than accurate recall. Indeed, all memory systems are inherently future-directed, rather than oriented toward the past, as one might assume. Take the classical conditioning first described by the physiologist Ivan Pavlov. He found that a dog learns to associate the sound of a bell with the arrival of food. The next time the dog hears the bell, it starts salivating before the food arrives. Although people typically think of this as memory, the dog is not really salivating at the memory of food, but in anticipation of, and preparation for, food. The sound of the bell makes the dog predict that food is about to arrive. Similarly, the memory system for facts must have evolved for its future-directed benefits. For instance, knowing where one's hideout is can

---

2 Sometimes people store more than the gist. A photographic memory, or eidetic memory, refers to a capacity to maintain access to an event as if one were still perceiving it. This is not to say that eidetikers remember everything they experience in this way, but they may for instance view a picture for a minute and then replicate it in detail. Other astounding feats of memory, such as the rapid learning of lists, numbers, or names, are typically the result of carefully deployed, deliberate memorization techniques.

quickly become fitness relevant. A warthog that knows this location has a better chance of surviving a lion attack than one that does not. Memory matters because of what it can do for you now and in the future. How might this work for episodic memory? The main benefit of memory for past events may be that it allows us to imagine future events.

Perhaps the Red Queen was right. It would be a poor sort of memory that only works backwards. Are our capacities for mental time travel into the past and into the future two sides of the same coin? Amnesic patients such as Clive Wearing who have lost their episodic memory have similar problems imagining future events; they draw a blank when asked about what they will do tomorrow. We have found that young children's capacity to answer such questions is linked with their ability to report what they did the day before. Introspectively, there are some similarities. For example, we tend to conceive of fewer details of events the further they are removed from the present—whether into the past or into the future. In old age we tend to report fewer details of both past and future events. Suicidally depressed and schizophrenic patients have difficulties recalling specifics about past events and show similar problems imagining the future. Brain imaging studies have found that when participants are asked to recall past events and imagine future situations, the same areas of the brain (including the hippocampus and regions in the prefrontal, parietal, and temporal cortex) are involved. Although there are some important differences between mental time travel into the past and mental time travel into the future—after all, one has happened and the other hasn't—many commonalities have been documented in recent years. Substantial evidence supports the proposal that episodic memory and episodic foresight are fundamentally linked in mind and brain.[3]

OUR EPISODIC MEMORY IS NOT limited to working backwards. One way for it to go forward is simply for it to project into the future. After all, the best predictor of future behavior, generally, is past behavior. Your dog's reaction the last time you attempted to steal his bone provides a reasonable guide about what to expect should you try to do so again next week.

---

3 *Science* magazine considered new evidence for this one of the top ten scientific breakthroughs of 2007.

However, you can do much more than predict reoccurrences of events. You can imagine situations you have never experienced before. For example, you might mentally simulate how you could distract the dog before going for his bone. You can imagine a virtually infinite number of future scenarios. You may consider a few options before pursuing the most promising path—no one likes being bitten. You do not need to suffer all the real-life consequences of trying out each possibility. We can test most things in our mind and assess how likely or pleasant they might be. For instance, I suspect you can imagine whether you would enjoy hot mustard mixed with your vanilla ice cream without having tried the combination before. To imagine new events you need an open-ended system capable of combining old information into new scenarios. If mental time travel evolved for this purpose, then the price of this flexibility is that we may at times reconstruct past events creatively rather than faithfully—which explains some of the typical errors of episodic memory.

Shakespeare wrote, "All the world's a stage," and this is also a useful metaphor for our mental world. Just as a theater production involves certain roles, so too does our mental scenario building. Imagine you have to prepare a speech for a function or a wedding. To create such a situation in your mind you need to disengage from the present situation and imagine (stage) the scenario. You need to have some idea about your self (actor), perhaps about your strengths and weaknesses in such situations. You may also need to consider who you are addressing and what their expectations might be (other actors). Depending on the particulars you imagine, this may make you feel confident or anxious. You might frame your mental event in some context, such as the venue (the set), and consider the obstacles it presents. For instance, do you need to project your voice? Do you have a podium?

Now imagine what you are actually going to say. Is it better to start formally or with a joke? Perhaps you need to consider different introductions. These scripts have to be generated (playwright) so you can cycle through the possibilities. You might choose to conduct various mental rehearsals and assess each scenario's appeal (director).

Imagined situations trigger emotions much like their real counterparts, and we can evaluate their desirability. Generating and comparing these imaginary scenarios requires the capacity for embedded thinking that we came across in the previous chapter as a unique aspect of human language. You need to reflect on your thoughts. While going through sufficient scenarios to develop your presentation, you may need to put on hold other activities. At some point you will need to make an executive decision about when you should stop such mental simulations and start putting your plan into action (executive producer).

Of course this is not to say that there is a little theater in your head (or that the mind is populated by homunculi). The metaphor simply highlights what component capacities are involved in episodic foresight. Your successful simulation of scenarios depends on a capacity for imagination and for thinking about things other than the present. It also requires understanding of yourself and others; knowledge about how the physical world works; a creative faculty that allows you to combine imaginary actors, acts, and objects in novel ways; an ability to rehearse and reflect on options; and the power to override immediate urges to pursue more remote rewards. Mental time travel is complex, resource-intensive, and error-prone. But it is a key to how our minds conquered the world.

MENTAL TIME TRAVEL UNLOCKED A new realm of possibilities for our species. We can hatch plans and make decisions that drastically increase our chances of future survival and reproduction. By foreseeing events we can seize opportunities that lie ahead and take steps to avoid approaching disaster. We can imagine the consequences of what we are going to do before we do it—and berate others for not doing the same. We can also benefit from the past in novel ways. We can mentally revisit past events, reflect on them, and draw new conclusions. Surprise visits of a family friend will no doubt be remembered and interpreted quite differently if you subsequently find out this person had an affair with your spouse. These replays allow us to learn new lessons for the future. In this case, you may become more attuned to early signs of infidelity.

Mental time travel radically increases our opportunities to be prepared. Even foreseeing that the future is difficult to foresee can prompt us to prepare for common eventualities. Consider what is in your pocket or your handbag. Why do you carry keys, money, cards, condoms, cosmetics, or whatever is in there? You do this, for the most part, because these things may come in useful in the future. There can be little doubt that this has long been a crucial and quintessentially human survival strategy that has enabled us to thrive even in previously inhospitable habitats. Take Ötzi, the five-thousand-year-old Iceman found in the Alps over twenty years ago. He carried dozens of items (such as an axe, a dagger, a tool for working flint, a bow and arrow, medicinal fungi, and a fire-making kit) that prepared him for many eventualities. As it turned out, it did not prepare him for the arrow that struck him in the back, but having these tools while crossing the Alps surely increased his chances of surviving the perilous journey.

Many everyday behaviors are guided by foresight, from planning next weekend's barbecue to the pursuit of a long-term career goal. Because we can consider many different future paths and their consequences, we gain a (perhaps somewhat fanciful) sense of free will. Humans can elect to pursue dramatically different skills, knowledge, and expertise. We engage in deliberate practice and study to improve our future selves. You may not think of yourself as an expert, but chances are you have become an expert in various areas—be it at your job, crosswords, sports, housekeeping, music, matchmaking, or rocket science. People differ in innate talents and the pleasure they derive from certain activities, but much of the diversity of human expertise comes from our having the potential to decide what to spend time on, what we want to get better at, and what goals we aim to achieve. Of course, our freedom to do so varies vastly in different times and situations, but the potential exists in all of us.

In some cases we can even get better at certain things by merely practicing them in our mind. As a soccer player I always loved back-flip bicycle kicks, but, given the likelihood of hurting yourself, it is best to train for them primarily in your mind. It is not hard to foresee that practicing them on concrete is probably not a good idea. People get better at moving their bodies, making music, solving problems, and generally dealing with a variety of situations through repeated mental rehearsal. Practice, whether in mind, in play, or in action, accounts for much of our varied skills.

We differ also in how much we worry about the future and the past. Some psychologists now consider this an important variable that distinguishes our personalities. Yet all of us think about the future. Sometimes we think about it so much that we forget to notice the present. As John Lennon sang, "Life is what happens to you while you're busy making other plans."

Neuroscientists suggest that mental time travel may be our default mode. Brain-imaging studies of subjects at rest reveal that the brain regions that light up are associated with thinking about past and future events. We seem to have to go to great lengths, such as dedicated meditation retreats, to train ourselves to be mentally in the here and now. We often look forward to times when we are not stressed about all the things we still have to do. On holiday at the beach, little could be more enticing than the idea of not having to plan a thing—although even here you may be quite happy to plan to have a few drinks after your swim and buy tickets for that gig tomorrow night. Mental time travel is a pervasive part of the human condition.

Some of our most peculiar behaviors only make sense (at least partly) in the light of mental time travel. It can illuminate such apparent biological paradoxes such as celibacy, when a person is expecting a greater reward in an afterlife, or suicide, when the future outlook is particularly bleak. It helps explain why we may acquire much more than we can currently consume, because we are keen to secure not only our present needs but our imagined future needs, too. Whereas a pride of full-bellied lions is no threat to nearby buffaloes, a group of full-bellied humans might very well be. Our at times appallingly voracious greed may stem from this concern for our future. On the brighter side, the unwavering pursuit of one's dreams is one of the most remarkable examples of the power of the human spirit. It wasn't only Martin Luther King Jr. who had a dream: we all, in our own way, pursue visions of the future. To explain our behavior, we need to consider our minds—minds filled with expectations and plans, hopes and fears.

Even mundane activities such as going to work reflect complex goals and our attempts at managing our future. To drive such behavior, long-term goals have to compete with more immediate temptations, and we have had to learn new ways to control our motivation. For example, my desire to play, eat, and have sex will not stop me from completing this paragraph, because I think it more important for my longer-term happiness to do my job and concentrate on writing this book—at least for a little while longer. You may strike an inner bargain by rewarding your progress with the promise of satisfying a simpler wish. In my case, I have earned a fine Belgian beer tonight. Deal! This capacity to manage, or mess with, our own drive system gives us enormous behavioral flexibility. Unfortunately, it also causes us tremendous stress. We worry even about things we can do little about right now. What's worse, we are often wrong.

WE ARE NOT CLAIRVOYANTS. THE future frequently turns out to differ from what we thought it would be. Vivid examples of plans that appeared to be a good idea at the time but turned out to be spectacular miscalculations can be found among the annual winners of the Darwin Awards. We will never know what exactly a wheelchair-driving 2010 winner was anticipating would happen when, after missing a closing elevator, he decided to impatiently ram the door until it broke down—only to fall into the now-empty shaft. Most of our foresight errors are minor by comparison, leading to inconvenience or embarrassment. You may fail to usefully imagine the future because of shortcomings in any one of the components in the theater metaphor. Stage: you may fail to disengage from the present to imagine the future—perhaps like that wheelchair driver. Actors: you may miscalculate how others will feel or act—as so often happens in pranks. Set: you may misjudge physical relations—say, when you think the boat could surely take a much heavier load. Playwright: you may fail to generate the relevant scenarios—and you later have to admit that you didn't think of this or that. Director: you may not have practiced for the future sufficiently—leading you to look distinctly underprepared. Producer: you may end up selecting the wrong plan—d'oh! There are countless ways in which our attempted foresight can let us down.

However, we have radically improved our chances of getting it right through a wonderfully effective trick: we share our plans and predictions with others. We can transmit our mental plays and reflections to audiences around us and, in turn, consider their thoughts. In preparing a speech, it can be helpful to rehearse it not only in our mind but also in front of a friend. We can learn from others' memory and foresight, and listen to comments on ours. Indeed, we have a deep-seated drive to broadcast our minds and to read what is on the minds of others—to foreshadow the next chapter. And we have an extraordinarily effective way of exchanging our mind travels through language—to remind you of the previous chapter. Language is ideally suited for this mental exchange, and much of human conversation is indeed about past events (who did what to whom, and what happened next) and future events (what will happen to whom, and what we are going to do about it). By exchanging our experiences, plans, and advice, we have vastly increased our capacity for accurate prediction. In *Stumbling on Happiness* the psychologist Dan Gilbert discusses errors and biases in our foresight and argues that the most reliable way to predict a situation is to ask people who have experienced something similar. Indeed, for much of our past the stories of our fellow tribespeople would have been all we had to go by.

Note that we can share mental scenarios even without language—through mime, dance, and acting. While there are limits to what can be communicated in these ways, this could have been a start. The more we relied on our time-traveling minds for survival, the more we stood to benefit from language as a more flexible, open-ended communication system to link with other minds. As noted earlier, the evolution of new thoughts likely preceded the evolution of the means to communicate these thoughts.

Even our young offspring are driven to understand others' minds, and we are compelled to pass on what we have learned to the next generation. As an infant starts on the journey of life, almost everything is a first. Young children have a ravenous appetite for the stories of their elders, and in play they reenact scenarios and repeat them until they have them down pat. Stories, whether real or fantastical, teach not only specific situations but also the general ways in which narrative works. How parents talk to their children about past and future events influences children's memory and reasoning about the future: the more parents elaborate, the more their children do.

From as early as the age of two, children begin to talk about past and future events. Yet it takes time to understand time—parents will understand the struggle of explaining to children that they will get something later. My kids, at age one and three, would have had no chance of surviving on plans of their own if stranded on a remote island. We parents pack their lunches, fetch their jackets, and prepare the weekend's activities. Adults have a wealth of experience to draw on. Our earliest memories, however, tend to go back to only about age three. Freud referred to the lack of earlier memories as infantile amnesia, and he believed that we repress traumatic events of our early psychosexual development—those embarrassing things to do with diapers and breastfeeding may not be palatable for the sensibilities of the adult mind. The latest evidence suggests that the explanation has more to do with the maturation of essential cognitive factors of memory and mental time travel, as well as social instruction from others.

Nonetheless, even infants have some memory and anticipatory capacities. A few weeks after birth they can learn to kick a hanging mobile and do it again in a new context (procedural memory). A few months later they can imitate an action such as the making of a rattle out of three parts and retain this knowledge (semantic memory) to make rattles out of similar

FIGURE 5.1.
My children, Timo (age three and a half) and Nina (age one and a half),
on the North Island of New Zealand.

objects. Yet there is little indication that infants can explicitly recall particular episodes of their past (episodic memory), let alone plot events in the remote future. Although children start to talk about past and future events from age two, they initially appear to have fundamental problems in their understanding. When three-year-olds learn something new, say a new word for a color, they go on to insist that they have always known it, even if they just learned it that day. In one study we told children stories about two people, one who acquired something yesterday and one who will acquire the same thing tomorrow. When we then asked them who has the object right now, even four-year-olds struggled.

In studies less dependent on words, we have found that at least by age four children can remember a problem they only experienced once and act in prudent ways that secure its future solution. We presented children with a curious puzzle in one room and later, in a different room, gave them the option to take one of several objects back to the puzzle. Children at thirty-six months of age selected an object at random, but by forty-eight months they picked the object that would allow them to solve the puzzle.

Without any prompts they figured out which object would be a handy future solution to a problem they remembered.[4]

By age four they may well think a lot further into the future. The day Timo passed this task, he later sat next to me, put one hand on my leg, and said, "Papa, I don't want you to die." Gulp—"Neither do I," I said. He then elaborated: "When I get bigger, I will have children, and you will be a grandpa, and then you will die." He had clearly started to ponder the existential questions that mental time travel confronts us with.

By this age children have also been shown to be able to place past and future events appropriately on spatial time lines, such as a picture of a road that recedes into the distance representing the future. The psychologist William Friedman has described how children gradually improve at judging the relative temporal positions of past and future events, such as whether last New Year's Day or their last birthday was more recent. It takes several years before children have acquired culture-specific temporal concepts such as days, months, and years that allow them to communicate about events in a precise, shared framework.

Temporal concepts, timekeepers, and calendars have been developed to aid our orientation and planning. They allow us to coordinate our activities in ever more sophisticated ways. We pursue complex shared goals and subdivide the required labor according to our expertise. We can agree on plans, review progress, and make flexible adjustments as required. Predictably, many of our ambitious endeavors do not work out—just think of the attempts of detailed five-year plans for the entire communist economy of the former Soviet Union. (Yet cooperative plans allowed that same country to send the first person into space.)

Humans are constantly scheming to achieve better control of the future. Today, much of the human population is waking up to the realization that our actions have produced immense pollution and have drastically reduced biodiversity. Let us hope this realization will lead to efforts to combat this trend. Norway recently demonstrated tremendous farsightedness in constructing a "doomsday" seed bank to protect all known varieties of the world's crops for future generations. Future planning has gone global.

---

4   When the problem and solution were presented in the same situation, both age groups can solve the puzzle. In a follow-up, children could select a solution and then had to wait for five minutes before they could implement the solution. Again, children from about age four can do this task.

It seems safe to say that mental time travel is a significant human attribute, without which we would hardly have been able to change the face of the Earth—let alone control many of its other inhabitants.

The idea that our mental access to the future is crucial to the human condition is not entirely new. In Greek mythology Prometheus stole fire from heaven and gave human mortals some of the powers of the gods. The name Prometheus means "foresight." In some versions of the myth, Prometheus not only brought fire but also taught humans the arts of civilization, such as writing, mathematics, agriculture, medicine, and science. Foresight has given us unheralded powers.

> What's time? Now is for dogs
> and apes! Man has Forever!
> —ROBERT BROWNING

ANIMALS HAVE NOT TAMED FIRE, nor have they mastered the arts of civilization. There is no obvious evidence that animals have ever agreed on a five-year plan. Although children's films often feature animals solving a whodunit or thwarting the bad guys' evil plot, Lassie, Flipper, and Babe have no real-life counterparts. Their behavior is the product of careful conditioning by trainers who use immediate rewards to get the animals to act as if they understood the narrative. On the farm or in the zoo there is little to suggest that real animals plot to take control. There is no evidence animals have invented bags to carry a variety of tools in case they might need them. They do not choose a career path and do not seem to deliberately practice in preparation for anticipated events.[5] They do not show the same diversity of expertise as humans. But just because they do not act the way we do does not necessarily mean they cannot think about past and future events.

Animals are by no means insensitive to temporal matters. There is evidence that some of them detect the time of day (some dogs will get ready

---

5 Juvenile mammals tend to engage in play behavior that may function as preparation. They can, for example, practice the moves that later will allow them to hunt or fight. Such behavior is universal and limited to a few capacities. It thus seems to be an instinctual behavioral predisposition, rather than the result of the individual animal anticipating a particular future event for which it must train.

for the mail carrier to arrive) and can track short time intervals (if you feed a dog every half an hour, it will start to salivate just before the next feeding). Yet, as the psychologist William Roberts showed, these competencies can be achieved through basic mechanisms, such as simple associations with states of the natural body cycle, rather than anything akin to mental time travel.

Our most direct evidence for episodic memory in humans comes from people's verbal reports. As already discussed, some apes have been trained to communicate using human symbol systems, but they have so far not acquired tense and have not told stories about the past or visions of the future. These projects have provided some compelling evidence of semantic knowledge, however. After all, these animals have learned which symbols go with which objects or actions. The chimpanzee Panzee has even used a symbol board to announce what food was hidden outside her enclosure and pointed to get a human to retrieve it for her. This shows she knew where the food was hidden, but it does not necessarily mean she remembered the hiding event itself. One can know things without knowing how one has come to know them. For instance, you may know that Mount Kilimanjaro is the highest mountain in Africa but probably not the occasion on which you learned that fact—unless you learned it just now.

Although it is safe to conclude that animals have procedural and semantic memory systems, there is no obvious demonstration that they have episodic memory. Rats appear to use their hippocampus to create cognitive maps of their environment. Most species, even insects, demonstrate sophisticated navigational skills. But do they mentally reconstruct the particular events that shaped their knowledge? Do they reminisce about days of yore?

RECENTLY, RESEARCHERS HAVE MADE THE case for certain animals having something like episodic memory. Psychologists Nicola Clayton, Anthony Dickinson, and their colleagues at the University of Cambridge have produced some intriguing results that they say may reflect mental time travel in animals. Scrub jays hide food for later consumption. Clever experiments on caching and retrieval capacities indicate that these birds know what was cached where and when. For example, they adjust their search differently for cached worms, which rot quickly, and peanuts, which keep fresh longer, depending on how long ago they had stored them. They do not bother searching for the worms if they had been stored a long time ago because they would be rotten. Jays show this search pattern even when there are no cues, such as smell, to guide them. The researchers conclude that the birds have memory for the past occasion on which they stored these foods. They refer to this as "episodic-like" memory, leaving open the question whether the birds are conscious of the past.

This approach has stimulated headlines, debates, and a profusion of such studies on other animals. Although several species have failed to pass similar tests, some species, such as mice, rats, and chimpanzees, have passed them. These successful species are therefore also said to have a capacity for episodic-like memory. But how like episodic memory is episodic-like memory? Despite the cautious terminology, romantic proponents of the what-where-when approach often imply if it walks like a duck and quacks like a duck, then it probably is a duck. In other words, these species probably can travel mentally into the past. But is this really a duck hunt or a wild goose chase?

I have argued that evidence for episodic-like memory is neither necessary nor sufficient for mental time travel. You can know that something happened and yet not remember the event at all. For instance, I know what happened on November 24, 1967, in Vreden, Germany. I was born. Yet I have, of course, absolutely no recollection of the event. Conversely, you may recall a particular episode and yet be factually wrong about what precisely happened where and when. Recall (if you can) that episodic memory is notoriously unreliable. Thus evidence that animals can draw on accurate information about the what, where, and when of a particular event does not show that they travel mentally in time.

How then to explain the scrub jays' clever retrieval behavior? It is possible, and quite plausible I think, that scrub jays know what food is where and whether it is still good to eat without having to remember the caching episode itself. Here is a simple alternative explanation. As time passes, memories fade. Scrub jays may learn when it is still worthwhile to search for a particular type of food by associating the strength of their memory of the food location with whether it is still good to eat on recovery. They then simply apply a rule along the lines of: worms aren't worth searching for once the memory of their location has weakened beyond a certain point. The experience with nuts leads to a rule that they are worth searching for even when memory is much weaker. Such rules can effectively provide use-by dates for different types of stored food (without requiring conscious recollection of the caching event).

If this what-where-when approach does not demonstrate episodic memory, how could one show animals have episodic memory, if indeed they have it? Without language they cannot tell us about their time travel. As we saw, humans can also express past episodes through mime or dance. However, there is no suggestion that animals do this too. If evolution selected for episodic memory in animals, then this capacity must have benefitted survival and reproduction. Evolution could not have selected for it as a private indulgence without a tangible effect. Given the evidence of close links between episodic memory and foresight, and the fact that foresight offers clear fitness benefits, one would expect that animals that have mental time travel capacities should be able to control their future prudently. They should hatch plans and plot their way to future happiness.

> Long-term planning . . . is something utterly new
> on the planet, even alien. It exists only in human
> brains. The future is a new invention in evolution.
>
> —RICHARD DAWKINS

RICHARD DAWKINS AGREES THAT THERE is something unique about the human capacity to think about the future. Human long-term planning appears to have no obvious rivals in the animal kingdom. Yet many species act in ways that improve their future: animals construct nests for breeding, they migrate at the right time to warmer climates, and they search for food where it is likely to become available. Learning about recurring patterns in the context of mating, food, and predation has obvious evolutionary benefits. Over long periods species have acquired ways of taking advantage of what is regularly recurring. Even bacteria demonstrate future-directed capacities in this sense. Right now, Esherichia coli are moving through your digestive tract from an environment rich in lactose to one rich in maltose—and they have prepared for this by turning on genes necessary for digesting maltose. This does not mean each individual bacterium looks ahead and decides to prepare as it goes down your gut. Evolution has selected for this order of events over many generations of bacteria: E. coli that happened to show this pattern of preparatory gene activation survived and reproduced better than those that did not. Many species have evolved innate mechanisms that take advantage of long-term regularities.

These innate behaviors look clever, but the lack of foresight involved becomes clear when circumstances change. A classic example is the digger wasp. The wasp always inspects the nest before dragging its prey inside to feed its larvae. If in the meantime a mischievous human moves the food a few centimeters, then the wasp will regather the food, and repeat the sequence again by dropping it at the entrance and inspecting the nest. This can be repeated again and again, without the wasp breaking out of its behavioral program. Although provisioning the young appears to be a complex, future-directed behavior, the wasp executes it without any apparent thoughts about seeing them grow up. If the entrance is destroyed, the wasp will not feed its larvae but trample over them in its frantic search for the entrance. Similarly, various animals hoard food for the winter without necessarily understanding why they do it. Young squirrels, for instance, will hoard nuts even if they have never experienced a winter. These behavioral solutions to recurring seasonal changes may not be all that different from physical adaptations to the same problem—such as storing food for winter in body fat.

Innate mechanisms are reliable but not very flexible. Memory, by contrast, enables an individual, rather than a species, to learn about potentially recurring events during its lifetime. All memory, you may recall, is in a sense future-oriented, allowing an individual to adapt to its environment. If a stimulus reliably predicts a situation, like the bell predicted food for Pavlov's dog, animals can learn to take advantage of this association. A pigeon, for example, may learn to peck on a button for food only when a light indicates food is available. Associative learning can account for some complex-looking behavior, as we saw in the case of Clever Hans. Behaviorists have documented the rules that govern such learning, and one of their main findings has been that two things have to happen together or nearly together if they are to be associated. The link between an action and a consequence is typically learned only if they are separated by no more than a few seconds.

There are rare exceptions. Perhaps the most extreme is that rats can learn to associate the taste of food with feeling nauseous some hours later. Given that rats keenly explore novel food sources, learning to avoid foods that will make them sick has obvious survival value. So this appears to be a special capacity. The rats learn only that a taste predicts later sickness; they cannot learn that a sound or a sight has the same predictive relationship. Yet instinct and learning may combine to create sophisticated future-directed behaviors, some of which are not yet well understood. For example, how do gray squirrels learn to bite out the seeds of white oak (though not red oak) acorns before storing them, thereby preventing their germination? Curious though they are, in these cases, as in the food-caching scrub jays, the future-direct competences seem limited to a particular type of problem and show none of the open-ended flexibility evident in humans.

We saw in Chapter 3 that our closest animal relatives have a basic capacity to imagine other possible worlds. Can they use their minds to plot future actions? Maybe they can for some problems that lie in the immediate future. For example, when presented with a treat that is out of reach, great apes can go around the corner, where the treat can no longer be seen, and select a stick of the appropriate length to solve the problem. I had to learn this the hard way, when the chimpanzee Ockie noted that a stick at one side of his enclosure could be used to reach a TV I had put up earlier on the other side—with smashing consequences. They also carry tools over short distances in the wild. One of the most impressive examples of apparent forethought comes from chimpanzees in the Tai forest of the Ivory Coast. These apes use stones to crack open nuts and sometimes carry stones for a hundred yards or so to the site of the nuts. Presumably they pick up the rocks as a result of developing an appetite for nuts and a plan of where to get them.

In spite of these signs of short-term foresight, animal mental access to the future may be restricted in several important ways. Recall the range of processes involved in a theater production of the future. Shortcomings in any of them may limit capacity. Chimpanzees may have some capacity to imagine alternatives, but they may be fundamentally restricted by other requirements. One possibility is a limited understanding of one's self as an actor in a future scenario. The psychologists Norbert Bischof and Doris Bischof-Köhler have claimed that animals may not be able to imagine drives and wants they do not currently experience. You can easily imagine being thirsty even when quenched, and hence may want to secure future sources of drink. Recall the comparison between full-bellied lions and humans—we often try to get things we do not (currently) need.

Such a limitation could explain various curious animal behaviors. Take the case of laboratory monkeys that were fed biscuits only once a day. William Roberts recounts how in the 1970s the cebus monkeys in Michael D'Amato's laboratory would hungrily eat until they were full and then throw the remaining biscuits out of their cages. To their dismay, they would find themselves hungry again some hours later. You might wonder why they did not learn to guard their food to satisfy their future hunger. If you cannot imagine being hungry again, then perhaps biscuits' utility may lie in their quality as fine projectiles. There is no point in acting now to secure a future need you cannot conceive of. This Bischof-Köhler hypothesis is much debated. In some sense it cannot be quite right, because any hoarding of food is behavior that secures future needs. What we don't know is whether hoarding animals are thinking about future hunger and burying their treats in anticipation of satisfying those future cravings. Despite several attempts to disprove the hypothesis,[6] it still has currency. Animals do not seem to keep and refine their tools for repeated use, and there is no evidence of the kind of greed present in humans.

Another essential capacity for foresight is the ability to forego current temptation in pursuit of a more desirable future outcome. When monkeys are given the choice between a small reward now and a larger reward later, they, like most other animals, find it extremely difficult to wait if the delay is more than a few seconds. Great apes can do better and have been shown to delay gratification for several minutes. For a reward forty times larger than the immediate reward option, chimpanzees may wait up to eight minutes. This is impressive, but it is considerably less dramatic than humans delaying gratification for months, years, or even a lifetime. Still, our closest relatives are much better at waiting for something than other animals.

Perhaps the most prominent case for ape foresight comes from a study in which three bonobos and three orangutans were trained to use a tool to obtain a treat from a feeder apparatus. The animals were then ushered into a waiting room, and the experimenter removed all remaining tools left in

---

6 One high-profile study suggested that scrub jays can adjust their caching of food according to what they could expect to desire in the future. Unfortunately, the rich interpretations were not convincing for a variety of reasons, one of which being that the birds did not actually increasingly store the relevant food. Very recent work on Eurasian jays looks more promising, but new killjoy accounts have also emerged. In another study, two squirrel monkeys were given a choice between one and four pieces of thirst-inducing date. When they selected the larger reward, they did not get any water for three hours. The monkeys gradually changed their choices, preferring the smaller quantity, leading the authors to conclude that the monkeys anticipate their future thirst. However, the gradual change suggests associative learning, and it is unclear why, if they did understand, they then did not simply select the four pieces and only eat one, leaving the others for a time when enough water was available. An attempt to replicate the results with rhesus monkeys failed.

the experimental room. After an hour's wait, the apes were allowed back into the experimental room. On almost half of the trials the apes carried a tool with them from one room to the other and back—so they could use it to get more treats. Some apes did a lot better than others, and two animals even succeeded to hang on to their tools when they returned to the experimental room after an overnight delay. Unfortunately, because it was always the same tool the apes had to select on all trials, it remains unclear here, as in subsequent studies, whether the apes anticipated a specific future situation or had merely learned to associate the tool with rewards. In another study ten chimpanzees were taught that they could exchange a token for a food reward and were then given the opportunity to collect these tokens in anticipation of an exchange session an hour later. Over several experiments the chimpanzees failed to take tokens more often than other useless objects. They just did not think ahead.

An unusual report made headlines in 2009 when it suggested that a chimpanzee at a Swedish zoo may have spontaneously planned its projectile attacks for hours in advance. Three zookeepers reported that in the late 1990s a male chimpanzee tended to collect rocks and concrete into piles early in the morning so he could excitedly hurl this ammunition at zoo visitors a few hours later. Nothing quite like such planning has yet been reported from the wild or, for that matter, from any other captive chimpanzees. If chimpanzees have this kind of foresight, one would perhaps expect to hear about many more such examples—it's possible that this is what the future will bring.

So what are we to conclude? Although some research suggests that our closest relatives may have some, albeit limited, capacity for foresight, we have seen that other studies suggest they are profoundly shortsighted. Animals clearly share with humans some procedural and semantic memory capacities. However, there is little evidence that they have episodic memory. The best evidence for episodic memory should come from signs of episodic foresight, given that mental time travel in both directions is intimately linked. Yet animals do not overtly express any of the obvious manifestations of such a capacity. Only in Orwell and other fiction do they conspire to rebel. Animals can learn that one thing predicts another if the events are separated by only a few seconds. Many species have also evolved instincts that equip them to act in preparation for the future. However, evidence for flexibility in such future-directed behavior is scant. The cases that exist all deal with the immediate future only. There is no compelling evidence that animals flexibly generate mental scenarios of remote future events the way humans do, or communicate mental scenarios to one another to obtain feedback or to coordinate actions. In the next chapter, we will consider whether animals can even appreciate that others have minds in the first place.

Mental time travel is essential for explaining a vast array of characteristics of the human mind, ranging from emotions (such as hope and regret) to motivations (such as plans and revenge). It allows us to understand how things got to where they are and to wonder where everything is headed. It has given us unheralded powers to control plant and animal life to our advantage. But it also confronts us with that most unwelcome of all realizations that I remember pondering as I lay in my childhood bed staring at the ceiling: our mortality.

S I X

Mind Readers

Of all the species on Earth, only
humans possess . . . the ability to
infer what others are thinking.

—CARL ZIMMER

WHEN I TELL PEOPLE I am a psychologist, they sometimes respond
warily, suspicious that I can peek directly into their minds. Alas, I cannot.
In spite of attempts to demonstrate telepathy, there is no proof that anyone
can communicate using only their minds. During the cold war the United
States and the Soviet Union invested considerable effort into attempts to
harness purported psychic abilities for their military programs—with the
only positive apparent outcome being satirical movies, such as *The Men
Who Stare at Goats*, which lampooned these earnest projects. As much as I
like the idea of parapsychological powers, there simply is no compelling
evidence for them. Minds are private; we can never be entirely sure what
goes on in somebody else's head. On a fundamental level I cannot know
what it is like for you to see the color green, to want something badly,
to know your limits, to feel lonely, to anticipate something, or to have a

toothache. Yet, just as we can time travel without time machines, we can read minds without telepathy.

Indeed, we are avid mind readers. We think, and often worry, about what others feel, desire, and believe. In conversation we customize what we say based on what we think the other person wants and does or does not know. We care about making others happy and empathize when they are sad. We may try to put them at ease, to tickle their fancy or blow their minds. We regularly interpret even the simplest acts in terms of mental states.[1] A brief rolling of the eyes may be interpreted as disdain, contempt, or frustration. If you had seen me get up and go to the fridge a moment ago, you could have explained that I wanted a drink and believed I might find cold water in the fridge. Knowing what others desire and think is immensely useful for predicting what they are going to do. We live in a world of minds aware of other minds, and mind reading is absolutely fundamental to our social lives.

Cognitive psychologists call this ability "theory of mind." "Theory" here refers to the fundamental fact that we can only theorize about the mental states of others. There has been considerable debate about how we do this. Some scholars argue that we reason about others' minds much as we do science. We develop a commonsense psychology—ideas about desires and beliefs and how they influence actions—which we adjust and fine-tune in the light of evidence we encounter. Others argue that, because the only direct evidence we have is our own mind, we understand others' minds by imagining their situation and simulating their experiences. This requires mental scenario building, as in the theater metaphor I discussed in the previous chapter. We can stage others' situations and therefore think about their mental experiences.

It is likely that both answers contain some element of truth. We seem to be able to put ourselves into another's shoes and imagine how we would feel or what we would think if we were in their situation. With sufficient experience, we may also develop shortcuts that allow us to rapidly infer what is likely to be on someone else's mind. In other words, you can engage in both instant recognition and elaborate simulation of others' mental states. For example, you may immediately recognize that someone who was cheated is likely to be upset, but you can also pause to imagine what this must have really been like, and so more thoroughly appreci-

---

1 Daniel Dennett calls this the "intentional stance" and contrasts it with explanations and predictions based on function ("design stance") or on other forces ("physical stance").

ate that person's perspective. We seem to read each others' minds in both these ways. We have a fundamental urge to link our minds: to understand and to be understood.

EVEN CHILDREN APPEAR TO HAVE a basic drive to wire their minds into the social network of other minds. Babies have a special affinity for social stimuli, such as eyes and faces. Mothers typically try to establish prolonged eye contact as soon as possible, and newborns appear to be prepared for it. When given the option, they prefer to look at open eyes rather than shut eyes. As adults, we use eye contact to make apparent mind contact, for instance, when inviting interaction.[2] As the proverb goes: the eyes are the window to the soul. Through extensive face-to-face interactions, parents and infants build deep bonds.[3]

From two months onwards, infants begin to smile when their parents smile at them. Over the next months they learn to follow gaze, at first to objects within their visual field and later to points of interest behind them. Infants and adults begin to attend to the same objects and mutually interact with them—for instance, in give-and-take games. The developmental psychologist Chris Moore has highlighted the fundamental importance of these three-way (parent, infant, object) interactions for children's social-cognitive learning. Infants start to check the facial reaction of their parents to glean information about how to handle ambiguous situations.

By the end of the first year infants start to point to objects. It was a memorable day when my son, Timo, suddenly began to point—as if the penny had dropped, he pointed out everything he was excited about. He had gained the power to direct my attention to objects in the world. And he could get me to fetch them. Pointing is more than a tool children use for manipulating parents to get what they want, however. They draw others' attention to things because they are interested in them, often without receiving any reward other than shared attention for these efforts. They are strongly motivated to keep making links with minds around them.

With the development of language, the opportunities for making such links grow. Language turns mind reading into mind telling. We tell

---

2 For this and some other reasons, congenitally blind children are typically delayed in their development of theory of mind.

3 There do appear to be cultural differences, however. For instance, in one study Israeli parents showed much more face-to-face interactions with their five-month-old infants than Palestinian parents, who, in turn, created more affiliation through touch.

each other our experiences, opinions, wants, and needs. When an infant points to an object, adults typically name it. By the end of the first year infants articulate their first words, and adults increasingly point out new words during bouts of joint attention. In a sense, words themselves provide the opportunity for more effective sharing in attention. You say a word, and the attention of people around you is drawn to what it stands for. Koala. One amazing property of words, of course, is that they allow us to share attention not only to what is around us but to things that are not there. We may not be able to point to last week's guest, tonight's sunset, or cute, eucalyptus-munching marsupials, but we can raise them in conversation. Language allows us to jointly consider stuff that we only represent in our minds. You, dear reader, have just had your attention drawn to that very fact.

More than simply drawing attention to things, language allows us to comment on them. We can transmit potentially useful information about absent objects and events from one mind to another. Even toddlers can do this: "floor wet—naughty dog." Of course, they do not always have enlightening information to contribute. Yet they often express a ferocious appetite to get involved in these exchanges. When four-year-old Timo tells my spouse, Chris, and me something, two-year-old Nina frequently, and excitedly, interrupts. She calls out, "Mama, Mama, Mama; Papa, Papa, Papa," until she finally gets our attention, only to repeat what Timo has already said—or to inform us again of the fact that "I a girl; Timo boy." Nina has no new information, but she won't stop until her contribution is properly acknowledged. Timo has long figured out the importance of adding new information and contributes accordingly (rolling his eyes at Nina's repetitions). He, like most adults, takes particular delight in sharing a surprising morsel of information or even a secret. He wants to know how you feel and makes sure you know how he feels.

OUR URGE TO LINK OUR minds permeates much of what we do. We spend a lot of our social lives exchanging gossip, opinions, and advice. We listen to stories, read books, or watch shows that let us see the world from someone else's perspective. Technological advances, from radio to the internet, allow us to wire our minds together in ever faster and more efficient ways. Today, what is on someone's mind can spread to millions of other minds across the globe within minutes.

Nested processes are also involved in mind reading. I am currently thinking about what you might be thinking about my thinking about thinking about thinking. Recursion and our open-ended capacity to imagine different scenarios appear to be crucial not only for language and mental time travel but also for theory of mind. Indeed, one of the main points in Michael Corballis's and my original paper on mental time travel was that the same mental machinery we use to simulate past and future events may be used to simulate other people's minds. I long thought this was an original idea, but Nick Humphrey pointed out to me—and I need to embed this further, because since then Chris Moore told me that Nick learned it from him—that the essayist William Hazlitt had linked mind reading to foresight a couple of hundred years ago when he wrote: "The imagination . . . must carry me out of myself into the feelings of others by one and the same process by which I am thrown forward as it were into my future being, and interested in it."

We may like or dislike our thoughts or desires and, to embed this further, evaluate our own evaluations. Because we can imagine past and future scenarios, we can reflect on our own past mental states (e.g., I should have known) and imagine future states (e.g., I will not get upset) just as we can about those of others (e.g., she will be happy about that). Being interested in our future well-being, we may decide to change present behaviors, even if they are currently enjoyable. For instance, you may decide to skip the next schnapps and start drinking water instead, to save your self of tomorrow from a wicked hangover. It can be difficult looking after not only your present desires but also those of your future self, as anyone who ever tried to give up smoking can testify. But we can aim to change. Reading our own future minds, we may realize that we will be, say, embarrassed because we are completely underprepared for what is ahead—unless we do something. So we may choose to practice skills we anticipate needing to be better at or seek information that we foresee will be useful later. This adds to our sense of free will as, to some extent, we can attempt to deliberately shape our future self. Of course, as noted earlier, people vary considerably in how much they take their future into account.

People also vary in how much they worry about the minds of others. Males and females, on average, tend to differ in the time and effort they put into mind reading. (No prize for guessing which sex is less inclined to think about others' minds.) It has been suggested that some mental disorders are extreme versions of these female and male tendencies. People with paranoid schizophrenia tend to spend a lot of time pondering complex ideas about what other minds might be up to, whereas people with autism characteristically think very little about what is on others' minds. There has been a lot of research on disorders of theory of mind in recent years. Autism research, in particular, has frequently been driven by an interest in autistic people's peculiar limits to making mental connections with those around them.

In spite of some diversity in people's particular beliefs about minds,[4] basic mind-reading capacities are otherwise universal. Across cultures, for instance, people can recognize the facial expressions of our basic feelings; you can recognize fear, anger, disgust, surprise, sadness, and happiness in people from any culture as easily as they can recognize these emotions in you. As mind reading is fundamental to human interaction and cooperation, it is another good candidate for a uniquely human characteristic. Indeed, several prominent comparative researchers have argued that other animals do not read minds. The assumption is, although they probably have minds of their own, they might not make inferences about the not directly observable minds of others. A provocative killjoy review, for example, was titled: "On the lack of evidence that non-human animals possess anything remotely resembling a 'theory of mind.'"

THE NOW POPULAR STUDY OF theory of mind actually began with a paper on chimpanzee cognition. In 1978 the comparative psychologists David Premack and Guy Woodruff published an article in the inaugural volume of *Behavioral and Brain Sciences* that kick-started this entire research area. They reported results of studies with a chimpanzee called Sarah that suggested she was reasoning about minds. Sarah was presented with short videos of a human actor facing a problem such as trying to get out of a cage, and then with a selection of photos of which one depicted something crucial, such as keys, to achieving the man's goal. She reliably selected the correct solution. This behavior had three curious apparent implications. First, it seemed to imply that the chimpanzee could make sense of both videos and photos, and was able to relate the two (as we saw in Chapter 4, this competence has since been substantiated). Second, it suggested that

---

4 For example, some people believe and others do not believe in telepathy or in a vigilant divine observer of one's mind.

the chimpanzee knew a thing or two about problem solving (we will examine that issue in the next chapter). Third, and most important to the authors, it appeared that she attributed an intention to the actor in the video: she seemed to infer what the actor was trying to achieve. Intention is a mental state, and Premack and Woodruff therefore suggested that chimpanzees have a theory of mind.

Behavioral and Brain Sciences is an unusual journal in that it publishes dozens of commentaries on a target article as well as a reply from the initial authors. The commentaries to this article raised various problems with Premack and Woodruff's experimental design, making leaner interpretations possible. Nevertheless, they underscored the importance of the core question: If even chimpanzees might reason about minds, how can human scientists—particularly the remaining hard-core behaviorists—ignore the mind in their theories of behavior?

Three commentators scratched where it itched the most. They independently spelled out a solution to the problem of how one could unequivocally demonstrate that an animal or a child is reasoning about the minds of others. They argued that one needs to show that the individual in question understands that others act according to how they see the world, regardless of whether these views are factually true or not. In the case of true beliefs, one's views and reality are by definition identical, and there is therefore no empirical way of distinguishing whether somebody's actions are based on observable reality or on an inferred mental state in the other. In the case of false beliefs, however, mind and reality diverge, and someone who realizes this can predict the misguided actions of others who hold false beliefs. To understand that someone falsely believes something, one needs to be able to think about beliefs and their relationship to the world. This requires similar nested processes to the ones we encountered in Chapter 4 with four-year-old Rory painting herself painting a picture. Rory demonstrated an understanding of the representational relation between a picture and what it depicts. Reflecting on this relation (metarepresentation) allows you to ask to what extent your picture correctly depicts the real world and what you got wrong. Similar reflections allow you to ask to what extent the beliefs of other people represent or misrepresent reality. If we can show that a child or an animal expects someone else's behavior to be based on false beliefs, they demonstrate mind reading.

In 1983 the developmental psychologists Heinz Wimmer and Joseph Perner published their seminal studies on false-belief understanding in children. They told stories about a character, Maxi, who had put his chocolate somewhere, only for his mother to move it elsewhere during his absence. The children were then simply asked to predict where Maxi would look for the chocolate. Wimmer and Perner's study, and the hundreds of variations that followed, found that young children persistently claim that Maxi will look where the mother had moved the chocolate. Older children, however, can put their own knowledge of the facts aside and realize Maxi will first unsuccessfully search for the chocolate where he had put it. They appreciate that Maxi's search will be based on his false belief about where the chocolate is, rather than on what the observing child knows to be its true location. At this stage, then, children understand that people generally act according to how they represent the world—whether these views are correct or not. They can now take this into account in their conversations (as we saw in Chapter 4). They are certified mind readers.

Extensive research on false-belief tasks has found that this developmental pattern is quasi-universal across cultures. By age three and a half children tend to pass these tasks. The capacity develops slightly earlier in children who have older siblings from whom to learn and in children who do better on language tasks. This suggests that the social communicative environment in which children grow up influences the development of theory of mind. In line with this, it has been found that deaf children of hearing parents who only start to learn sign language late are delayed in passing false-belief tasks, whereas deaf children born to deaf parents, who are exposed to sign language from the start, develop the capacity to pass these tasks like their hearing peers.

Developmental psychologists have identified a whole battery of links between the emergence of false-belief understanding and different concepts. One important finding has been that at about the same time that children first become able to attribute false beliefs to others, they also begin to attribute them to themselves. That is, when asked what they thought was inside a candy box before they were shown that there were, say, pencils and not candies inside, younger children tend to state that they thought all along there were pencils inside. They do this in spite of having been excited about the prospect of candy just a minute ago.

Young children do not seem to grasp how they come to know what they know. If you put an object in a bag and ask toddlers how they would find out the object's color—by putting their hand in it or by peeking inside—they choose arbitrarily. Very young children do not seem to appreciate how perception and knowledge are linked. This is why they happily "show" you their new toy when you are on the phone with them (with video phones, of course, this example may not work anymore). In studies in which children were either shown or told new information such as the content of a drawer, children later failed to accurately explain how they know what they know. When you play hide and seek with them, you have to accept that they may want to go back to the same hiding place repeatedly. They do not yet appreciate that the aim is to get out of mind, not just out of sight.

The developmental psychologist John Flavell showed that children around this age also fail to distinguish between the appearance of an object and reality. We generally take for granted that when milk is poured into a blue glass, it does not change the milk's color but only our perception of it. However, young children have difficulty appreciating that something can look like one thing and still be another. They find it hard to consider two contradicting views of the same thing. Here the child needs to integrate two opposing thoughts about the milk—"is blue" and "is white"—and may need to tag these thoughts as "looks like" and "is truly."

Being able to concurrently consider contradicting interpretations of the same object or event is required in a lot of mind reading. Without it there can be no lies. You may say something that is not true—we all make mistakes—but to actually lie, you need to know that what you say is not true, and, furthermore, you need to want someone else to believe that it is true. In other words, to lie is to knowingly implant a false belief. Without mind reading there can be no pedagogy, for deliberate transfer of knowledge requires some understanding of what is or is not known by the pupil to devise a way in which future knowledge can be acquired. In short, theory of mind is essential for normal human cultural and social interaction.

ALTHOUGH RESEARCHERS HAVE LONG FOCUSED on three- and four-year-olds, there are both earlier and later developments in this arena. For a while, developmental psychologists were nearly obsessed with the classic false-belief task as the watershed. The fact that passing such tasks demonstrates mind reading was understandably enticing. However, failure on such tasks does not necessarily entail absence of a capacity for mind reading per se. Nor does passing the task mean that a child has achieved all there is to achieve.

For example, even five-year-old children have problems with increased embedding, such as I *expect* that you *think* that everyone *knows* that. In fact, even adults can typically only juggle five or six such nested ideas before we lose the plot. I *think* that you *suspect* that I *intend* you to *believe* this. Joey, in the TV sitcom *Friends*, for instance, struggles desperately to follow when Phoebe, on discovering that Chandler and Monica know that she and Rachel know that they are dating, succinctly states: "They don't know that we know they know we know!" She warns Joey not to tell the others, and he despairs: "I couldn't even if I wanted to." As Robin Dunbar pointed out, only exceptionally clever writers like Shakespeare make our minds stretch with five or six levels of nesting, as he, for instance, *intends* us to *believe* that Iago *wants* Othello to *suppose* that Desdemona *loves* Cassio, who actually *loves* Bianca.

Children have plenty to learn about embedding and the numerous intricacies of the mind. For example, it takes them some time to get to properly appreciate faux pas. Consider, for instance, a scenario in which, say, Jane visits Frank and then accidentally breaks his bowl. The bowl is a present that Jane gave Frank some years ago. Now imagine that when Jane apologizes for the accident, Frank, not remembering how he got it in the first place, tries to put her at ease by saying, "Oh, no worries. I never liked the bowl anyway." Young children do not understand such a violation of etiquette. To appreciate it, one has to combine several different types of mental states: Frank did not remember where he got the bowl from, he did not have an intention to hurt Jane, but hurt she is. It can all be rather complicated, as I suspect you know. Indeed, our learning about the sensibilities and workings of other people's minds might never stop. Just consider the misunderstandings characteristic of interactions between romantic partners or between different cultures. Many adults keep trying to learn more about minds. We may meditate, read self-help books, go to workshops, or study psychology.

Sometimes minds seem to be awfully complex and unpredictable, and we may even despair at not understanding one another. Just as we get our time travels wrong, we regularly misjudge what is on others' minds. And sadly we often feel misunderstood. At other times, we seem to see right through someone else's thinking and "read them like an open book," as the saying goes. We have meetings of minds and sometimes feel we thoroughly understand each other. We fall in love with each others' minds.

One reason mind reading is such an exceedingly complicated affair is that people frequently say one thing but mean quite another. I refer here not only to lies but to devices such as irony and metaphor. You may, for instance, say, "That's just wonderful; it's exactly what I needed," on hearing news about a strike at the airport—as you are packing your bags. In irony we may state the opposite of what we mean, and if we do this in a harsh tone, it turns into sarcasm. We may exaggerate or understate, make fun through satire or parody, make vague suggestions, or employ double entendres. Not only actors can pretend to have one mental state while in fact being in quite a different one. People can cry and yet not actually be sad, and vice versa. We have some control over what we want to reveal and what we want to conceal about our minds. We can manage the impressions we give. Strategic games unfold as we try to influence each other's minds. It is the stuff of our lives—or at least of our soap operas.

While there is much left to learn once children pass false-belief tasks, failing these tasks need not mean younger children are clueless about minds. Indeed, as we have seen, newborns already have a special interest in social stimuli and over the first few months at least appear to develop some appreciation of the minds of others. They form expectations about the goals and intentions of animate objects—even of cartoons on computer screens. Joint activities, shared attention, gaze following, and declarative pointing suggest that minds increasingly matter to them. In the second year they show some competence at appropriately interpreting emotions, desires, and intentions. They copy what someone else intends to do, even if that person fails to achieve the goal. They get upset if you do not join in and attend what they attend to. "Look at me, look at me," my two-year-old daughter is yelling in the background, as if to illustrate the point.

In fact, even toddlers may have some understanding of false beliefs. When examining spontaneous responses, such as looking direction, toddlers and young children seem to have an expectation of where someone who holds a false belief about the location of an object will look first. When you ask them, however, they will insist that this person will search for the target where it actually is, not where it is mistakenly believed to be. This suggests that they have some problem inhibiting their own knowledge of the facts to reasoning about others'. But they may already develop a basic capacity for mind reading a lot earlier than previously thought. There continues to be debate about the nature of this early understanding.

The psychologists Ian Apperly and Stephen Butterfill, for instance, argue that humans have two different systems that track belief: one that is implicit and evident in toddlers, and one that is explicit and develops later in preschoolers.

Researchers are increasingly finding evidence that theory of mind develops gradually through social scaffolding and not with the big bang once imaged. The developmental psychologists Henry Wellman, Candi Peterson, and colleagues, for instance, have documented how mind reading typically progresses from an early understanding that individuals can have diverse desires to an understanding of diverse beliefs. Children pass standard false-belief tasks before they appreciate that someone can feel one thing and yet display a different emotion.

Do other animals show any of these capacities? Do they consider the minds around them? Do they stare deeply into each other's eyes and link their minds? Since Premack and Woodruff's paper, extensive efforts have been made to find out.

IF YOU LOOK STRAIGHT AT a rhesus monkey you might well be attacked. For primates, staring into another's eyes is typically a threatening gesture. Therefore, primates largely avoid eye contact, and face-to-face interactions are surprisingly unusual. Even chimpanzees look into each others' eyes only on rare occasions.[5] When you stare into the eyes of chimpanzees, one thing you will notice is that their eyes do not show any white. Human eyes physically differ from those of other primates in that our sclera is white and more of it is exposed in relation to the eye outline. Human eyes signal gaze direction. We advertise where we look, and we read where others are looking. The eyes of other primates, if anything, seem to camouflage gaze direction. They do not roll their eyes to express disdain or shed tears to express their sorrow.

We use the eyes extensively in mind reading without words. For instance, soccer players often rely on this when taking penalty kicks. In my youth I could reasonably reliably convert the shot by simply peeking briefly at one corner of the goal, running up, and then slotting the ball in

5   Chimpanzee mothers and infants sometimes show some brief bouts of gazing in each other's eyes. For the most part, however, primates, even mother and offspring, do not spend much time in such gazing.

FIGURE 6.1.
Ockie and me. Chimpanzees on the odd occasion do make eye contact.

the other corner. This technique relies not on the accuracy or velocity of the shot but almost entirely on fooling the keeper. Eventually clever keepers began to pick up on this simplest of tricks and tried to thwart me by jumping in the opposite corner to the one I looked at. Some even moved closer to the other post and hence "offered" one side. The battle became increasingly more difficult, as I had to size up the keeper's ability to read my intentions and do the opposite of what I thought he thought. This is "theory of mind" in action.

Following Premack and Woodruff's paper, field studies added enthusiasm for the possibility that other primates read minds and demonstrate this in their actions. The work of primatologists like Jane Goodall suggested that primate societies are much more complex than had been previously assumed. The notion that social pressures drove the evolution of intelligence gathered momentum, and it was fitting to suppose that primates may have evolved some mind-reading capacities to understand and control the behavior of others. For instance, reports of apparent tactical deceptions in primate societies suggested that they can reason about each

others' desires and beliefs—and that they may be able to deliberately implant false beliefs.[6] In a classic example, a baboon was observed mating
with a female behind a rock in a position such that the alpha male of the
group could see his head but not what he was doing. Was the baboon in-
tending to mislead the alpha male into thinking he was doing something
else? As we have seen, leaner explanations can also account for such observations. For instance, the baboon may have been punished in the past
by the alpha male when mating in the open but not when mating behind
rocks. So he may mate behind rocks without necessarily reasoning about
what the alpha male can or cannot see and know. Nevertheless, early laboratory experiments supported the richer interpretation of mind-reading
in at least our closest living relatives. In the early 1990s the comparative
psychologist Daniel Povinelli reported studies that suggested chimpanzees,
though not monkeys, might be able to reason about others' knowledge
and take their perspective.

Following these initial findings, Povinelli set up his own chimpanzee research center at the University of Louisiana and examined their
understanding of seeing, pointing, intention, and knowledge. To general
surprise, he failed to find additional support for chimpanzee theory of
mind. Instead, he discovered numerous reasons for lean interpretations
of their behavior. In scores of studies, he obtained consistently negative
results from his group of young chimpanzees. For example, they would
beg from someone who was wearing a bucket over her head just as often
as from someone who could see what was going on. When one trainer
saw where food was hidden and another could not see this hiding—
because they had left the room, looked away, or wore a blindfold or a
bucket—the apes were equally likely to follow the advice of the ignorant
as of the knowledgeable human. Chimpanzees that had been trained on a
cooperative task, such as pulling a box with two ropes, ignore that a naïve
chimpanzee lacks the relevant knowledge to get the task done and do not
teach them. Povinelli championed the killjoy cause and concluded that
chimpanzees only reason about behavior, not about minds.

Povinelli's research suggests that theory of mind is uniquely human.
Great ape behavior may be driven by more basic calculations. Indeed, we

---

6 An alternative possibility is that the need to understand other species drove the evolution
of mind reading. Predators, for instance, could benefit from better prediction of prey behavior
and prey species from being better able to predict predators. This may have led to an evolutionary arms race of wits. In support of this theory, note that there is evidence that mammalian
predator and prey brain sizes have increased in tandem over geological time.

might be misguided if we think our own behaviors are caused by theory of mind simply because we explain them in such terms. Perhaps we often merely *reinterpret* behavior in mental terms. For instance, to keep with the theme of soccer that I kicked this section off with, when an offensive player tries to dribble past a defensive player, a common trick is to make the opponent commit to one side and to delay one's own final move in order to capitalize on that commitment. When we explain these actions afterwards, we may say that we wanted to fake out the defender, misleading him into thinking we were going to go the other way. Yet it is not clear such thoughts drive our actions. We have time before a penalty to plot deception and counterdeception. In the heat of the action, however, we can't stand there and formulate explicit notions of each other's intent. We may automatically go through all the motions and only afterwards interpret them in mental terms.

Povinelli and colleagues argue that ape behaviors that look as though they might be driven by mind reading, such as apparent deception, empathy, or grudging, may not be. When we see chimpanzees engage in a chase of rapid side-to-side movements, we may be misinterpreting their behavior in terms of faking a turn to fool the pursuer into thinking a certain way; the apes may merely be concerned with actions.[7] Only humans, they claim, have evolved a capacity to reinterpret behavior in mental terms.

There is an alternative explanation to this proposal, however. Andrew Whiten and I have suggested that in some cases the rapid actions of deception and counterdeceptions are only automatic because we have initially engaged in a lot of practice involving explicit cognitive considerations. Human skill development is replete with examples in which behaviors that had once been governed by slow conscious processes become automatic with training. Just think of the complexities of driving a car. Initially you need to carefully think about what your arms and legs are supposed to be doing to control the vehicle. With experience you can focus on having a conversation or listening to the radio without attending to how your autopilot, as it were, is driving the car. Similarly, strategies that allow one to get around defenders in soccer may become automatized only after extensive

---

7   Such pursuits can certainly look thoughtful, though. Consider the following example of a male chimpanzee aggressively chasing a female. As the female sought cover behind a tree trunk, the male moved to the left, which prompted the female to move to the right. In full motion, the male then threw a brick in the direction of her rightward path while continuing to move to the left himself. To avoid the projectile the female changed directions, only to find herself caught by the male.

training. It requires a lot of practice to become a good player. Perhaps theory of mind also develops like other skill acquisition, from effortful, explicit, and controlled to fast, automatic, noneffortful processing that comes with practice. Initial mental simulations can give way to quicker shortcuts. For example, we may tend to quite automatically follow the gaze of others not, as Povinelli suggests, because we have some low-level mechanism for this but because we had sufficient experience and practice with these situations.

In spite of Povinelli's considerable output, other laboratories—particularly the Max Planck Institute for Evolutionary Anthropology in Leipzig—gradually flooded the field with findings supporting richer interpretations of animal behavior. The comparative psychologists Michael Tomasello and Josep Call and their colleagues showed, for example, that gaze-following abilities in apes were more sophisticated than previously thought. Even dogs and monkeys appear to be able to follow gaze under some circumstances. Chimpanzees can project somebody else's line of sight geometrically beyond their own immediate visual field and do so even around barriers. They move in ways that allow them to see what another is looking at—even if the target is behind an obstacle. This suggests that chimpanzees may interpret gaze in terms of what the other person might see. They sometimes even glance back and forth as if to check what in fact the experimenter is finding so interesting.

In collaboration with Brian Hare, the Leipzig group produced some ingenious studies supporting the possibility that chimpanzees reason about minds. In one set of experiments chimpanzees had to compete with another, more dominant chimpanzee for food. The chimpanzees were found to preferentially head toward seizure of whichever of two food items was visually screened from the dominant competitor. This suggests that they do understand something about what another can see. When the opaque screens were made transparent, the preference for the "concealed" food vanished, presumably because they recognized that it no longer blocked the view of the competitor.

Rhesus monkeys similarly act as if they know what another sees. When given the option between a grape in front of a person who looks at it and a grape in front of a person who looks away, or whose sight is blocked, they consistently prefer the latter—in apparent recognition that "stealing" from someone who is not looking is safer. Chimpanzees in earlier experiments by Povinelli and others might have failed because the tasks involved cooperation, such as a human informing the ape about where food is. That is not a particularly natural situation. Unlike human toddlers, who constantly want to point out things to their parents and others, primates do not seem to have much inclination to inform others. They tend to compete for rewards instead, and so it may not be surprising that we can observe their competence better when tested in competitive situations.

Unfortunately, these newer results, in spite of their substantial publicity, do not actually prove that the primates reasoned about what the human could see. A much leaner explanation is available. It remains possible that the monkeys simply learned that a human facing the grape is more likely to interfere with their attempt at obtaining the treat than one that is not. The same is true of the chimpanzee example. The lower-ranked individual may have learned purely behavioral rules, such as: if a dominant faces the food, it is not safe to approach.

But other results suggest that chimpanzees appreciate not only what another sees but also what another has previously seen. In an extension of Hare and colleagues' previous experiments, the chimpanzees took into account whether a dominant competitor had observed the hiding of a food item or not. When the placement of food behind a particular screen was witnessed by the dominant, the subordinate was subsequently less likely to approach the food than when it was not witnessed. In another experiment, the dominant saw the hiding but was then replaced with a different dominant animal that did not see the hiding. Again, the subordinate chimpanzee was more likely to approach the treat when the competitor was ignorant than when he was knowledgeable about the food location. So it remains possible that chimpanzees, after all, have some capacity to reason about the minds of others.

Indeed, other studies support the possibility of competence in great apes. Like two-year-old children, some great apes appear to recognize what someone else is trying to achieve, even if the attempt failed. Some results suggest that they can distinguish accidental from purposeful actions. Other experiments suggest that they can discriminate between someone who is unwilling to do something and someone who is unable to do it. One study suggests that they may be able to distinguish appearance from reality, and they seem to be able to take advantage of when a competitor cannot see them.

There is some indication, though scarce, for such capacities in other species. Grey squirrels, for instance, space their caches farther apart when observed by other squirrels, presumably to avoid pilfering. They even preferentially cache while oriented with their backs to other squirrels. Similarly, scrub jays, when in the presence of a potential competitor, preferentially cache food further away, in darker and more occluded areas, than they do otherwise. Hence other species may also take into account what another can or cannot see. Then again, they may not.

Unfortunately, none of these behaviors need imply any reasoning about the mind of the competitor. Squirrels, jays, and apes in all of these cases may behave as they do simply on the basis of observable behaviors. Acting one way gets a reward; acting another way leads to punishment. Recall that to demonstrate an animal is taking the mind of another into account, one has to show attribution of false beliefs. In spite of several clever attempts at this holy grail of theory of mind research, so far no nonhuman animal has passed false-belief tasks. Even chimpanzees that have demonstrated some impressive competence at other components of the task fail when false beliefs need to be taken into account. It is therefore possible to maintain a lean interpretation, as many comparative psychologists do, and conclude that no other animal has anything like a theory of mind.

MY OWN HUNCH, ACTUALLY, IS that the truth lies between the extreme romantic and killjoy positions. The wealth of recent positive data from great apes suggest, though they certainly do not prove, that they have limited understanding of basic mental states. It will be interesting to see whether apes demonstrate any sign of the early (implicit) false-belief understanding that has recently been documented in infants. Researchers are trying to use eye-tracking devices to test chimpanzees on such tasks, but it is not easy. Perhaps results on such tests will become available by the time you read this book.

Given that great apes perform like two-year-old children on a variety of tasks that involve considering more than meets the eye (as discussed in Chapter 3), it would not be surprising if they did so here. They may have a limited, perhaps implicit, understanding of what another sees, believes, knows, attends, desires, and intends, but this possibility should not detract from the remaining gap between the mind reading in humans and the limited reasoning that on current evidence maximally exists in our closest animal relatives. Although Povinelli's group and Tomasello's group, the two most influential research laboratories in this field, have been at loggerheads about lean and rich interpretations of data on ape theory of mind, they both agree that there is no sign yet of any false-belief understanding. Thus we have consensus that there is something unique about human theory of mind. Povinelli argues that only humans have a theory of mind, period. Tomasello, Call, and colleagues believe that their findings are more parsimoniously explained by granting great apes some basic mind-reading capacities. Yet they, too, maintain that apes lack the most fundamental human socio-cognitive skills.

In fact, Tomasello and colleagues argue that great apes do not even show the basic social awareness that is typical of human infants when they point, show, or offer things because they want to communicate about them. These authors argue that the main difference is that humans have what they call "shared intentionality." As we have seen, humans have a fundamental motivation to share their own mental states with others. This inclination allows us to construct a sense of "we" that enables us to collaborate on unheralded flexible scales—socially constructing tools, meals, games, and theories (drawing extensively on language and mental time travel). Infants demonstrate an inclination toward this sharing early, well before they pass false-belief tasks. For instance, when one-year-olds are engaged in a collaborative task with an adult and the adult stops, the infants typically try to get the adult to reengage. Infant chimpanzees, on the other hand, simply try to do the task themselves. Chimpanzees may gesture to get someone to do something for them, but humans often gesture (and talk) just to inform.

Chimpanzees do not point to each other in the wild. This may be the case because it is pointless to point (pardon the pun) if the other chimpanzees do not give you what you want. In experiments, they are poor at using and providing social cues in cooperative tasks requiring communication. For a while experiments suggested that great apes, unlike dogs, could not even understand pointing by humans. More recent work shows that they only struggle when the human points to things that are close to each other; they can discriminate when the options are far apart. Great apes can also learn to point to humans but do so virtually only to request, rather than to declare. (Recall that only some 5 percent of utterances in "language-trained" apes could be categorized as resembling statements or declarations.) Children, on the other hand, constantly want to point out things to share information. My own children are quite insistent that I drop everything and join in the excitement. Humans may be uniquely motivated to establish links between their minds, through read-ing and telling, that allow us to create common mental worlds of goals, ideas, and beliefs.

Great apes may have some capacity to reason about basic states of mind. Even so, they do not seem to have a particularly strong drive to build connections between their minds, which may significantly limit what cooperation they can mount. A recent large-scale examination of over a hundred great apes found that they performed similarly to two-and-a-half-year-old children on a battery of tasks on physical cognition, but much more poorly on tasks of social cognition.[8] Of course, such com-parisons depend on what exact tasks are used: social tasks in which apes have to interact with human adults may not be comparable to social tasks in which human infants interact with human adults. Nonetheless, the re-sults add to the mounting evidence that humans' urge to link with other minds is unique. After over three decades of research, there is no strong evidence that apes understand the representational character of beliefs. It remains possible that they only reason about observables, rather than about minds.

Humans evidently are mind readers. Though we may frequently get it wrong, we can read and tell our minds sufficiently well to enable us to co-operate in a multitude of clever ways. We share ideas, advice, and goals. We can develop intricate plans and collaborate in bringing them to fruition. We teach and learn from each other's experiences. We set out to entertain each other and care about what others might find funny or pleasing. We come together to share attention in celebrations and performances. Our cultural inheritance, as we will discuss later, can be conceived of as an accumulation of cooperative exchanges between minds over many gener-ations. We reflect on the nature of mind reading. We even spend consid-erable effort trying to overcome the obstacles associated with establishing the nature of animal minds. Much of this scientific mind measuring has focused on animals' capacity for smart problem solving. And so we turn now to research on intelligence.

---

8 Note that such a coarse distinction between social and physical cognition is far from clear-cut. Social factors often feature in physical tasks (e.g., when presented by a social other), and social tasks often involve reasoning about physical components.

SEVEN

Smarter Apes

Man is most uniquely human when he
turns obstacles into opportunities.

—ERIC HOFFER

HUMANS HAVE OVERCOME MANY OBSTACLES on the way to planetary dominance. We have created light where we could not see. We have created warmth where we were cold. Our smarts have given us tools to do what could not be done before, whether it's hunting from a distance or curing illness. Increasingly, our technologies give us control over what we care about. We keep finding new solutions to problems and call this progress— all while tacitly ignoring the fact that most of our solutions have spawned new problems of their own, from hard labor to pollution.

We are a long way from creating worldwide utopias with sustainable supplies of what we want. Even basic necessities, such as sufficient food and clean water, remain out of reach for many human communities. But no matter to what embarrassing failures or glorious successes you might point, it is self-evident that humans can be outstandingly resourceful and

clever. We survive by our wit. We keep finding ingenious solutions to obstacles and turning them into opportunities.

Other animals also have effective ways to solve problems such as finding food, shelter, and partners. They generally cope well with whatever re-
current challenges they face. It could hardly be any other way. Just as their
bodies are adapted to their environments, so are their senses, cognitions,
and behaviors. On many tasks other species are clearly our superiors. Some
sharks, birds, and turtles navigate according to electromagnetic fields we
do not even perceive. Bats, dolphins, and some shrews use echolocation to
scan their environment. Bees use optic flow to control their flight in such
an elegantly simple way that humans are now trying to implement it in
aircraft navigation. There is a difference, however, between an individual
figuring out a clever solution and mechanisms that are universal to all
members of a species. Some clever-looking animal behaviors are hardwired. In these cases new obstacles may remain barriers they cannot turn
into opportunities. We saw in Chapter 5 that digger wasps, for instance,
always inspect their nest before dragging food inside and are incapable of
snapping out of this routine even when an experimenter keeps moving the
food away.

Yet some animals demonstrate flexible problem-solving capacities, and
this is not restricted to big-brained primates and cetaceans. For example,
some Australian crows have figured out a way of eating poisonous cane
toads. These toads were introduced to Australia in 1935 and have become a
great pest. The birds learned to turn the toads onto their backs and then
peck at their harmless bellies. Even invertebrates may have considerable
smarts. We have seen that some cephalopods show immense deceptive capacities (and octopuses are invertebrates that enjoy vertebrates for lunch).
Whether or not these examples can be explained by lean interpretations,
we should not be presumptuous about intelligence only existing in vertebrates, in mammals, or in primates. I suspect that so far comparative psychology has merely scratched the surface in its attempts at documenting
animal problem-solving capacities.

Nevertheless, human minds demonstrate intellectual flexibility that
appears to be unmatched. What might be unique about our intelligence
and creativity? We can start by examining what research has uncovered
about the nature of human intelligence.

RESEARCH ON INTELLIGENCE IS IN some sense one of the greatest success stories of the discipline of psychology. Millions of people are given intelligence tests every year. In the Netherlands, for example, for decades virtually all young males have been tested. Intelligence tests originated from two main sources. In England, Sir Francis Galton, a cousin of Charles Darwin, believed that intelligence was a combination of sensory acuity and effort, and he developed various tests to measure these factors. But people became disenchanted with Galton's approach, as scores on his tests failed to make good predictions about, for example, who would perform well at school. In France, intelligence tests derived from just such pragmatic concerns. The French government asked Alfred Binet to devise an objective test to identify children who would not benefit from normal classroom teaching with their same-aged peers. In response Binet developed a battery of tasks—including measures of memory, general knowledge, problem solving, and so forth—based on his view that intelligence was an aggregate of diverse abilities. He gave his test to many children to establish their competences and calculate the average score for each age group. These averages became his yardstick. If you performed like the average twelve-year-old, your mental age was twelve. Simple.

In 1912 the German psychologist William Stern used mental age to derive the infamous Intelligence Quotient. The IQ was your mental age divided by your actual chronological age times 100. If you are 10 years old and perform like the average 12-year-old, you have an IQ of 120. If you perform like the average 10-year-old, your IQ is 100. This quotient, however, only makes sense for assessing children. If you extend this logic into adulthood, then you would find that everyone would want to make comparisons to the aged. After all, if you perform like the average 90-year-old and are in fact 30 years old, you would have an IQ of 300. Today's measure of IQ is not really a quotient but your score's relative position in a standard population distribution.[1] Binet's test was revised in the United States into the Stanford-Binet test, which, together with the Wechsler scale, is still the most widely used intelligence test.

Chances are you have taken one of these tests at some stage in your life. They test your capacity to comprehend and define words; share knowledge

---

1   Individual scores are converted to a scale on which 100 still indicates the population mean. The distribution has a standard deviation of 15, and this entails that just over two-thirds of people's IQs fall between 85 and 115. If your IQ is 115, you score higher than 84 percent of the population; if it is 130, only 2 percent of people have a higher score.

of some facts; reason by analogy, deduction, and inference; solve arithmetic problems; repeat strings of digits backwards; put puzzle pieces together; copy a design; identify what is missing from a picture; put picture sequences in order; and translate signs from a novel symbol system. Being good at these things means you are intelligent. At least that is what many intelligence researchers believe, because a boring definition of intelligence is "intelligence is what the tests test" (as E. G. Boring wrote in 1923). This is circular reasoning, of course, but Boring noted strong findings supporting these tests. They unearthed individual differences, and children's relative rank order tended to be stable even though performance improved with age. Importantly, it has long been found that people who are good at one part of these tests also tend to be good at others. This points to a general intelligence factor, usually called "g," and this factor has been shown to make reliable predictions about real-life outcomes.

For the most part, IQ tests are given to assess someone's chances of success in training, job performance, and the like, rather than to measure their intellectual wealth per se. IQ tests predict various indicators of "success," from school drop-out rates to future income. During the twentieth century the tests therefore became increasingly popular in Western societies.[2]

Researchers have identified many variables that affect performance on the tests. Prenatal exposure to alcohol, for example, leads to reduced IQ in children, whereas having high IQ parents predicts high scores. IQ is highly inheritable, yet test scores have been increasing overall over the last one hundred years. This has stimulated debate about whether we are actually getting smarter or merely better at taking these tests.

Most importantly, what do these tests really tell us about the nature of intelligence? A basic consensus in the IQ testing community is that intelligence involves the capacity to learn from experience, to adapt to the surrounding environment, and to reflect on one's own performance. Because there is such a great wealth of IQ data available, many intelligence researchers have examined the relationship between performance on various subtests for clues about the underlying structure of intelligence.

---

2   Especially in the United States, intelligence testing has been of immense influence. People who score low on IQ tests are more likely to be incarcerated, bear children outside of marriage, and be welfare recipients than those who score highly. These data have led to some debate about whether American society is increasingly becoming stratified according to intelligence. Note, however, that testing itself affects this result, as only people with certain scores will be given certain opportunities (e.g., to study).

Alas, the resulting theories of intelligence have deeply contradicted each other. For example, although many researchers highlight a single general intelligence factor (g), others have shown that we need to distinguish at least two factors: crystalized intelligence and fluid intelligence. The latter refers to processing capacities that decline with advanced age, and the former refers to knowledge of facts, which does not tend to decline. Other theorists have made distinctions between 7 abilities (verbal comprehension, verbal fluency, inductive reasoning, spatial visualization, number, memory, and perceptual speed) and even between 150 (too many to list here). While some researchers subscribe to a hierarchical structure, others see discrete components. Worst of all, there is no clear way of deciding which of these theories is correct. Although intelligence testing has been a resounding success in some sense (such as in terms of predictions and money made), the research on millions of test scores and their correlations has not established consensus on the structure of human intelligence.

The IQ testing approach to studying intelligence certainly has its critics. One long-standing criticism is that the tests reflect specific Western values of intelligence and measures these only with a restricted range of artificial tasks. Consider, for instance, the fact that tests are often timed. Although speedy decision making may be a hallmark capacity of smart stock market traders or air traffic controllers, in other cultures and contexts speed may not be valued in the same way. In fact, for many cases of intelligent decision making, speed is quite unimportant compared to, say, getting it absolutely right. Consider weighty questions such as whom to marry, what house to buy, and whether to go to war.

IQs are established with paper and pencil in a quiet testing room. Yet the real world is noisy and often lacking in the luxury of quiet desk space. Universities are full of people with high IQ scores, and yet among them are some who—as an Australian might say—"couldn't organize a piss-up in a brewery." As the psychologist Robert Sternberg suggests, practical intelligence is quite distinct from the analytical intelligence measured in IQ tests. You can score low on IQ tests and be very smart in your practical life and vice versa. I suspect that some of the most successful people in life—certain politicians spring to mind—do not score unusually high on a standard IQ test.

There are in fact a few alternative accounts of intelligence. One scheme recognizes multiple intelligences, including linguistic, logical, musical, spatial, kinesthetic, naturalist, interpersonal, intrapersonal, and existential intelligence. You may also have heard of emotional intelligence as a popular addition. These proposals go beyond the standard tests and acknowledge the manifold capacities that people may have. It is often said that everyone has a talent—you just need to find it. In fact, the term "talent" may be more appropriate for many of these purported intelligences.

Whatever your view on IQ testing, the tests are not all that helpful for our purposes. We want to know how humans might differ from animals, not how we differ from each other. Since the tests all involve verbal instructions, we cannot simply give them to animals, though I foolhardily tried.[3] In order to compare intelligence in humans and animals, we must return to the essential foundation of what intelligence is. We can all recognize it when we see it, but researchers have been so preoccupied with individual differences that many have overlooked what intelligence we have in common. Steven Pinker offers the following definition: "intelligence . . . is the ability to attain goals in the face of obstacles by means of decisions based on rational (truth-obeying) rules."

This definition draws attention to two crucial points. First, intelligence is practical: it enables the overcoming of obstacles in pursuit of goals. To judge an act as intelligent you need to take into account what it is the individual wants to achieve. Someone may superficially appear a total fool (e.g., dropping things, forgetting others, making costly mistakes) but may still be acting intelligently. Given our capacity to reason about others' minds, we may intend to be perceived as stupid—for instance, if you want someone else to believe you are not cut out for a task you do not wish to do. Without a goal an action can hardly be intelligent.[4] Second, to intelligently achieve a goal the action must be based on reasoning by *rational* rules. If you get what you want by chance alone, you can hardly take the credit.

---

3 I once made a large version of a test called Raven's progressive matrices (a series of pattern completion tasks, highly correlated with IQ) and tried to test the chimpanzees Cassie and Ockie. Alas, they did not understand the basic premise, and I quickly abandoned the attempt.

4 In this light, the term "artificial intelligence" may be a misnomer. Computers have better memory than we do and may calculate numbers more quickly and accurately, but as long as they do not want to achieve anything, they may not be regarded as intelligent. This "want" refers not just to a goal (which is easily programmed) but to what William James called "having an interest." Computers do not care if you turn them off—as far as I can tell.

Man is a rational animal—so at least I have been
told. Throughout a long life, I have looked diligently
for evidence in favour of this statement . . .

—BERTRAND RUSSELL

ALTHOUGH ARISTOTLE PROCLAIMED THAT HUMANS are rational animals, we often fail to live up to expectations. The psychologists Amos Tversky and Daniel Kahneman documented numerous biases and heuristics that people commonly use to reach decisions. For example, we frequently base judgments on how easily we can call to mind relevant information and make a decision as soon as we have a satisfying answer. We therefore frequently fail to act optimally given available information. Yet we tend to be supremely (over)confident about our judgments and generally resist evidence that demonstrates we are wrong. In hindsight we are sure we would have predicted what we now know to have happened. Some researchers (as well as some fictional characters such as Spock and Dr. Sheldon Cooper) take great delight in pointing out the logical shortcomings of human thinking, and many studies support them. Suffice it to say, I am often irrational—and so are you.

In spite of this confession, humans evidently are capable of rational thinking. Bertrand Russell certainly was. We can try out potential solutions in our minds. We can infer and deduce, even if we prefer shortcuts. We can reason, even if we are often guided by emotions. We can think scientifically, even if we might prefer mystical explanations. A common bumper sticker in Australia reads, "Magic happens," and I had to smile when I saw ABC Science retorting with a sticker of its own: "Logic happens." It does.

A fundamental capacity involved in any form of reasoning is the ability to store and process information in one's mind. Differences in this storage capacity explain a lot about differences in reasoning and intelligence. Short-term memory has to be distinguished from long-term memory, because one can be intact even when the other fails. Most information is only briefly held in mind and then is lost forever. Try to recall the biases I listed two paragraphs ago. You may recall the gist, but you have probably lost much of the detail. Yet to follow a written passage, such as this one, you need to keep information in mind long enough to meaningfully link what you are reading with what you read before. Early research suggested that we can only hold up to seven (plus or minus two) chunks of information in short-term memory. When more information has to be considered, some of the earlier encoded information will be lost from short-term memory (unless it is transferred to a more long-term store). If I give you a number to read and you have to close your eyes and repeat it backwards in your mind, you should find it easy to do with five digits (48372) but much less so with ten (3747297497).

This is a consistent finding—unless you cheat. One way to cheat is to chunk information together so that the task of remembering, say, the ten-letter sequence AC DCA BCL OL turns into a three-chunk sequence ACDC ABC LOL that is easier to remember because it occupies only three slots by linking together familiar letter sequences. People who excel at memory tasks usually employ a host of such mnemonic strategies to increase performance. When it is made impossible to use these strategies and to rehearse, usually by asking participants to do a distracter task in parallel, recent research suggests that human short-term memory capacity is limited to a mere three to five chunks. Short-term memory is limited indeed.

Psychologists these days are inclined to speak of "working memory" rather than short-term memory, because the system is not merely a passive information store.[5] Working memory is our capacity to hold and manipulate chunks of information in our minds. We use working memory in all manner of mental activity, from simple tasks such as rehearsing a telephone number to creative endeavors such as designing a house. It is the workbench for our conscious mental operations. We can disengage from perception and imagine alternative scenarios, such as those we need for mind reading and time traveling. In a sense, working memory is the stage in the theater metaphor of mental scenario building. It allows us to reason offline, as it were. With sufficient working-memory capacity we can temporarily bind several concepts and reflect on their relationships. Thinking about thinking and other embedded processes are only possible when one can juggle several chunks of information in working memory.

Working-memory capacity constrains the number of relations one can consider together. This accounts for major differences in intelligence. In-

---

5 The psychologist Alan Baddeley proposed the term "working memory" and suggested that it comprises distinct components: a phonological loop and a visual-spatial sketch pad that operate independently from each other. Furthermore, he proposed a "central executive" that controls how these components are used. In a later version Baddeley added an "episodic buffer," a limited capacity store in which information can be integrated, combined, and manipulated.

deed, it has been established that how well people do on working-memory tasks predicts how well they score on reasoning and intelligence tests. As much as half of the variability in IQ can be explained by variability in working memory.

Children increase their working memory capacity steadily between ages four and eleven, and these increases have been linked to the kind of tasks they can solve. My colleague Graeme Halford has made the case that toddlers only have the capacity to bind two concepts in working memory and can hence only understand simple relationships, such as the concept "smaller," as in one thing being smaller than another. Preschoolers develop a capacity to process the relationship between three variables such that they can compute formal additions (e.g., 4 plus 5 equals 9). Only later still do they become able to consider four items and so can compute complex relations such as proportions (e.g., is 2 to 3 equivalent to 6 to 9?). Halford and colleagues have argued that many changes in reasoning capacity during a child's development can be explained in terms of the growing capacity to deal with processing load.

One persistent problem with this theory, however, is that processes and concepts may be chunked, just as numbers and letters can. Halford gives the example of the concept "speed," which can be represented as distance traveled divided by time, but turns into a single variable when simply read as a pointer on a dial. So a three-year-old might talk sensibly about speed without considering the relation between distance and time. But she cannot answer questions such as "How does speed change if we cover the same distance in half the time?" until she can entertain these relationships in working memory. Limits in working-memory capacity constrain reasoning.[6]

Temporary storage and processing space are important for our ability to imagine multiple mental scenarios, to integrate them into a larger narrative, and to compare and evaluate them. It is essential for creating any kind of nested, recursive thought. Therefore, sufficient working-memory capacity is critical for language, mental time travel, and theory of mind. It is now widely discussed as a potentially crucial factor in human cognitive evolution. Yet there is more to our smarts than a simple capacity increase.

---

6 Recent research suggests that capacity for integration of relations and for storage and processing of chunks are related but distinguishable concepts.

One way in which humans radically improve their imaginative capacity is better chunking. We can treat mental scenarios themselves as single chunks of information and embed them in more complex trains of thought. In this way we can use the limited working-memory platform to reflect on scenarios and consider their respective likelihoods and desirability. We can hierarchically organize them and construct higher-order (meta) scenarios. For example, the idea of, say, "getting a degree" consists of numerous scenarios involving lectures, study, and exams. By chunking them under this one heading—represented, for example, by an image of a framed diploma—we can reflect on the conglomeration of all these activities, without all the details. The image acts as a placeholder, allowing us to reason about the value of the achievement and the opportunities it would bring, without having to simulate the day-to-day activities that would get us there. Thus we are able to use placeholders to represent (symbolize) complex propositions and treat them as one mental chunk.

Clever chunking and embedding allows us to *decontextualize*: to think abstractly, without the clutter of the concrete. Because this thinking is no longer closely tied to specifics, we can apply what we learn in one context to any other. Cooking, as you may recall, affords us innumerable metaphors. I am not going to *mince* words: this capacity is an essential *ingredient* in the *recipe* of the human mind. Such decontextualized thinking allows us to use metaphors, infer and deduce unseen forces, build general theories, and consider logical coherence. So we come to be able to form, and reason about, abstract concepts such as the economy, nouns, or evolution. This system gives us supreme flexibility and potential.

Much of our thinking is abstract rather than episodic. Yet it has its roots in our capacity to generate scenarios, substitute them with placeholders, and recursively treat them as chunks of information.

THERE IS YET ANOTHER PERSPECTIVE on human intelligence and scenario building that requires attention. Robert Sternberg suggests that in addition to analytical and practical intelligence, we need to acknowledge that people differ in their imagination and creativity. These are essential aspects of our intellect. Indeed, one of the most famously smart people, Albert Einstein, once said, "Imagination is more important than knowledge."

We can mentally build scenarios of things that are not real (yet). We use imagination to design and innovate in numerous domains, such as architecture, art, fashion, literature, science, and technology. You do not need to be a genius to be creative in all sorts of ordinary contexts, such as cooking, gardening, playing sports, and fixing your car. In sheds and workshops around the world countless functional and aesthetic objects are created every day. As we have seen, when you speak, you easily generate entirely novel sentences. Although some individuals seem more creative than others, every one of us has immense mental power to conjure up ideas, stories, and solutions to problems.

> The imagination is one of the highest prerogatives
> of man. By this faculty he unites, independently of
> the will, former images and ideas, and this creates
> brilliant and novel results.
>
> —CHARLES DARWIN

A RECURRING THEME HAS BEEN that recursion is a key mechanism that unites "former images and ideas," allowing us to produce novelty in language, music, technology, and art through recombination. Generating novel content is not enough, however, unless we want to grant creativity to a random number generator. Creativity also requires a capacity to assess what is generated.

We sometimes disagree with each others' evaluations, of course. Indeed, objectively assessing creativity is notoriously difficult. What I think is creative may seem derivative to you and vice versa. Researchers have developed simple tests in their attempts to quantify creativity. In so-called divergent-thinking tasks, for instance, participants are asked questions such as, "Tell me all the things you can do with a newspaper," and the researcher records the number of appropriate answers a child comes up with. Sometimes they also give scores for the originality of responses (e.g., if no other person in the sample came up with, say, "make a paper hat," then that answer receives an originality point). To generate appropriate answers, children need to search their own knowledge base and assess options. This thinking about knowledge may require similar capacities as theory of mind tasks. Indeed, in a couple of early studies Claire Fletcher-Flinn and I found associations between children's theory of mind and divergent-thinking scores.[7] Once children passed false-belief tasks, they generated more answers as well as more original answers.

Human generativity coupled with our ability to mentally project ourselves into future scenarios enables us to prudently design aspects of our environment. Designing is the capacity to imagine a new object or situation with a specific function or aesthetic in mind. Design is not limited to professional architects or couturiers but includes everyday activities such as arranging a flower bouquet or one's living room in a premeditated fashion. When we design objects, we combine and recombine basic elements recursively and appraise their imagined constellation in terms of the desired function. Rather than adapting to the environment, we have increasingly used this design capacity to flexibly shape our world to meet our fancy. We like a challenge, and we even invent novel problems to solve. Sudoku, anyone?

ANIMALS, LIKE HUMANS, PRODUCE ARTIFACTS that significantly change their environments. Termites create mounds, spiders spin webs, and beavers build dams. Yet even the most impressive of these constructions, such as the elaborate bowers of Vogelkop bowerbirds, may not be based on a reasoned plan. All members of these species (or of the relevant sex) build the objects in question. Furthermore, they all seem to build only one or a few types of items. There is no evidence of the open-ended flexibility that characterizes human design. But perhaps this underestimates their competence. Some animals use tools, and a few species even make tools. Great apes, as we have seen, have demonstrated at least some capacity at imagining alternative worlds. Various creatures act in ways that seem intelligent and creative. Consider the Queensland jumping spider, *Portia*. It hunts for other spiders, taking detours to abseil on top of them or moving across its prey's web only when wind and other disturbances offer a smoke

---

7   Children who failed theory of mind tasks produced few correct solutions. They often searched the test room for answers, and a salient idea seemed to stay in their focus of attention. When asked to name things that are red, for instance, they might say, "fire engine," and then keep listing related items, rather than to note that books, balls, or almost anything else could also be red. Flexible scanning of one's own knowledge for potential solutions may require executive capacities to disengage from an answer and meta-representational skills to generate and evaluate potential solutions.

screen. Many examples of rather clever-looking behavior exist in a variety of nonhuman taxa. Is this not intelligence?

I noted that the IQ testing community regards three components as essential to intelligence: to learn from experience, to adapt to the environment, and to reflect on one's own performance. Many nonhuman animals meet the first two parts of this consensus definition of intelligence: they learn and adapt. Predation, for instance, would have exerted strong selection pressures on these capacities; consider orcas or lions hunting in packs and their prey trying to avoid being caught. On the other hand, the third component, reflection, may point to something uniquely human. Embedded thinking, or to think about thinking, may well set humans apart.

There is little to suggest that other animals reflect on their minds. One line of research, however, raises the possibility of some degree of meta-cognition in nonhuman animals. The comparative psychologist J. David Smith and colleagues found that a dolphin could discriminate between a high-pitch and a low-pitch tone but increasingly hesitated when the two tones were similar in frequency. Given an option to decline trials, the dolphin did so preferentially when the stimuli were similar and the risk of error was high, suggesting some awareness of uncertainty. Subsequent studies, primarily with monkeys, have confirmed that following extensive training on simple discrimination tasks, some animals can eventually learn to avoid trials they are likely to fail.

One way to describe such behavior is to say these animals know when they do not know. Several leaner interpretations have been ruled out through careful experimental study, but this is not to say that a rich meta-representational interpretation is necessarily correct. Smith and colleagues try to claim the middle ground by arguing that some animals are capable of uncertainty monitoring and so deliberately select to decline difficult trials.[8] This is more than associative models predict but need not involve deep reflection on one's inner mental life. We saw in the previous chapter that animals have so far failed to provide compelling evidence for representing others' mental representations. Absence of nested thought, of

---

8 It should not be surprising if some animals monitored uncertainty. Many species have to track indicators of, say, whether to attack or to flee, and it could be highly beneficial to recognize uncertainty about what one can or cannot handle. A monkey traversing the canopy needs to take into account how far it can jump.

meta-representation and recursion, might severely constrain the flexibility of animal intelligence.

As a hunt illustrates, nonhuman animals can pursue goals in the face of obstacles and so also meet part of Pinker's definition of intelligence. The complexity of goals may be restricted (especially in light of limits in mental time travel), but they evidently can pursue certain objectives. What is less clear is whether they employ reason in their pursuits. Most cases of apparent reasoning have attracted both rich and lean interpretations; the resulting debates are complex and multifaceted. I will not provide a comprehensive review of this large literature, but I have selected key examples in the hope of conveying the current state of science on the following question: Can animals solve problems rationally?

Perhaps the most famous case of animal intelligent problem solving comes from Wolfgang Köhler's classic experiments on chimpanzees during World War I. Köhler, the German gestalt psychologist, was on Tenerife at the time—it has been rumored that he might have been a spy—and conducted experiments he described in his influential book titled *The Mentality of Apes*. In his studies he presented a group of captive chimpanzees with various puzzles. For instance, he attached bananas to the ceiling of their enclosure and observed how the chimpanzees stacked boxes to reach the treat. He would put bananas out of reach outside the enclosure, and they would connect sticks to rake them in. Köhler's star pupil was a male chimpanzee called Sultan. Köhler argued that Sultan considered the situation until hitting on a solution, solving the problem through insight rather than associative learning.

Given these early successes, it is surprising that subsequent research has produced mixed results. Though there are some remarkable cases of problem solving described in the literature, ape behavior frequently appears to be much more haphazard than the word "insight" suggests. Great apes do not typically sit still and then rapidly enact a flawless solution. Instead, there is often a great deal of trial and error. There are reports of astounding and persistent failures to solve simple problems, and there seem to be a lot of individual differences in the ability to find solutions. Still, the basic observation that some chimpanzees, as well as other great apes, can figure out solutions where most other animals are stumped has frequently been reported.

In one study gorillas and orangutans selected tools of the required length to reach a reward. They picked the correct tool even when the problem was out of sight, suggesting that they mentally represented what was required. In another experiment these same apes also could obtain one tool in order to get at another tool that was subsequently used to reach the food. Such "meta-tool" use may be a precursor to using tools to make tools.

It used to be thought that only humans make tools. This skill is rare indeed, but research has established that at least a few other species manufacture tools: great apes, elephants, woodpecker finches, and New Caledonian crows. I recently visited the research station on the island of Mare, where Gavin Hunt, Russell Gray, Alex Taylor, and their colleagues were conducting their research on these crows. There I saw a crow inspect a baited hole, fly to a nearby pandanus bush, cut and rip a sliver of the barbed leaf, and then insert it into the hole to retrieve the food. When the treat could not be reached, it flew back to make a longer tool. The researchers showed that these birds can also use a tool to obtain another tool. Crows had to retrieve a short stick tool to reach a longer stick tool, which they in turn deployed to get food.

Crows are part of a family of birds known as corvids, which also includes magpies, ravens, and jays. The comparative psychologists Nathan Emery and Nicola Clayton have argued that in a range of different domains corvids have capacities quite similar to those of great apes. Ravens, for example, are capable of solving problems such as pulling up strings to get whatever is attached to them. When faced with parallel slanted strings, they consistently pull up the one that has food on it, and ignore the other. Some researchers argue that they therefore have demonstrated insight into the causal relations of connectivity.

Yet there remains some lingering skepticism about these cases of animal problem solving. As we have seen repeatedly, behavior that looks smart need not necessarily be the result of intelligent thought. The Clever Hans effect is a persistent problem for rich interpretations of animal behavior. For instance, the string pulling behavior of corvids may mean that they understand the connection of the string and the food, but their behavior can also be explained in terms of simple associative learning, since with each pull the bird is rewarded by the food moving closer. Taylor and colleagues recently conducted experiments that manipulated what the crows can see when they pull up food. When they cannot see the food getting closer with each pull, they stop pulling. When they can see it, even through a mirror, they continue to pull.

Disrupting the visual feedback disrupts the action. This suggests that the crows do not have insight into the problem but rather act on immediate reinforcement.

Although one may question how powerful associative learning really is,[9] we always need to ask whether leaner explanations might be responsible for observed behavior. Daniel Povinelli, the major advocate of killjoy interpretations of ape theory of mind–like behaviors, failed to find support for insight and causal understanding even in chimpanzees. He found that, in spite of making and using tools in the wild, apes have limited understanding of the functional properties of tools. In Povinelli's studies chimpanzees were equally likely to choose to pull on a string that lies on top of a banana as to choose a string that was tied around a banana. They performed poorly when given the option between a rake with a floppy end and one with a solid end in their attempts to rake in food. They made the most elementary mistakes, and Povinelli concluded that they simply cannot reason about abstract causal forces such as gravity or support. Instead, they learn associations between observable events.

But when presented with choices between natural stick tools, orangutans recently solved some of Povinelli's connectivity problems. Several other results suggest that his conclusions were premature. In one ingenious study, orangutans used water to solve a problem creatively. Presented with a tube that had a peanut inside that they could not reach directly, the orangutans took water in their mouths and returned repeatedly to spit it into the tube until the peanut floated to the top and became accessible. Several other apes tested subsequently had problems finding this solution. Rooks, corvids that are not known to use tools in the wild, have recently been shown to be able to manipulate water levels in similarly clever ways as described in Aesop's fable of the crow and the pitcher. Presented with floating worms in a vessel they spontaneously put stones in, thereby raising the water level until the food could be reached. New Caledonian crows can learn to do this as well; they select functional objects among distractors. Perhaps, then, some corvids and great apes can solve prob-

---

9 While big consequences can lead even to one trial learning (you do not need to burn your hand on a stove repeatedly), minor ones may not have much power in shaping behavior. Consider, for example, whether you turn your key left or right when opening a car. This is an action that you might have done thousands of times, and you get either rewarded when moving the right way (the door opens) or punished when moving the wrong way (the door is still shut). I still keep getting it wrong.

lems through insight after all—at least some of the time and under some circumstances.

The so-called trap tube task is particularly revealing. The animal has to insert a stick into a Plexiglas tube to push out a morsel of food. The trick is that there is a trap on one side of the tube, such that when you push the tool in, say, from left to right, the food falls in the trap; when you push it from right to left, it comes out of the apparatus and can be consumed. After ninety trials only one of four capuchin monkeys learned to push the food away from the trap. Yet when the tube was turned around, even this monkey failed, suggesting that these primates had no understanding of the simple causal relations involved. Chimpanzees fared slightly better. However, they pushed in the same way regardless of whether the trap was at the bottom or top of the tube, even though when the trap is oriented upwards, gravity dictates that food will not fall into it. Follow-up studies suggest that the majority of great apes fail these tasks, although some can learn to solve them. Nevertheless, even the successful ones fail when there are small changes to the setup—such as when they need to push out rather than rake in the food.

Povinelli and his colleagues Penn and Holyoak therefore maintain that great apes do not comprehend the analogical similarity between perceptually different but functionally equivalent tasks. In fact, these authors propose that this is what essentially separates human from animal minds: only humans form "higher-order relations between relations." The idea is that only humans construct theories about the underlying causal mechanisms that govern the world. This lean account of animal analogical capacities—you guessed it—has been challenged. When presented with a new version of the trap tube task that did not involve a tool, some chimpanzees can avoid the trap and can transfer competence to analog versions in some circumstances. In fact, even New Caledonian crows were recently shown to pass the standard task and to transfer it to other versions that had distinct perceptual cues (although they had trouble when the trap itself was altered).[10] Thus great apes and these crows, at least, appear to have

---

10   Another recent study suggests that these crows can consider hidden humans as causal agents. When the crows saw a human go into a hide, then saw a stick moving through a hole in the hide's wall, and finally saw the human leave again, they were more willing to approach the hide than when they just saw the moving stick. This suggests that the birds attributed the movement in the former condition to the hidden human.

some capacity to reason about causal relations and to transfer their insight. There is other evidence that chimpanzees may, after all, reason by analogy.

The psychologist David Premack taught chimpanzees to place a plastic symbol "same" between two oranges and "different" between a banana and an apple. The animals could then transfer this to other pairs of objects that were either like each other or not. The chimpanzees could solve analogies such as small triangle is to large triangle as small square is to large square.[11] One study also found that a chimpanzee understood functional analogies such as "a can opener is to a can as a key is to a lock." The link here is not a perceptual equivalence but the sharing of an equivalent goal: opening. This argues strongly against the argument by Povinelli and colleagues about what makes human cognition unique. This result has not yet been replicated, so debates between rich and lean interpretations continue.

Note that even those researchers who argue for analogical reasoning in great apes report stark individual differences and limits in capacity. The problem of inconsistent performance also emerges in other research on reasoning. Consider one more series of recent experiments. Imagine I had one treat and put it in either my left or my right hand. If I show you that my left hand is empty, you can infer, assuming there is no trickery, that the treat therefore is in the right hand. There is some evidence that great apes can make such spontaneous inferences. Josep Call put food in one of two tubes. When the tested apes looked in one tube and discovered it empty, they sometimes instantly retrieved the food from the second tube without looking into it (see Figure 7.1.). But they did not do it often, possibly because there is little cost in peeking into both tubes. A better test, therefore, may be to give animals a forced choice.

Call placed food in one of two cups. He then shook both cups in turn, with the baited cup making the telltale sound. Great apes, when subsequently given the choice, tend to select the cup with the food. But surprisingly, only a minority of the tested apes (nine of twenty-four) did so

---

11   Paula Irving and I once tested dolphins on a similar task at SeaWorld. At first, we rewarded them for touching with their nose one of two boards that displayed a symbol they were shown beforehand on the other side of a pontoon. We then wanted them to match items that were perceptually different but represented an analog relationship. Alas, the dolphins were poor at this task. When they got it wrong, instead of examining the problem, they would try to hit the board harder or do a flip first or some other fancy trick. As we have seen, negative results are difficult to interpret. One possible explanation for our findings is simply that their interactions with humans typically involve fish rewards for acrobatics rather than for correct choices.

FIGURE 7.1.
Andrew Hill playing the tubes task with the female orangutan
Punya *(photo Emma Collier-Baker)*. One orangutan and two
chimpanzees showed signs of spontaneous inferences by exclusions
in this replication of Josep Call's study.

reliably. The others often selected the cup that did not make a sound. The
successful animals were then given follow-up tests to examine if they can
reason by exclusion. The first test was again very simple. One of two cups
is baited, but this time only one cup is shaken. If it makes a sound, it
obviously contains the food. If it does not make a sound, reasoning by
exclusion indicates that the other cup must have the treat inside. Three of
the remaining nine apes reliably selected the other cup when the empty
cup was shaken.

Did at least these three apes understand, or did they simply use the
sound as an associative cue without understanding? To rule out such a
simple explanation, Call introduced a series of clever tests. He held a tape
recorder over the cups and pressed play to produce the sound when held
over the baited cup. In this study, most apes did not select the baited cup
more often than expected by chance. Their choice was therefore not driven
by a simple association between sound and food. This suggests that in the
previous study, they were not simply acting on the basis of such an association. Nonetheless, of all of the animals tested, only one ape, a gorilla,
performed across the board in a way consistent with inferential reasoning;
the others did not. So again, while there is evidence for more than trial-
and-error learning, performance is inconsistent.

My PhD student Andrew Hill followed up this finding by testing twenty chimpanzees, orangutans, and small apes. Again, most of them did not perform well. Yet two chimpanzees selected in ways that were entirely consistent with an inferential reasoning explanation.[12] On current evidence, then, reasoning by exclusion is not a uniquely human trait. Still, the difficulties most apes have with these simplest of inferences highlights substantial differences between humans and our closest relatives. It remains unclear what exactly the nature of their capacities are.

The classic debate between rich and lean interpretations of animal behavior typically boils down to the two options of either associative learning without insight or insightful logic and reasoning like humans. In reality, as these examples strongly suggest, this is a misleading simplification. Showing that a species behaves in ways that cannot be explained by trial and error learning does not mean that they necessarily reason like a human being. We have seen how limited and inconsistent their solutions are. Conversely, if animal behavior is not driven by humanlike reasoning, it does not immediately follow that it must be the result of some "mindless" associative learning alone. Species differ in their problem-solving capacities, and these accounts fail to explain why this is so. In fact, the same animals are often good at learning one thing but not at all good at learning another.[13] Such findings suggest that animals do not simply share an all-purpose learning machinery, as was once imagined by behaviorists.

Different animal species have evolved a variety of cognitive means to solve problems. Even if they do not reason exactly like humans, they may still have a range of solution mechanisms over and above simple trial-and-error learning. They may be prepared to learn some causal relationships but not others. They may be able to attend to critical informa-

---

12   It had been argued that Call's tape recorder condition does not address the possibility that an association between the compound of shaking and sound was learned. To examine this, Andrew indicated the location of the treat through the sound of shaking a duplicate cup. The apes performed at chance in this condition, but clearly above chance when the target cup itself was shaken.

13   For instance, when we recently gave New Caledonian crows the choice between two boxes with a food reward, one with a stick poking out with the meat skewered at the end and another where the food was either not attached to the stick, attached to a broken stick, or otherwise not functional, the birds acted randomly. When given the option between functional and broken tools to be used to rake in food, they immediately made the correct choice. Though avid stick tool users, it took them many trials to learn what to do when the stick already had food on it. Some things are easier to learn than others.

tion while doing one thing but not another. And so forth. The challenge for comparative psychologists is to move beyond the simple dichotomy of rational agent versus associative machine, and try to map the diversity of cognitive abilities that exist in the natural world. My task here is to look at what sets human minds apart. But that does not mean that all animals have the same capacities—they do not.

On current evidence we can conclude that across all the areas of problem solving discussed, the blanket statement that animals cannot reason can be rejected. Some animals can reason sometimes, under some circumstances. Nonetheless, there appear to be profound limits to their reasoning abilities. Even in the most convincing demonstrations, their performances are inconsistent. There is no sign yet of the construction of explicit theories that describe the relationship of forces. Some animals make tools, but none so far seem to design and refine tools by assembling various components and with various functions. Without embedded scenario building, without the benefits of human mental time travel, theory of mind, and language, it would not be surprising that their capacities for reasoning are limited even in the simplest of tasks. A key potential constraining factor is working-memory capacity.

AS WE HAVE SEEN, MUCH of the individual differences in human IQ seem to relate to differences in working-memory capacity. The comparative psychologist Tetsuro Matsuzawa and colleagues have conducted innovative studies on chimpanzees at Kyoto University that suggest surprising memory capacities. In these experiments the chimpanzee Ai faced a computer touch screen and was trained to press numbers in ascending order that appeared in random places on the screen. In one version, when the first number was pressed, the other numbers were masked with white squares, and Ai had to press them in order of the numbers that were no longer visible. She performed about 65 percent correctly when there was a total of five numbers on the screen. It has therefore been suggested that Ai had a working-memory capacity of five. Though impressive, this performance may only require recall of three, not five, numbers. The first number press need not involve memory (one can point straight away), and the last number is always whichever square is left.

A more recent study reported that some chimpanzees can do this with up to nine numbers (suggesting a memory of seven chunks). Ai's son

Ayumu could even beat humans on a version of the task when five numbers were presented for merely a fifth of a second on the screen—too quick
to explore the screen through eye movement. This is a very impressive performance when you see it in action. The numbers appear and disappear,
and then Ayumu quickly presses the five locations in ascending order. At
a conference we were shown a video of a chimpanzee being interrupted
on a trial, looking away, and then returning to complete the sequence in a
flash. The researchers now argue that the chimpanzees solve the task with
something akin to photographic memory. The findings that the chimpanzee did better than adult humans caused significant attention in scientific
and nonscientific circles alike. The comparison was perhaps not entirely
fair: humans did not receive anywhere near the practice on the task that
the chimpanzee did. In follow-up research humans were given practice
and subsequently outperformed the chimpanzee.

What, then, is the working-memory span of chimpanzees? No conclusive tests have yet been conducted that are directly comparable to
human working-memory measures (which typically involve distractor
tasks). Analysis of various task performances in the wild and in the laboratory led anthropologist Dwight Read to suggest that chimpanzee
working-memory capacity is actually limited to between two to three
concepts. He examined, for instance, the number of intelligible word
combinations produced by Kanzi and Nim Chimpsky and the number of
object combinations in natural tool use. Such a working-memory capacity
might account for a lack of embedded thought and could represent a
fundamental limit. A gradual increase of working-memory capacity in
human evolution could explain a lot about the qualitative changes that
characterize the human mind. This is an intriguing hypothesis. But since
there is no established nonverbal measure of working memory in nonhuman animals, we will need to wait for further research before drawing
a firm conclusion.

I NOTED PREVIOUSLY THAT ANIMALS can learn some things better than
others. For instance, rats can learn that a taste, but not a sound, predicts
later nausea. Evolution may have shaped what learning matters for a species. Given that rats are generalists that frequently explore novel food
sources, learning to link taste with sickness is important. Many species
show clever behavior in specific contexts but not in others. According to

the primatologists Dorothy Cheney and Robert Seyfarth, animals show "laser-beam intelligence." David Premack concurs and points to teaching in cats as an example of a clever but restricted ability that serves one goal: teaching hunting. Human teaching, by contrast, is domain general and can serve many goals, as I will discuss in the next chapter. Premack argues that what sets human intelligence apart is our flexibility. The abilities we have discussed thus far—language, foresight, mind reading, reasoning—are not fixed to a particular domain but can be employed to virtually endless varieties of goals.

Species differ in the way they can interact with the world. Some species have few means of responding to environmental challenges (e.g., a snail's defense is to hide in its shell), and others have a diversity of options at their disposal (e.g., a monkey may threaten, hide, recruit support from its group, or climb to safety). The philosopher Kim Sterelny refers to this as "response breadth." Humans can react to situations in flexible, diverse ways. We innovate new ways of responding to situations. We are also correspondingly curious. We seek out novel information and prefer situations that are likely to yield new insights. The German word for curious is *Neugierig*, which literally means "news greedy." Indeed, we generally crave new information, and endorphins are released as we comprehend—they give us pleasure by activating the same opioid receptors that are activated by certain drugs. We all know that reading a good book can be rewarding. Irving Biederman has called humans "infovores" to highlight our innate hunger for novel, interpretable information. (He does not, however, restrict the term to humans.)

A classic study on zoo animals presented sets of wooden blocks, dowels, chains, and rubber tubing to over one hundred species. Primates and carnivores were found to be more than twice as curious as rodents and other mammals. Furthermore, great apes spend twice as much time observing the objects than the other primates tested. Our closest relatives are not only curious but also quite innovative. Andrew Whiten and I noted that chimpanzees who were unable to poke out a bolt necessary to obtain a treat from inside a puzzle box tried thirty-eight different ways to solve the problem. They employed one hand and two hands; they used their lips, feet, and a tool. They pushed, pulled, and poked. They grasped, gripped, and hit. They did not give up easily. Baboons used far fewer ways of exploring the same apparatus, in spite of the fact that their hands, including the thumb to finger ratios, are more like ours than those of chimpanzees.

One way to examine response breadth, then, is to offer animals objects and record the diversity of responses they produce.

One study recorded the diversity of actions primates directed at a knotted rope that had one end secured outside the cage. There was no food reward to be gained; the researcher examined playful rather than functional manipulations. Great apes employed significantly more combinations of body parts and actions than any other primates. These results have been replicated: great apes appear particularly inventive. To put it a different way, their behavior is the least predictable. This result is in line with field observations of innovation and creativity with objects and social others,[14] as well as with the diversity of socially maintained traditions we will discuss in the next chapter. Thus there appear to be some signs of gradual change in the evolution of intelligence and creativity. Our closest relatives are particularly adept at flexibly interacting with their environment.

Nonetheless, human response diversity is exponentially larger. Our inventiveness appears to know no bounds. We can generate virtually infinite combinations of elements, creating novelty in behavior, tools, and sentences. With language we can learn from the responses of others, even if we did not witness them ourselves. With mental time travel we can test consequences of potential actions, even if we do not physically try them out. We can therefore overcome obstacles and discover opportunities in our mind. We can treat scenarios as chunks of information and use placeholders to construct higher-order relations. We can decontextualize these relations and reason about entirely abstract concepts. We can construct elaborate theories about the forces that govern this world and systematically test whether they are correct. Only humans do science.

In a sense Cicero may have been right in asserting that "before all other things, man is distinguished by his pursuit and investigation of TRUTH." The acquisition of knowledge is a goal driving many human endeavors, and we derive happiness from gaining understanding. We can build on others' insights and observations to accumulate knowledge. We have established cultures that permeate virtually everything we do and help us act smart in our environment. To culture, then, we turn next.

---

14 Rates of behavioral innovation in animals are associated with brain size measures as well as with frequency of tool use and social learning.

EIGHT

A New Heritage

The primary difference between our species and all others is our reliance on cultural transmission of information and hence on cultural evolution.

—DANIEL DENNETT

WE ARE DEEPLY CULTURAL BEINGS. That does not mean that we are all connoisseurs of classical music, literature, and fine arts. Culture in a broad sense comprises everything enduring that we learn from others; it includes the commonplace—even banal—customs, values, knowledge, and objects our societies have invented and propagated. Shoes, for instance, are deeply cultural. Someone realized that adding a sole to one's foot is a good thing, and ever since, people around the world have created new versions. You and I benefit from this knowledge, even if we did not conceive of the idea, obtain the raw material, design shoes, nor produce shoes—all we had to do was buy them. This cooperation is extraordinary. No monkey has any shoes—at least not any made and sold by other monkeys.

What is most powerful here is the accumulation of knowledge, skills, and artifacts over time. We benefit from what others have done long ago.

We do not have to reinvent the wheel, as the saying goes. Someone invented the wheel some six thousand years ago or so, and the idea spread rapidly. From the original uses as potters' wheels, such as in the Mesopotamian city of Ur, to chariots, mechanical clocks, pulleys, and the hula hoop—there are many thousands of uses to which this basic idea has been applied. We build upon the cultural achievements of others.

Daniel Dennett argues that cultural objects make us smarter. They allow us to do things we could not do before and hence enable us to explore new ways of intelligently interacting with the world. The day someone invented the boat, oceans of possibilities opened up before all humankind. Such broadening of horizons occurs for nonmaterial aspects of culture as well. Culture provides us with mind tools. Words, for instance, are tools not only for communication but for categorizing, thinking, and reasoning. We do not have to reinvent concepts and symbols anew but can acquire those of our group. The word "shoe" itself is a part of your cultural heritage. As we saw earlier, you may know tens of thousands of words, but few, if any, are words you invented yourself. Most of the concepts you have are imported from others. The ideas of "software" and "evolution" are recent cultural inventions, and you can use them without having had to lay the theoretical groundwork. A mind with words is profoundly different from one without, and different words may influence your thinking in different ways.[1]

Although our individual understanding is often flawed and our foresight misguided, by linking our minds to those of others we have enormously increased our predictive capacities and powers of control. With theory of mind and language we are able to wire our scenario-building minds into much larger networks. We teach each other and copy each other, allowing us to pass on what we have experienced, abstracted, innovated, or learned from another. Thus are populations able to socially maintain and accumulate knowledge, customs, and survival strategies.

Culture penetrates most of what we do. We are part of a larger matrix that links us to the cultural achievements of our forebears and contempo-

---

1 This may be the case, regardless of what one might think about "linguistic relativity"—the idea that language determines thought. The popular example of Inuit having many different words for snow has been debunked. Nonetheless, it is clear that experts in any discipline have more words and hence conceptual categories for the things they study. Where I see some fish, my spouse, Chris, who is a marine biologist, sees dozens of different species and complex ecosystems.

raries. Our minds are shaped by the cultural heritage of our group. Alone we may be weak, but together we are strong. Human culture has led to civilizations that, for better or worse, have changed much of the planet. This system is built on extraordinary levels of cooperation. In this chapter I will discuss culture first in terms of the fundamental problems of widespread cooperation that had to be overcome and then in terms of the key mechanisms through which culture is transmitted and changed.

WHATEVER YOU MIGHT THINK ABOUT humanity, we are a remarkably cooperative lot. People habitually collaborate with friends and family members, communities, teams, clubs, companies, societies, associations, and national or international institutions. With the ability to read minds and tell each other what we are thinking, we can coordinate our actions in an unprecedentedly flexible manner. With foresight we can construct and pursue long-term collaborative plans. We even cooperate with people we do not know. I banked on the truth of this claim when I hitchhiked through various parts of Europe, Asia, and America. Wherever I travelled, I benefited from the kindness and knowledge of strangers. You can cram us in a bus or a football stadium, and only rarely does mayhem ensue.

We also cooperate economically. For the right price most people will trade goods and services with almost anyone. Most of your possessions were likely made by others: your clothes, furniture, music, spices, art—certainly, the book you are reading right now. Though I am writing this down under, you may be reading it on the other side of the world. You are benefiting from my labor (whether or not you agree with what I write).

The extent of our dependence on others' ideas and labor becomes obvious when you imagine trying to recreate your belongings stranded alone on an island. Who knows how to build a bicycle, let alone a car, and even if you did, how would you obtain the raw materials to do it? Even growing your own food depends on prior knowledge of principles discovered by your ancestors. Our scenario-building minds draw heavily on the ideas and experience of other minds to guide our own future. Our modern means of communication allow us to cooperate with virtually anyone anywhere on the planet.

Other animals cooperate, too. Symbioses are widespread. For example, you are host to millions of bacteria that could not do without you, nor could you do without them. In fact, bacteria outnumber human cells in your body. They do a job for you, and you provide them with a fine habitat. Symbiosis is common, as long as the benefits outweigh the costs. Some animals even take what appear to be considerable risks in their co-operation. Cleaner fish eat parasites out of the mouths of larger fish, which in turn do not swallow them. Some ants protect aphids (plant lice) from predators and even store their eggs. In return, the ants get to milk the aphids for honeydew.

Cooperation between members of the same species is similarly wide-spread. Ants, like other social insects, demonstrate large-scale cooperation. The reason they do this, ultimately, is that they are all closely related. A bee may sting an attacker and die, but its individual sacrifice increases the chances of survival and reproduction of its hivemates (and genetic line). Such species act, in some sense, as one super-organism. Naked mole rats are a mammalian species that follows a similar strategy. A queen repro-duces with the help of a couple of males, whereas workers do all the other jobs around the burrow. Though the workers do not reproduce themselves, their actions increase the fitness of their reproducing relatives.

The evolutionary biologist William D. Hamilton suggested it is not the number of viable offspring per se that matters in evolution, but the genes that make it into the next generation. Since we share genes with our relatives—you share at least 50 percent of your genes with each of your siblings, children, and parents; 25 percent with half siblings, grandchil-dren, aunts/uncles, and nephews/nieces; 12.5 percent with first cousins; and so forth[2]—the frequency of our genes in the next generation is deter-mined in part by the reproductive success of our relatives. If your behavior can aid your relatives' fitness, it may be selected even if it is costly to you. Hamilton proposed a rule about when costly behavior is compensated by benefits to relatives.[3]

Cooperation between relatives can therefore be explained by kin selec-tion. Consider how much pain you would be prepared to suffer for a given family member. To do this, sit as if you were sitting on a chair while lean-ing against the wall, with no chair, legs bent at 90 degrees. Try to hold this

---

2   The actual genetic similarity is often significantly higher. When there is little migration, the baseline average relationship of group members can be quite high, and so the similarity of close family members must be even higher.

3   Hamilton's rule is $rB > C$. Individuals may act altruistically if the fitness cost to the in-dividual (C) is less than the benefit (B) to the recipient times the degree of relationship (r) between the individual and the recipient.

position for as long as possible. You'll find that it is easy at first but will become painful the longer the position is held. Some enterprising, and possibly cruel, researchers used this exercise to quantify how much people were willing to hurt for their family, by promising participants more money the longer they could hold the position. If the money was for the participant, they ended up holding on the longest. If the money was for a relative, people held on according to how closely related they were. They held the position longer for a parent or sibling than for an aunt or grandparent. They held on less still for a cousin and the least for an unrelated person. Such behavior is exactly what would be predicted from the kin selection perspective.

You may well object that there are exceptions.[4] For instance, you may have decided not to suffer at all anymore for a particular family member. You may well be willing to help close friends much more than you would be willing to help any of your relatives. The experimental results above, as well as corroborating ones, are only averages, so exceptions may be accommodated. Still, it is clear that not all human cooperation can simply be explained in terms of kin selection because we evidently cooperate extensively with people to whom we are not particularly closely related. It is this cooperation with nonrelatives that appears to be unusual in the animal kingdom and is essential for explaining human societies and cultures.

The sociobiologist Robert Trivers proposed that we cooperate with nonkin largely in expectation that the favor will be reciprocated in the future: I'll scratch your back if you scratch mine. The return of the favor need not be immediate, and payment may be indirect—for instance, through money. In other words, we might have evolved to be so helpful because we ultimately get something back in return.

But is all our helping so selfish? People sometimes give generously without expecting anything in return. This is a virtue of which we are typically quite proud. I was not related to any of the people who gave me lifts

---

4   One glaring exception, at first sight, seems to be the fact that the highest number of child abuse and murder cases comes from within the family. However, the evolutionary psychologists Martin Daly and Margo Wilson showed that the perpetrators of these crimes are typically not blood relatives. Stepparents are much more likely to harm the children of their spouses than biological parents are, in what has become known as the Cinderella effect. Of course, such abuse is rare, but among the parents who abuse their children, stepparents are over-represented. For instance, in Canada between 1974 and 1990, less than 3 fathers per million fatally beat their child, whereas 321 per million stepfathers did.

as a hitchhiker. When a natural disaster strikes, people donate time and money—with little potential for reciprocity. During the process of writing this book, our house in Brisbane, like more than twenty thousand others, was severely flooded. Friends, neighbors, and countless strangers helped us shovel the mud out of the house, tear down the soaked walls, and clean what was salvageable. Someone even gave us a washing machine. It's extraordinary how humans can pull together in times of crisis, creating a sense of indomitable community (and faith in the power of the human spirit). I do not recall a single argument or conflict, and we had some laughs in spite of it all. History is full of compassionate, selfless, and heroic actions. People are capable of sacrificing time, effort, and sometimes even their lives—all for the sake of helping others.

Nonetheless, many philosophers and economists question whether true altruism exists. They argue that the apparent altruists always benefit, or they think they benefit, in some way or another. Such a view may appear uncharitable, but it is certainly true that helping others may secure you future support. Your house may be flooded or hit by an earthquake in the future. People may not consider this factor at the moment they decide to help and simply act out of compassion. However, the tendency for selfless giving may have evolved because those who did act altruistically on average ended up benefitting (or at least their relatives did), even if they did not consciously seek benefits.[5] For example, people who help a lot gain a good reputation, which in turn can bring advantages, and not only from those who were helped directly.

Many people believe in some sort of karma—a general cosmic rule that what goes around comes around. Various religions promise some personal account keeping and payback or reward, if not here, then in an afterlife. This encourages helping behavior but also turns even the most generous, selfless acts into something that is fundamentally self-serving. I'd like to think that humans are not merely selfish, yet I do not know of conclusive evidence against such ultimate explanations, and my liking or disliking the idea is, sadly, irrelevant. Be that as it may, we can probably agree that in

---

5   The ethologist Niko Tinbergen clarified that we need to distinguish the *proximate* explanations for behavior, the mechanisms involved, and how they develop from the *ultimate* explanations for behavior, how the behavior evolved, and what its functions are. The proximate explanations of feeling empathy and having been raised to "do the right thing" may serve ultimately more selfish functions.

the long run most of us expect some kind of return for our giving. When the balance between giving and receiving gets too out of balance, most of us become rather irritated.

A notorious problem for reciprocal altruism, then, is that individuals may cheat. There are always takers who do not give anything in return. Whether you call them leeches, parasites, freeloaders, or some more offensive term, we tend to bristle at people who take advantage of us—and we may seek retribution. Extreme freeloaders may be called sociopaths. They are characteristically irresponsible, unreliable, and egocentric, taking advantage of others' goodwill by exploiting their prosocial tendencies. In small societies people quickly learn to mistrust and stop cooperating with those with a reputation for cheating. In today's large, mobile societies, however, sociopaths may sometimes move from one group to another and start afresh. Yet freeloading is not only a trait of people you dislike. In fact, humans often cooperate in one context and cheat in another. For example, many otherwise prosocial people will proudly state how they avoid paying taxes (which, of course, is cheating everybody in the country). Violations of reciprocal altruism may range from small, simple oversights to large-scale exploitation—from not doing the dishes to thieving, from assault to invasion.

Because of the problem of cheating, it has been hotly debated how cooperation among nonrelatives could have possibly become common practice in the first place. As Dawkins argued so persuasively, from a genetic perspective we are hosts to selfish genes that try to replicate themselves either directly or via our kin. If this is correct, then cooperation should only work in the long run if it benefits our genes. Cooperative systems are always under threat from freeloaders because they get the benefits without having to pay the cost. So freeloaders should win, and their genes should spread more than those of the cooperators—leading to the ultimate collapse of cooperation. Human societies managed to overcome this problem, even if they are evidently not immune to abuse. We developed means to detect, punish, and deter cheating.[6] Groups established effective ways of encouraging and enforcing cooperation. We managed to cooperate reasonably consistently and, eventually, on a large scale.

---

6   The evolutionary psychologist Leda Cosmides has even argued that we evolved innate cheater-detection mechanisms. The evidence for this, however, is controversial.

With our already discussed capacities for problem solving, mind reading, time traveling, and exchanging our minds, humans can agree on shared sets of rules in everybody's mutual long-term interest. For instance, we may agree on the rule that if you want to take certain things from someone, you need to ask permission. Today, in most societies governments have turned cooperative rules into written laws. Police enforce these rules, judges determine punishment for transgressions, and legions of lawyers argue about what you can get away with. In principle, the important thing is that violations of rules need to be identified and addressed. Repeated or severe transgressions typically attract harsher consequences, such as exclusion from the group through banishment or death. But gossip and public shame are usually powerful enough deterrents, especially in small groups, because cooperators, for obvious reasons, would rather interact with other cooperators than with cheats. Who wants to hang out with, and rely on, someone who will cross you? People therefore work to clear their names if wrongfully accused, and one's honor is of paramount importance. Reputation matters. It allows for "indirect reciprocity," in which other group members bestow benefits on those who cooperate consistently and costs on those who do not.

We internalize rules of conduct and monitor and evaluate our own and others' behavior accordingly. Our morality is fundamental for effective cooperation with nonrelatives. We will look at this in more detail in the next chapter, but for now I'll simply note that through cooperative rules humans have been able to establish flourishing, cooperative societies. The big prize this persistent cooperation brought is that our ancestors could complement genetic evolution with a powerful new way of rapidly meeting adaptive challenges: cultural inheritance.

YOUR SCHOOLING WAS DESIGNED FOR the acquisition of cultural knowledge, which was accrued over many generations and deemed important to teach the next generation. Even without formal schooling, every human group passes on cultural heritage. Obviously, groups differ in their traditions (e.g., Australians play the didgeridoo and Austrians the alphorn), but they all discovered or invented thousands of tunes, symbols, technologies, and customs passed on from one generation to the next. Each new generation then builds on this heritage. The capacity for this kind of cumulative culture appears to be universally human. Michael Tomasello argues it is also uniquely human:

> *New forms of cultural learning created the possibility of a kind of ratchet effect in which human beings not only pooled their cognitive resources contemporaneously, they also built on one another's cognitive inventions over time. This new form of cultural evolution thus created artifacts and social practices with a history, so that each new generation of children grew up in something like the accumulated wisdom of their entire social group, past and present.*

We all stand on the shoulders of giants[7]—or, rather, we stand on the shoulders of millions of ordinary, mostly dead people from whom we inherited our culture. We have evolved a fast and flexible way to pass on information to the next generation. What worked well in the past is maintained until something more suitable comes along. Cumulative culture has a role in most of what we do: it shapes our minds and is essential for explaining how we have transformed the Earth. It is evident in architecture, arithmetic, ceremonies, clothes, conversations, crafts, cuisines, customs, dance, games, infrastructure, mating rituals, music, philosophy, public performances, rites of passage, science, spirituality, stories, and technologies— among other areas. If the mechanisms of cumulative culture are unique, then they would explain much of the apparent uniqueness of the items on this list.

The most important characteristic of human culture, then, is that it acts as a second inheritance system, in addition to genetic inheritance. Like genes that were selected because they gave individuals an adaptive advantage, cultural information can have distinct survival and reproductive benefits. One obvious advantage of cultural evolution is that it enables us to adapt much faster than we ever could biologically. And this might have given us an additional edge over other creatures. In response to a sudden onset of an ice age, people could simply make warmer clothes, whereas gradual biological selection for thicker fur takes many generations—and much chilly suffering. Richard Dawkins suggests that cultural evolution is

---

7   Isaac Newton famously declared, "If I have seen a little further it is by standing on the shoulders of Giants."

based on the replication of what he called "memes" (in analogy to genes), such as ideas, behaviors, or tunes that spread from one person to another. While there are heated debates about the precise similarities and differences between cultural and biological evolution,[8] the important point is that cultural knowledge can make a difference to survival and reproduction and hence can influence biological evolution itself.

Cultural knowledge evolves in response to local demands. Aboriginal Australians living in the dry center of the continent managed to eke out a living because they knew how to find water and food. Even the wisest among them would probably perish if suddenly transported into the northern arctic. The Inuit, by contrast, have managed to hunt with the limited materials available in the Arctic and devise ways to stay warm. Yet an Inuit would probably fare no better than our transported Aboriginal person were she to suddenly find herself alone in the Australian desert. (I would no doubt perish in either place.) Relevant local cultural knowledge is essential for human survival in diverse habitats.

Whereas culture used to be extremely localized, written language and modern information channels enable us today to rapidly exchange knowledge across the globe. Perhaps it is now possible for a single person to accumulate the tools and knowledge necessary to survive in all manner of harsh environments. Before such opportunities for mass distribution of "memes" existed, however, they must have been transmitted person to person. Cultural forms fit local functions (just as Darwin observed that biological form fits function—to raise a parallel to biological evolution). Each generation must have learned exactly how the solution to a problem worked and passed it on—if not, it was lost.[9] One generation not speaking the language of their parents can mean, and has often meant, that the language dies out. Unlike recessive genes, social learning could not lie

---

8   Like biological evolution, cultural evolution clearly involves variation and differential replication. But this need not mean that there are other parallels, such as cultural versions of genotype, phenotype, RNA, sexual selection, and so forth. Whether the concept of memes (the "meta-meme," if you like) is helpful in the study of culture is unclear. Some prominent scholars find the very idea totally misguided.

9   Once the island of Tasmania was physically separated from mainland Australia some ten thousand years ago, the Aboriginal Tasmanians could not share in subsequent mainland inventions such as boomerangs and in fact even lost technologies they previously possessed, such as bone tools. This is not to say that their cultural heritage was not exceedingly rich and successful in the environment that they lived in—at least until a radical change occurred with the arrival of Europeans.

dormant for a generation (to raise a difference from biological evolution). It was crucially important that all the significant "memes" are transmitted reliably and with sufficient accuracy.

To achieve high-fidelity social learning in the absence of written language, humans had to rely on two processes. Information is passed on either with the intent of the possessor or with the intent of the receiver. *Teaching* and *imitation* are the two recognized pillars of human cultural inheritance. Each new generation acquires the material, social, and symbolic traditions of its group in these ways.

IMITATION IS WIDESPREAD. YOU MIGHT not admit to it, but you copy what others say, what they wear, and what they do. For culture to be transmitted across generations it is essential that children reliably imitate. In fact, humans demonstrate the first signs of a capacity for imitation from birth. When you stick your tongue out at a newborn, there is a chance that the baby will do the same back to you. It is unclear how this earliest copying is related to later imitative abilities, but it certainly encourages adults to copy their babies in attempts to socially engage.

By nine months infants can copy novel actions and thus acquire new skills. For instance, through observation they may learn how to combine a couple of objects to make a rattle. As we saw in Chapter 5, they retain this new knowledge and can later make rattles from similar objects. By one year, infants begin to imitate rationally: when they copy actions, they appear to take the model's situation into account. In one study infants saw an adult use his head rather than his hand to turn on a light switch, and they copied the act. In another condition in which the model had his arms tied behind his back, and there was hence a rationale as to why he did not use his hands to turn the light on, the infants instead used their hands to turn on the light. They seemed to understand what the model was trying to achieve and what he could do in the circumstances. From about eighteen months onwards, imitation often becomes a favorite past-time. Toddlers engage in sustained imitation games with older children and adults, taking turns copying and being copied.

When my colleague Mark Nielsen showed twelve-, eighteen-, and twenty-four-month-old children how to open puzzle boxes with a tool to get at a reward, they could all do it. Yet while the younger children simply used their hands to open the boxes, the older children imitated the use of the tool—even though they could have more easily opened it with their hands. When twelve-month-olds first see a model unsuccessfully attempting to open the box with their hands and then resort to tool use, the infants also choose to use the tool rather than their hands. Thus they imitate tool use when there is a rational reason to copy the exact actions. Only the older toddlers, however, copy even when there is no obvious reason to act in the more complex way. In a sense the younger infants' strategy appears to be smarter: they copy the most efficient way to solve a problem, whereas the older toddlers "overimitate"—that is, they also copy superfluous actions. Yet there must be other reasons driving the older toddlers' imitation. Why else would they put in the extra effort?

One possibility is that they desire to be like the person they are copying. Such identification may be important for effective cultural transmission. Copying actions even if they do not yet understand why it was done or modeled the way it was enables children to faithfully acquire proven cultural traditions. It ensures high-fidelity transmission and explains why, in addition to useful behaviors, bizarre, irrational, and superstitious practices survive and even flourish in our cultures. More importantly, it explains how culture can be maintained over generations without short-sighted youngsters abandoning all the hard-learned lessons.

Imitation is critical to normal social and cognitive development. The psychiatrist Justin Williams and the psychologists David Perrett, Andrew Whiten, and I proposed that autism is a disorder that may have its origin in imitation problems. Underlying this problem may be cells in the brain known as mirror neurons. These neurons fire both when you see a particular action, say someone tearing a piece of paper, and when you do the action yourself. In other words, the observation of another's behavior activates the brain mechanism for doing that same action. The discovery of this mirror system has caused great excitement among neuroscientists because it suggests a neural mechanism involved in imitation and a range of other important capacities including theory of mind, language, and empathy—abilities that are characteristically impaired in autism.[10]

Humans often copy each other without being aware of it. When we are with close friends, we tend to unconsciously mimic their postures, move-

---

10 Our proposal has garnered much research interest in recent years, which has produced both challenges and support for the idea.

ments, and the way they speak. Monitor yourself, and you will see what I mean. Such copying is associated with greater mutual liking and has come to be known as the chameleon effect. Research suggests that when you imitate someone (and they do not notice this), they tend to act more prosocially toward you.[11] In one study, for instance, people who had been copied by an experimenter for a few minutes would instantly help pick up pens the experimenter "accidentally" dropped, whereas only a few participants who had not been copied helped her. But beware—the power of this finding has not escaped astute salespeople and politicians.

TEACHING IS THE FLIP SIDE of imitation: while imitation is based on the ignorant trying to learn from the knowledgeable, in teaching the knowledgeable party is trying to impart information to the ignorant. Parents regularly interfere in their offspring's play to instruct. For instance, they may make important aspects of a task salient, partially solve a problem, or select the easiest versions of a behavior to get their child started. Parents provide opportunities for children to learn and practice skills without being harmed. Though psychologists have extensively studied learning in children and pedagogy in adults, rather little is known about the development of teaching in children. An exception is a study on Mayan children in which researchers found that from age four onwards children initiated teaching situations. By age eight they modeled actions and corrected the learner's attempts, using language to describe and explain their actions.

Language is crucial for our teaching of facts. In teaching skills, however, such as how to play an instrument, a teacher may model and instruct the pupil to imitate. In such cases the two pillars of cultural learning are frequently combined. Learning may be encouraged by drawing attention to the crucial steps, perhaps by slowing down, by breaking down the process into more manageable chunks and teaching them one by one, or by repeating the sequences with the teacher highlighting the value of the reward it can bring. Teaching can be subtle or quite hands-on, as the teacher interferes with the pupils' attempts, adjusting their body or their mind in the desired way.

---

11   More generally, behavioral synchrony is associated with increasing levels of oxytocin, the hormone associated with bonding and affiliation. Lack of social mirroring has been found to be associated with more stress and increased cortisol levels.

Without teaching, cultural transmission would no doubt be significantly limited.

Schools and curricula are relatively recent institutionalizations of teaching and learning. Even without schooling, traditional societies transmit important knowledge incidentally as well as in carefully managed situations. For instance, rites of passage may impart knowledge and inform potential teachers that an individual is ready to learn about a new stage of life. Stories are told that harbor diverse lessons for the listener. Though there is great variation in nature and extent, teaching appears to be a cross-cultural universal.

Reflecting on teaching and learning may be less common but can bring many benefits. The teaching of complex skills and knowledge often involves some mental time travel, as a long-term plan may be required to overcome numerous obstacles to future expertise. Considering what is or is not known by a pupil helps devise a method by which the pupil can acquire more knowledge. In this sense, pedagogy must also require some theory of mind. Indeed, teaching competence in young children is associated with performance on theory of mind tasks. The pupils too benefit from a capacity for mind reading, as it can be useful to appreciate the teacher's purposes. They also crucially benefit from mental time travel. Choosing to practice to get better at something in the future is essential for many types of complex learning.[12] As we saw, differential practice must be partly responsible for the great diversity of human expertise. Of course, language immensely improves teaching and learning, as it enables us to directly exchange ideas. Thus all four domains of the mental gap discussed in the previous chapters contribute to our capacity to teach effectively.

BEFORE ANYONE CAN BENEFIT FROM imitation and instruction, someone first needs to invent something worth transmitting. On occasion we might stumble on something by sheer luck, but we often actively pursue new solutions. As we saw in the previous chapter, we have a knack for turning problems into opportunities. This need not mean that each human group

---

12 In some as yet unpublished studies we found that children between ages four and five begin to select to practice those problems that they have reason to anticipate they will need to master in the future.

depended on geniuses, like Leonardo da Vinci, who single-handedly make considerable contributions to cultural heritage. Minor improvements accumulate. Because cultural inheritance is not limited to transfer from parents to children, adaptive information can spread from any member of a cultural group to any other. Human groups also typically have some contact with other groups and may hence also incorporate inventions and practices from their neighbors. Much of the rapid transfer of ideas was likely due to trading, migration, and war.

A serious concern that arises with today's interconnected cultures is that we are rapidly losing diversity by importing all the same "memes." Languages are disappearing, and with them much of the cultural heritage of their speakers. One might argue that barriers are worth breaking down. Globalization and large-scale cooperation works best when we can all communicate with the same tongue. Nevertheless, it may turn out to be critically important to maintain our diverse cultural heritage, so we have variation on which selection can work in the future. Besides, it makes for a more colorful world.

Rather than wholesale exchange, cultures traditionally import subsets of the cultural heritage of other groups and then modify them to their own liking or circumstances. We pass on not only solutions but *potential* solutions. When people from a preliterate culture saw that an outsider used symbols to account for his goods, this observation may have been sufficient for the group to devise its own symbol system. Although Leonardo da Vinci had no way to successfully construct a helicopter, his idea for such a contraption survived him and was eventually put into action. We can pass on questions, not just answers. Obviously, not every so inherited problem will be solved (e.g., a perpetual motion machine), but as knowledge increases, past visions can become tomorrow's reality (e.g., Jules Verne's submarines or solar-propelled spacecraft).

Finally, cultural innovation can be deliberate and goal directed. People set out to find a solution to a problem. With mental time travel, we can even start working on problems that do not yet exist. For example, what are we going to do if a large meteorite heads our way? Indeed, some spectacular attempts have been made to socially engineer cultures themselves. Communism was a deliberate attempt at changing cultures for what was believed to be a fairer, more cooperative future. Yet as we all know, plans do not always work out. Although it is probably the case that cultures primarily evolved from the bottom up rather than the top down, as it were, we can certainly develop some of our culture by choice.[13]

However it was shaped, culture is critical to our minds. What would you be without cultural input? Your mind would no doubt be unimaginably different. Other animals raised in human cultural environments, even the language-trained apes, do not quite import our culture the way our children do. We appear to be predisposed to acquiring culture; I will elaborate on this later. First, however, we must challenge human arrogance again and ask whether other animals have not evolved their own cultures.

COOPERATION EXISTS IN THE ANIMAL kingdom. As we have seen, much of this is mutually beneficial symbiosis or can be explained in terms of kin selection. Cooperation with unrelated individuals is unusual. Researchers have proposed that it is rare because effective reciprocal altruism requires sophisticated cognitive capacities, such as a numerical capacity to keep account of the giving and taking in interactions, and an ability to detect and punish cheaters. Nonetheless, a few cases of apparent reciprocity have been noted. Vampire bats, in spite of their sinister image, were famously reported to kindly share blood with weak and unsuccessful unrelated bats. Cuddlier primates depend to a large extend on reciprocal grooming as a social glue. Among chimpanzees, having groomed another increases one's chances of getting a share of the other's food. As noted earlier, chimpanzees also form coalitions and reciprocally support each other in fights. These coalitions are the basis of their political struggles. Like us, they are more likely to help those who have helped them.

Furthermore, chimpanzees sometimes seem to know whom best to work with. In one study chimpanzees were presented with a problem in

---

13   Social engineering need not be based on totalitarian edicts but can be democratically agreed on and scientifically tested. For example, the 2009 Nobel prizewinner of economics, the late Elinor Ostrom, produced research that provides guidance on how groups can effectively manage common resources and avoid overexploitation. In particular, she suggests that there need to be clearly defined rules of entitlement, an individual's duties should be proportional to allocated benefits, and adequate conflict resolution mechanisms must be in place. Sanctioning of free-riders should be done by the community itself, should start mildly (as we have seen, threats to reputation are often enough), and gradually increase for repeated transgressions. Governance should involve collective decision making. Given that this was found to work for others, perhaps it works for your group. People may hence deliberately choose to organize themselves along such lines.

which they had to decide when to recruit a partner to pull in food. The treats were put on a tray outside of the enclosure and could be obtained only when two chimpanzees pulled on the two ends of a string in synchrony. A chimpanzee could let in either of two other chimpanzees by opening their cage doors. Chimpanzees generally recruited whoever was more effective in cooperating previously. Another recent study found that chimpanzees learn preferentially from high-status individuals. We are not the only species to have figured out ways to effectively cooperate with non-relatives. (Whether other animals have morality to support such cooperation will be discussed in the next chapter.)

Many animals also change their environment and so pass on more to their offspring than merely genes. Though no deliberate planning needs to be involved, a beaver's dam, an extensive burrow, or a termite mound can significantly alter the lives of future generations—perhaps not entirely unlike our old houses and infrastructure. Yet is there social inheritance of behavior? Ethologists have documented how during a critical period young animals acquire certain characteristics from their parents, such as which traits are desirable in a mate. Such imprinting does not offer much flexibility or room for knowledge accumulation. Do animals teach and learn from each other the way we do? Do they exchange ideas with unrelated others? Is there anything we could justify calling "animal culture"?

Perhaps the most famous case of purported animal culture comes from monkeys. In 1953 a Japanese macaque, Imo, was observed washing sweet potatoes, which had been provided by researchers, to rid them of sand. This behavior spread through the group and so suggested social learning by the monkeys. Some people went so far as to suggest that the behavior spread more mysteriously, and this case comes up in writings of New Age spiritualists. A closer look at the actual spread of the invention, however, suggests little need to appeal to culture—nor to any psychic connections. Imo's mother learned the trick about three months after Imo. Two years later, seven members of the group did it, and three years later it had spread to eleven individuals. By 1962, thirty-six out of forty-nine animals did it.

The speed of transmission was not exactly rapid, nor did the rate increase with time. If the spread had been the result of teaching or imitation, one should expect it to accelerate as more models become available for the uninitiated monkeys to learn from. Furthermore, monkeys generally brush sand off food. I once saw an ibis, a common wading bird in Australia, wash a potato chip for apparently the same reason. Thus acquisition of the behavior may not be that unusual. It is possible that each monkey acquired the skill individually through trial-and-error learning, though it is quite likely that some social learning played a role.

There is, in fact, mounting experimental evidence that behavioral patterns can socially spread in groups of various species. When pigeons trained to peck through the covers of food were released into a flock of naïve birds, the behavior spread more quickly than in flocks that did not have a model. Such so-called diffusion experiments have become more sophisticated over recent years and examine how the behavior of animals can socially spread in populations of fish, birds, and mammals.

In one study researchers trained one captive chimpanzee to get food from an apparatus in one way and another to solve the same problem in a different way. Each was then returned to their native groups to see how the seeded technique spread. Thirty of thirty-two chimpanzees mastered the technique introduced into their group, whereas none of the control group learned to solve the problem by themselves. In a recent study on orangutans a similar result was obtained. Chimpanzees and orangutans, at least, are capable of socially transmitting these techniques (or memes, if you prefer).

Do our close relatives therefore possess a second inheritance system? There are some indications that they do. The long-term studies of chimpanzees in Africa have documented behaviors of different groups in great detail. Andrew Whiten brought the directors of these projects together to pool their data on potential cultural variations. They mapped behavioral patterns that were common in at least one site but absent elsewhere. For example, chimpanzees at Mahale in Tanzania often groom each other while holding hands in the air, whereas chimpanzees at Jane Goodall's research site at Gombe, only about 150 kilometers away, do not. At several sites chimpanzees use probing tools to fish for termites and ants. At Gombe, chimpanzees dip for ants by inserting a relatively long wand into a nest and then wiping the ants off with their hand to put them into the mouth, whereas in the Tai forest chimpanzees insert a short stick and so gather fewer ants they eat straight off the stick. Only at a site in the Republic of Congo have chimpanzees been observed to use two sticks, one to thrust into the mound to create an opening and a more delicate one to extract the termites. Chimpanzees at Bossou in Guinea crack nuts with stone hammers, whereas at the Tai forest in the Ivory Coast they also use wooden hammers. At Gombe they use neither. The systematic comparison has yielded between one and two dozen of such traits per group, and an overall current total of thirty-nine such different behaviors.

There are no simple ecological or genetic explanations for these differences. Given that we know that behaviors can spread socially in captive populations, it is now generally accepted that this diversity in behaviors between wild groups is due, at least in part, to socially maintained traditions. In other words, chimpanzees have a kind of culture. Some traits may have been passed on over many generations; recall from Chapter 2 that there is evidence that chimpanzees in the Tai forest already used stones to crack nuts over four thousand years ago.

Subsequent work on Sumatran orangutans identified two dozen such behavioral traditions. They too may hence qualify as having culture. Cetaceans, as well, appear to have multiple social traditions. At Shark Bay in Western Australia, for example, dolphins break off sponges and wear them over their snout when probing the sea floor. Such behavior appears to have no genetic basis and is likely maintained socially. Furthermore, there is evidence they learn foraging strategies from their mothers. Socially maintained behaviors of a single variant have now been documented in various others species—from the foraging techniques of insects and rats to some dialects of birds.

New Caledonian crows, it has been argued, may even have cumulative social traditions. Gavin Hunt and Russell Gray have demonstrated that different groups of these crows cut out three distinct tool designs from the barbed edges of pandanus leaves. The tools are used to extract grubs from crevices; the designs vary from a simple strip to a complex stepped design. The design made at a particular location stays stable over decades. Hunt and Gray made the case that the more complex designs arose from the simpler design. If that is correct, it would be the first recorded instance of accumulation of technology in a nonhuman animal. Subsequent studies on the learning of the manufacturing skill, however, suggest that individual trial-and-error learning plays a pivotal role. Social transmission may be limited to providing the young with exposure to tools and tool use.

In sum, some animals are capable of behavioral traditions. Admittedly, the number of socially maintained traits is low. The most prolific traditions are found in our closest relatives, but in comparison to the tens of thousands of "memes" that characterize each human culture, even the number of traits chimpanzees maintain socially is exceedingly small. Scholars disagree over whether to call animal traditions "cultures" at all. I do not mind either way. In either case, there is a big quantitative difference between humans and our closest animal relatives. And it seems likely that a qualitative difference is responsible for this. Other animals do not appear to fully exploit the powerful potential of a flexible, cumulative cultural inheritance system. They do not show anything like the ratchet effect by which numerous solutions are continually refined and improved. Given the limits in language, mental time travel, theory of mind, and innovation discussed so far, this may not be a surprise. A key reason may be that animals' cultural transmission mechanisms are not appropriate for the spread and accumulation of vast amounts of information.

COMPARATIVE PSYCHOLOGISTS HAVE IDENTIFIED DIFFERENT types of social learning that may be involved in behavioral traditions. On the perhaps most primitive level there is *contagion*. As we all know, yawning is contagious. Seeing someone yawn substantially increases the likelihood that you will yawn as well. Such contagious behavior occurs in other primates, too. Yawning might have evolved as a means of a tired individual influencing the group to stop moving and rest. On a slightly more complex level, any interaction with an object can draw others' attention. The effect is regularly observable when you have more than one child in the house. As one of them starts to play with a toy, that toy can suddenly acquire a new degree of attraction for the other child, no matter how long it has been ignored before. Animals that see another animal intently exploring something can, by the same token, also become interested in examining that object. Even an octopus that sees another octopus selecting one of two objects is afterwards more likely to pick the same object. It makes evolutionary sense to attend to what others find rewarding—after all, there might be a reward for you, too.

Learning by copying others is a more complex form of social learning.[14] There has been considerable debate about what qualifies as imitation learning among comparative psychologists. To rule out previous

---

14 Some comparative psychologists differentiate this further. For instance, the copying of another's goals is distinguished from the copying of the means to achieving those goals. Both may involve an understanding of a model's intention, but the former, often dubbed "emulation," need not involve the copying of the exact actions of the model. It has been suggested that apes emulate rather than imitate, but this distinction is not always clear.

associative learning, some have advocated considering only copying of novel behaviors as evidence. Others are happy to refer to imitation whenever any behavior is copied. Like human newborns, chimpanzee and macaque infants are more likely to poke their tongue out when a human faces them and pokes the tongue out than in control conditions. One explanation for this behavior is commonly assumed to be a functioning mirror neuron system, which, as discussed earlier, links what is seen with one's own action. In fact, the mirror neuron system was first discovered in monkeys. Yet, surprisingly, there is little evidence to suggest that monkeys imitate in other ways, in spite of their reputation. Monkey see; monkey don't do.

Yet other animals do—at least in some contexts. Imitation of sound is widespread in the animal kingdom. Many songbirds imitate the songs of their parents, and thus song traditions are maintained. Imitation of sound may be substantially easier than imitation of action. In copying sound, one needs to match one's own sounds to those heard, whereas one's own and observed actions look different from each other. Observation of another's, say, dance move, cannot be directly compared to doing it yourself. To imitate actions you may need to mentally assume the perspective of the model. Nonetheless, sound imitation can be quite complex. The most impressive sound imitators I know are the lyrebirds in Queensland. Male lyrebirds sing and dance to attract females. They can copy the sounds of other birds and animals but are also proficient at imitating the sounds of chainsaws, didgeridoos, camera shutters, and even the opening of beer cans.

Some mammals copy elaborate songs to attract the opposite sex. Male humpback whales are renowned for their songs. All males in a population sing the same song, although it changes over time. My colleague Mike Noad observed how the song sung on the east coast of Australia underwent a radical transformation. In 1996, two of eighty-two whales were recorded singing an utterly different song from the other whales. Their song was typical of another population of whales migrating up the western coast of Australia—as if this pair had taken the wrong turn on their way back from the Antarctic. The next year, some 40 percent of males adopted the new song. It was a hit. On the subsequent southward migration, virtually all males sang the new song. This rapid spread of the new song must have depended on imitation. Such transmission of songs between unrelated individuals has since been documented in other humpback populations.

Humpbacks may copy other actions. Off the coast of Brazil they have been observed displaying a strange behavior: prolonged floating vertically in the water with tail in the air. These observations have increased over time, suggesting possible social transmission. In captivity, some cetaceans have indeed demonstrated a capacity for bodily imitation. The comparative psychologist Louis Herman has shown that dolphins can imitate on command and even copy the behavior of human models. When the human turns and flaps his arms, the dolphin must somehow map its different body plan onto that of the model. These marine mammals are the exception, not the rule. There is little evidence that other mammals so readily imitate.

The other notable exceptions are the great apes. Orangutans at a reintroduction center, for instance, have been observed copying human activities such as hanging a hammock, applying insect repellent, and sweeping with a broom. One, you might recall, was even observed trying to imitate the lighting of fire. Great apes, moreover, appear to recognize when a human copies them. We first found evidence for this by asking Emma Collier-Baker to mirror everything the chimpanzee Cassie did. He would repeat behaviors more frequently in this than in various control conditions, and he even appeared to test her resolve with unusual sequences of actions. These responses to copying have since been documented in other great apes. Furthermore, the ape language pioneers Keith and Katherine Hayes taught the chimpanzee Viki to copy whatever they did on the command "do this." Viki understood the idea sufficiently to later imitate entirely novel behaviors. This "do as I do" paradigm is perhaps the most direct way to assess understanding of imitation, and great apes have been shown to do well. Similar attempts with monkeys have so far remained fruitless.[15]

Do great apes use imitation to learn new tricks from one another? Andrew Whiten and colleagues have conducted numerous studies on this question. For instance, they presented chimpanzees with a puzzle box—an artificial fruit with a treat inside—that could be opened in two ways. Participants in one group saw a human demonstrator pull and twist out bolts, turn and remove a pin, and then turn a handle to get at the treat.

---

15 One study did find that when capuchin monkeys were copied by a human, they displayed something similar to the chameleon effect: showing more affiliate behavior toward the imitator than toward another who did not copy them.

FIGURE 8.1.
The chimpanzee Cassie may recognize it when his
postures are being copied.

The other group saw a model poke out the bolts with a finger, twist and remove a pin, and then pull the handle up to release the lid. Chimpanzees opened the box primarily in the way they were shown, suggesting that they acquired their knowledge from the social model. However, in other experiments chimpanzees did not learn socially, acted inconsistently, or demonstrated social learning other than imitation. The emerging picture suggests that chimpanzees can but often do not imitate.

Not surprisingly, there were debates about what all this might mean. In a seminal study Victoria Horner and Andrew Whiten found data that go a long way toward an explanation. They used a version of the artificial fruit task and found a profound difference between the behavior of chimpanzees and that of human children. A model demonstrated first poking a stick into a hole at the top part of an artificial fruit and then inserting it in another hole further below. Both chimpanzees and children copied both actions of the model to retrieve the reward. In a second condition, the apparatus was made of transparent material. It became patently obvious that the first action, poking into the top, had no causal role in the opening of the box. You could simply poke the stick into the lower hole to obtain the reward, and this is what the chimpanzees subsequently did. They stopped imitating the first action and instead immediately proceeded with the

FIGURE 8.2.
The young male orangutan Putu trying to open
one of Andrew Whiten's puzzle boxes *(photo courtesy
of Mark Nielsen)*. Like chimpanzees, orangutans
showed no signs of overimitation.

second. Three- and four-year-old children, on the other hand, continued
to copy the superfluous first action before putting the stick into the lower
hole. Human children tend to "overimitate," whereas chimpanzees imitated only to the extent required to achieve their immediate goal.

Though the chimpanzees in this experiment acted more efficiently and
arguably more rationally, it is the children's overimitation that is crucial
for faithful transmission of cultural knowledge and for accumulation.[16]
As we saw earlier, by age two children tend to overimitate even when there

---

16 There are other "transmission biases" in humans, such as conformity, success, and prestige biases, that may be important to cumulative culture.

is a quicker way to the goal. Mark Nielsen recently examined this in children in the Kalahari Desert and found the same patterns as that described in European and Australian children. Great apes, on the other hand, do not seem to overimitate (see Figure 8.2). Perhaps faithful copying is a key difference between the cultural transmission of humans and our close relatives. It enabled our cultures to gradually ratchet up the repertoire of our second inheritance system.

THERE ALSO APPEAR TO BE differences in teaching. Mammals may provide opportunities for their young to learn in a safe environment. Adults encourage or discourage certain behaviors. But do they structure learning events in light of their assessment of the pupils' knowledge? There is no obvious evidence of animals having anything like a curriculum. Without sophisticated cognitive capacities such as foresight and theory of mind, flexible, tailored teaching would seem to be impossible. It was therefore long thought that other animals do not teach, period. However, it has become clear recently that animals can at least act in ways that function like teaching—even if the teacher might not understand it as such.

A simple definition of teaching is that a teacher modifies his or her behavior in the presence of a naïve pupil without immediate benefit, and that this behavior fosters learning in the pupil. Even using this definition there is only limited evidence of teaching in the animal kingdom. Cats can be observed throwing half-dead prey, such as mice, in front of their young in apparent attempts to get the offspring to learn a fundamental hunting skill. Similar behavior has been observed in some other carnivores. The most compelling evidence comes from meerkats. Young meerkats learn to handle dangerous foods such as scorpions within the first three months of life. Adults gradually introduce them to the prey by bringing first killed and then disabled prey to begging pups. Adults remove the sting of scorpions and so make it safe for the young to interact. As they get older, pups are increasingly given live and intact prey. This adult behavior clearly aids learning. It may justifiably be called "teaching" because the adults change their behavior in the presence of the learner, the behavior brings no immediate benefit to the teacher, and the pups can learn in ways they otherwise could not.

Yet it turns out the adult meerkat behavior is not based on assessing the skill level of the pups. Playback studies have shown what really drives the behavior is the pups' calls. If the begging calls of old pups are played to adults, they bring live prey even if their real pups are far too young to handle them. Conversely, playing back the begging calls of very young pups to a group with old pups increases the number of dead prey brought back. Although this behavior functions like teaching, it may have little in common with the mechanisms underlying flexible teaching in humans.

With more research I suspect that we will find functional teaching in a range of other species. Some killer whales use the dangerous technique of stranding themselves to hunt elephant seals. Adult whales have been observed pushing juveniles forward in their early attacks and, importantly, helping them get back off the beach. Functional teaching may be found in unexpected places. Indeed, perhaps the second most compelling case for teaching in the animal kingdom comes from ants. Ants that know the way to a food source guide others. They can run "in tandem," thereby slowing down progress for the leader but ensuring that the follower learns the route. Nonetheless, there are only a handful of such cases of functional teaching in the literature at present, and in each the teaching is limited to one type. They do not show the flexible, customized teaching of virtually indefinite content that is evident in humans.

Curiously, there is as yet little evidence of functional teaching in our closest relatives. As already discussed, they do not seem to declaratively point things out to each other in the wild. Indeed, there are few signs that the traditions of chimpanzees or orangutans are maintained through instruction. Where chimpanzees crack nuts with hammers and anvils, it takes the young several years to learn the process. The mothers do not typically guide the young or help them along. They usually allow the infant to observe closely and, at times, to indulge in some of the morsels of goodness that await the successful nutcracker. Active instruction could transmit this information far more quickly and effectively.

The primatologist Christophe Boesch, after years of observation, documented two instances in which the chimpanzee mother acted in a way suggestive of active teaching. One involved a five-year-old who failed to crack a nut after several minutes of trying. The mother, who had been resting, joined her, took the hammer, and slowly rotated it into a better position. She then cracked ten nuts, of which six were eaten by the offspring, before returning to rest. The daughter, adopting the better hammer grip, managed to open four nuts over the next fifteen minutes. The other instance involved a mother apparently helping her infant by putting a nut in a better position on an anvil. These behaviors look like teaching, although a skeptic might account for them in leaner ways. What is striking, though, is that this is so unusual. After decades of systematic observation, these are the only two instances of proposed ape teaching in the literature. Regardless of whether one takes a rich or lean view on the evidence, it is clear that chimpanzees do not commonly instruct each other.

Human teaching relies heavily on language, theory of mind, and mental time travel, so animal teaching may be limited for all the reasons already discussed. Just as great apes are not particularly motivated to share intentionality and experiences with others, their capacity and motivation to teach appears to be profoundly limited. In a recent study chimpanzees, capuchin monkeys, and human children were presented with a sequence of puzzle boxes. The children taught each other through language and gestures, imitated each other, and shared rewards. Children who received such social support did better on the tasks. By contrast, the chimpanzees and monkeys acted to secure rewards for themselves and showed no signs of teaching others.

Great apes have some capacity for cooperation and social learning through which they maintain behavioral traditions. However, without sufficiently sturdy pillars of imitation and teaching, neither the knowledgeable nor the ignorant are sufficiently equipped for a ratchet effect to build up an ever-increasing, cumulative cultural heritage.

Humans have a strong urge to link their minds, to overimitate, and to teach. In this way our inventions, skills, and knowledge spread and are adjusted to local conditions, fine-tuned, or optimized by others. Human social groups cooperate to accumulate cultural capital over many generations to an extent that has seriously altered the fitness of its members. As I will discuss further in the next chapter, we have developed moral norms and means of indirect reciprocity that critically facilitate reliable cooperation. With the right cultural knowledge we can live in the desert or in the Arctic. We can even leave the Earth and briefly live in space. Our second inheritance system has allowed us to gain powers vastly beyond what any previous organisms have wielded. This has confronted us with novel challenges because with great powers come great responsibilities (to paraphrase Voltaire—or Spiderman's Uncle Ben). In spite of our many failings, most of us try to do the right thing. To the eternal struggle between right and wrong, between good and evil, we therefore turn next.

NINE

Right
and Wrong

Of all the differences between man and the lower
animals, the moral sense or conscience is by far the
most important.

—CHARLES DARWIN

I GREW UP IN A small town in Germany in a house that my grandmother
built with her own two hands after the war. She was alone with her two
girls at the time—my grandfather had been killed on the eastern front. The
old house had been destroyed by bombing. My mother, seven years old in
1945, hid for days with her younger cousin in a hole in the ground as the
bombs rained down. She never liked to talk about the war, but as I grew
up, I started to question why anyone would have ever wanted to drop bombs
on our town. It is difficult to describe how painful it was to discover
that my people, even if not my family members, were responsible for arguably the most heinous genocide in history. Learning that the nice old man
down the road used to be an SS officer disgusted me and made me skeptical
about any elderly Germans. When I first set foot in a concentration camp as

a teenager and saw for myself the furnaces in which people were burnt, I cried in despair at humankind.

Morality is the division of thoughts and behaviors into right and wrong. We have a conscience with which we evaluate our choices and those of others. If the moral sense is the most important difference between human and animals, as Darwin proclaimed, how can people commit such atrocities? Of course, humans violate at times what their own conscience dictates. Most ordinary drafted soldiers, like my grandfather, simply had no choice but to follow orders, short of rebellion or desertion. Stuck in such a situation, many of them, no doubt, believed the propaganda, seeing themselves as the good guys—as difficult as that may be to imagine.[1] But what about those who committed the worst atrocities? Surely the guards at the concentration camps could not possibly have killed so many innocent, defenseless people and thought they were doing the right thing? Had they—and other perpetrators of such atrocities in history—lost their capacity for empathy and compassion, their conscience and morality?

In his diary, the concentration camp doctor Johann Kremer described the deep bond he felt with his poor little canary. When the bird died, he noted the endless sympathies he experienced as it succumbed to its agonies. In other entries, he outlines his gruesome work at Auschwitz, such as removing the organs of the executed. One of the most notoriously brutal concentration camp guards was one Hildegard Lächert, aka Bloody Brigitte, who also was a nurse and mother of three. It seems that even these people were not amoral—they had compassion, and they helped others, but they did not apply their morals to the humans in their camp.[2] Victims were construed as vermin, toward whom they could act in ruthless ways.

We argue a lot about what is moral and what is immoral. Yet it seems humans generally have some morals, however skewed they may appear to be.

---

1 Soldiers in the German army marching east, for instance, may have thought they were bringing God back to the godless Bolsheviks. After all, Hitler had signed a concordat with the pope, and Germans paid tax to the Vatican (and still do so to this day).

2 Within their group and at the time, such people could have had a reputation for high morals. As Alfred North Whitehead observed, "What is morality in any given time or place? It is what the majority then and there happen to like, and immorality is what they dislike."

WHATEVER OUR SPECIFIC MORALS ARE, it is clear that we take pleasure in acting morally, and it pains us when we think we acted immorally. In guiding our behavior, conscience and morality are crucial to human long-term cooperation and cultural evolution. What are their psychological bases? According to Frans de Waal, the foundation of morality can be subdivided into three broad levels: (1) the basic building blocks of empathy and reciprocity; (2) the group pressures that keep individuals in line; and (3) the capacity for self-reflective moral reasoning and judgment. I will discuss each of these in turn.

Excepting those, like psychopaths, who may be pathologically incapable,[3] human beings are all able to feel empathy and compassion for others. Empathy is sometimes equated with mind-reading capacities. However, humans can take another's perspective based either on inference or on complex empathic simulation (sometimes called "cold" and "hot" processes, respectively). Note that psychopaths may be good at inferential mind reading and use this reasoning to devise more sadistic tortures. What de Waal and others mean, then, when they refer to empathy as a moral building block is the more specific capacity for sympathetic concern for others' well-being. We can share in others' feelings and, as a result, be motivated to relieve their suffering or enhance their happiness. We try to avoid unnecessarily hurting other group members and are instead inclined to lend a helping hand. In turn, other people are compassionate and helpful to us.

As we saw, a key driver for cooperation with unrelated others is reciprocity: we support those who helped us and vice versa. We have a sense of fairness that monitors whether the giving and taking are roughly in balance. Because cultures are the product of persistent, long-term cooperation, it may not be far-fetched to suppose that human nature has evolved traits that support reciprocity and compassion. To kick off a cycle of reciprocal helping, someone has to first be kind.

Are humans inherently good-natured? Michael Tomasello and colleagues have offered evidence that human infants, well before they can be indoctrinated by moral teachings, have a fundamental prosocial urge.

---

3   In psychopaths lack of sympathy and remorse is associated with specific neurological problems (e.g., lesions in the orbitofrontal cortex). Some concentration camp guards may well have been psychopaths.

They share resources, help others achieve their goals, and provide useful information. At eight months my daughter, Nina, surprised me by starting to take a bite of a biscuit and then feeding me instead. Research suggests that toddlers help others even when there is no apparent reward or praise. They start pointing out where something is hiding by one year of age. They do so to help you, even if they do not want the object themselves, and they stop pointing when you have found it. If you have dropped something, they tend to fetch it for you. Once they acquire language, children tend to tell the truth and often contribute to the goals of conversations with useful bits of information. All these findings fit nicely with Tomasello's thesis: infants relate to other human minds in a deeply social way. By eighteen months they show clear signs of sympathy and console others in distress. Toddlers wave good-bye, smile, and laugh. They take turns smothering you, and each other, with hugs and kisses.

While such a picture of infant goodness may make you warm and fuzzy, it is far too rosy to be quite complete. Infants, just like adults, are not always nice. Young children can be awfully self-centered and lacking in compassion. The expression "the terrible twos" has its basis in reality. Toddlers often take what they want, throw a tantrum if they cannot get it, and subsequently refuse to cooperate at all. Children can be cruel, unhelpful, and downright annoying. The idea that they are good to begin with is therefore not entirely compelling.

Young toddlers help less than older toddlers, especially when it is at a cost to them. They initially require much prompting from adults to engage in prosocial behavior. Even three-year-olds, in research using games in which they can give rewards to themselves and others, turn out to be quite selfish. They are happy to take more than their "fair share." As they become older they increasingly make choices that avoid inequality, even when it is to their disadvantage to do so. Between ages three and seven children begin to increasingly direct their helping to those who (a) are closely related, (b) have reciprocated in the past, and (c) have been observed sharing with others. In other words, with experience they increasingly cooperate with those with whom cooperation ultimately makes the most sense: family, friends, and people with a good reputation for reciprocity. Although it remains possible that some prosocial inclinations have become part of our innate nature, these results suggest that experience shapes children's prosociality and therefore that these tendencies are socially, rather than biologically, inherited. Even by age one a child will have had plenty of experience with rewards and approval, either directly or indirectly, for prosocial behaviors. Similarly, they will have experienced plenty of examples of antisocial behavior being punished. Indeed, as soon as they can walk, they keep encountering the limits of acceptable behavior, as adults give frequent feedback.

NATURE IS CRUEL. THERE IS no escaping the fact that all animals have to consume other organisms for sustenance. Even vegetarians and vegans can only restrict which organisms they choose to eat, not whether to eat any. The Darwinian perspective on life, with its emphasis on survival of the fittest, appears to suggest that humans, like other animals, should be inherently selfish. In various situations an individual's interest is in conflict with the interests of others, and humans evidently are sometimes willing to hurt each other.

Thomas Hobbes observed that reasons for human quarrels typically fall into three distinct categories. The first is competition for limited resources. Those who are good at forcing their way past others to obtain food, territory, or a mate leave more offspring in the next generation than those who are not. Thus it is not surprising that this form of aggression is widespread and already common in infancy. The second is self-protection. When potentially attack-minded people are around, even a peace-loving person may be forced to defend himself. One form of self-defense is a preemptive strike. Therefore, even two parties that do not want to pursue aggression for the first reason may end up fighting because both are suspicious.

To stop the other from striking first, cold war politicians (and evolutionary psychologists like Steven Pinker) tell us the best strategy is credible deterrence: if you strike me, rest assured I will get you too. We reciprocate bad deeds just as we do good deeds. An individual who demonstrates a capacity and willingness to retaliate against any aggressor may increase her likelihood of being left in peace by others. For deterrence to be credible, it is important to demonstrate determination and capacity. You need to reliably settle your scores, as apparent lack of resolve may invite ruthless adversaries to exploit this perceived weakness. Pinker argues that the need for credible deterrence leads to Hobbes's third reason for human aggression: quarrels over "trifles" such as disagreements, insults, or other signs of disrespect. People fight tooth and nail over seemingly banal and inane things, from parking spots to flippant remarks about one's mother. The reason may be that backing down in these matters undermines one's reputation, one's honor. Indeed, people fight much more over trifles when there are witnesses than when there are none. Revenge, feuds, and so on follow naturally.

Hobbes thought uncivilized human life was not noble but "nasty, brutish and short." Only civilizations with government monopolies on violence offered an escape from these cycles of threat, aggression, and retaliation. Centralized powers can ensure that aggressors will be punished and so reduce the reason to pursue aggression for any of the three reasons. The risk of being caught by the police is a disincentive to use violence to take what you want. There is subsequently less reason to attack preemptively or to retaliate every affront in an effort to deter others. This way, Hobbes suggested, humans' aggressive tendencies can be pacified, and civilized society can flourish. There can be little doubt that law and order can reduce violence and increase overall cooperation, but this is not to say that humans are naturally bad.

The view that humans are inherently bad is probably just as misleading as the view that humans are naturally good. Infants are not simply born bad (and then are civilized) or good (and then corrupted) but have both prosocial and antisocial tendencies. We are a species of contradictions: we take others' blood, and we donate our own. Our distinct cultural evolution relies on our extraordinary capacity for cooperation, yet we can cooperate to commit genocide. We can be selfless and compassionate but also greedy and merciless. We all have what might be called angelic and demonic tendencies. From a moral perspective, only one is desirable, but from an evolutionary point of view, both can provide distinct adaptive advantages in certain circumstances.

SOCIAL PRESSURE, OR WHAT de Waal referred to as level 2 morality, is clearly not limited to state-enforced, institutionalized law. Cooperative groups have goals that all members have in common, such as territorial defense or sharing of certain resources. To contribute to these common goals, group members may be willing to put pressure on others, through reward and punishment, approval and disapproval. Many unwritten social norms guide how we ought to behave. For example, in some cultures we line up at a counter; in such places, people do not look kindly on those who cut into line. Whatever the norms, we all have a stake in maintaining a harmonious social environment that discourages cheating and encourages compliance to norms of cooperation. As Albert Einstein wrote, "In the last analysis, every kind of peaceful cooperation among men is primarily based on mutual trust and only secondly on institutions such as courts of justice and police."

Mutual trust is the belief that the other person is going to do "the right thing." Each human group has inherited rules about proper conduct: the obligatory, the forbidden, and the virtuous. They represent a social contract comprising rights and responsibilities: you must do X, you are not allowed to do Y, and it is especially good of you to do Z. Although there is much cultural diversity, certain moral rules appear universal. Obligations typically include reciprocating favors, keeping promises, and protecting weaker members of the group from harm. Most groups prohibit killing, stealing, and lying.[4] The reason for this is obvious: people who frequently lie, kill, and steal from each other will not build lasting cooperative societies. Finally, virtuous actions go beyond the call of duty. Successfully protecting the group and its interests at the risk of personal injury or death is generally regarded as heroic. Such acts enhance one's reputation and attract rewards. Together, moral codes are the user manuals that make human cooperative societies work.

Recall from the previous chapter that cooperation is always risky because people may cheat—there always exists the temptation to take the rewards of cooperation without paying the price. Or they may want rights without duties. If too many individuals give in to such temptation, cooperation collapses.[5] This explains why cooperation in animals is mostly based on kinship rather than on complex and vulnerable systems of reciprocity

---

4 Adultery, breaking promises, causing pain, and incest are other usually forbidden acts. Note that most cultures also specify exceptions to these rules. For example, causing pain may be acceptable when it is done to help another—such as when a dentist extracts your rotten tooth.

5 Extensive research on cooperative games has examined the strategies that lead to cooperation or collapses. Game theory has identified "tit for tat" as one potentially evolutionarily stable strategy. Individuals that follow this strategy are nice to begin with, retaliatory when the other fails to cooperate, but also forgiving, as they reengage in cooperation as soon as the other person does. This approach encourages cooperation without there being too much risk of being taken advantage of. Because it is not always clear whether another cooperated or not (e.g., they may claim that they tried), a strategy that allows the other person one slip-up before one stops cooperation is even more friendly and sustainable. But at some point people stop cooperating with those who do not cooperate—most people do not let themselves get burned three times in a row.

and reputation. Humans, on the other hand, handled this problem by creating norms that made fairness obligatory, cheating forbidden, and generosity virtuous. Such standards make others' behavior predictable and encourage trusting cooperation. Critically, as we saw in the previous chapter, compliance with these norms is enforced not only by the individuals directly affected but also by others in the group who reward virtue and punish violations (indirect reciprocity). The economist Ernst Fehr has shown that third-party punishment can lead to stable cooperation, whereas lack of sanctioning leads to a decline in cooperation.

Various studies show that people are willing to punish norm violations even when they do not benefit, it is costly, and they themselves have cheated before.[6] Punishment is typically based on strong moral conviction and has a profound consequence on general adherence to a moral code. Without it, the temptation to free ride and cheat is high. Throughout history people have gone to great lengths to detect and deal with immorality (often leading to zealous prosecution). Similarly, acts of valor are rewarded and recognized not only by those who directly benefited but by others as well. In general, a good reputation attracts future cooperation. Being honest, law-abiding, and generous can therefore be a rewarding strategy.

Indeed, studies on hunter-gatherer societies suggest that people have a tendency to act for the greater good, rather than for their own immediate personal interests, as was long assumed. Moreover, experiments in economics have demonstrated that people often prefer win/win situations over outcomes in which they win and another loses. People often make generous offers when they could be selfish, reject "unfair" offers even if it means losing resources, share when they do not need to, and contribute to public goods even when they could get away with not giving anything. We believe in a better world—and we tell others about it. Every day millions of sermons are delivered about how we ought to act.

Even young children are swift to inform others about the norms they have learned. When two-year-old Timo learned the rule that feet are not to be put on the table, he quickly started berating me in moments of lazy reclining. Guests were also duly reprimanded—he would not rest until all feet were back on the floor. Children adopt norms, such as how a game should be played, even in the absence of any explicit instruction by adults.

---

6 It is even in a selfish individual's interest to posture for altruism—which explains much hypocrisy.

And they are keen to teach them to others. This tendency is part of the general desire we have encountered so frequently: to link our minds. It enhances the spread and standardization of norms, making us support those who follow the rules and impose costs on those who violate them. Virtue, honor, and decency are central to most people's lives, and many invest heavily in the pursuit of nobility (or at least in the public perception thereof). In our groups, morals matter.

Cooperation with strangers from another tribe is riskier, as the same group pressures need not apply. People may murder or steal from outsiders, even when these acts are forbidden within their own group. There are hundreds of studies showing how humans treat members of their own group differently from those of another group. Even when group membership is arbitrarily assigned ad hoc (e.g., according to T-shirt color) people instantly become more prosocial to their *in-group* and more antisocial to the *out-group*. Although these days most of us are members of many different groups (your village, your sports team, your political party, or your assigned group in a social psychology experiment), throughout much of prehistory we would be primarily stuck with our immediate tribes. So rituals, ethnic signaling, and other indicators that groups share basic values and agree on a code of conduct were important in encouraging trust in interactions with other groups.

A key factor facilitating the standardization of moral rules within and across groups in human history has been religion. In most societies, fundamental cooperative rules are absolute and unquestionable by virtue of being presented as divine commands. God, religions promise, will reward adherers and punish transgressors. In a sense this is the ultimate form of indirect reciprocity. Religion reduces the need for policing because believers are to some extent policing themselves through their conscience—to avoid divine, rather than secular, punishment. Of course, people can derive and follow a moral code without, or in spite of, these threats and promises. Nevertheless, the religious approach has proven immensely successful in keeping people in line (although exceptions spring to mind). Followers of the same religion can assume that they share a basic code of conduct. If you have the same God, there is no hiding, and you will be judged by the same rules.

While helping and hurting are the most fundamental moral domains, norms frequently extend to questions of authority, loyalty, obedience, and purity, both bodily and spiritual. There is some debate about what qualifies as moral. A common distinction is made between moral and conventional norms. Morals are typically seen as prescriptive and universally enforceable because violations lead to harm, whereas conventions do not. For instance, there may be a norm about what clothes one wears to a particular occasion, but violating that norm does not harm anyone. Stealing the clothes from someone else, on the other hand, violates the owner's rights and is therefore morally wrong. Even preschool children make this distinction quite readily. However, in some cultures the most apparently arbitrary conventions can be moral by virtue of a spiritual logic that links the act to harm. For example, in one study the anthropologist Richard Shweder and colleagues asked Hindi children in Bhubaneswar to rank a list of breaches of conventions in terms of their seriousness. According to the children, the most serious was: "The day after his father's death, the eldest son had a haircut and ate chicken." These acts were considered worse than incest between a brother and a sister or a husband beating his wife. Norm violations, such as eating the wrong food, may be thought to cause immense harm in the afterlife. Thus spiritual ideas can create powerful pressure for people to conscientiously conform to social norms, and religions have accordingly proved to be great catalysts in the rise of civilizations, enabling ever-larger numbers of people to conform and cooperate.

Many guiding moral principles and norms advocate loyalty, trust, and caring—essentials for large-scale cooperation. One of the most famous principles is the Golden Rule: "Do to others as you would have them do to you" (or "Do not do to others what you do not want done to yourself"). This rule encapsulates the crucial relationship between empathy and reciprocity that is fundamental to human morality and cooperation. Versions of this rule can be found in the early writings of the civilizations of Babylon, China, Greece, India, Judea, and Persia. By spreading the same moral code across many tribes, people could increasingly work together in the building of civilizations. Moral communities expanded. Yet the flip side of in-group cooperation, as we have seen, can be antisocial behavior to the out-group. In fact, conflicts between followers of different religions have provoked some of the most abominable wars and persecutions in history.

With the Enlightenment, European societies started to adopt a more civil, more rational, and more compassionate stance than was evident in the Middle Ages. Torture and cruel capital punishment, for instance, became increasingly objectionable, and these moral norms spread. Changing views about cruelty did not end intergroup conflict and warfare. However, the circle of sympathy generally expanded to become more inclusive. For some people this is still restricted to immediate relatives; for others it extends to a select group of members of a gang, religion, nation, or "race." Darwin foreshadowed that civilization will eventually lead us to extending sympathy to all humanity:

> As man advances in civilisation, and small tribes are united into larger communities, the simplest reason would tell each individual that he ought to extend his social instincts and sympathies to all the members of the nation, though personally unknown to him. This point being once reached, there is only an artificial barrier to prevent his sympathies extending to the men of all nations and races. If, indeed, such men separated from him by great differences in appearance and habits, experience unfortunately shows us how long it is, before we look at them as our fellow-creatures.

Following the Holocaust, humans of all nations eventually sat down to agree on this. The United Nations published the Universal Declaration of Human Rights. The first article reads: "All human beings are born free and equal in dignity and rights. They are endowed with reason and conscience and should act towards one another in a spirit of brotherhood." The declaration is an appeal to extend our morality to all humans, to stop slavery and abuse, to give everyone equal rights—in other words, to treat all humans like relatives. Over the course of history, human cooperation has grown on a progressively larger scale. We are finally putting group pressure on all humans to follow the same basic moral rules to prevent harm and encourage helping. In spite of recurring conflicts, cooperation and respect among humans of all cultures is now, for the first time in our history, a real possibility. The declaration is of course silent on animal rights, but we shall get to that later.

DE WAAL'S THIRD LEVEL OF human morality is our capacity for self-reflective judgment and reasoning. We regulate our own behavior based on our moral assessments. We can reflect on why we do what we do and want what we want—and we can decide to change tack. We think about what "ought" to be the case. We can inform others about our views and judge them. We can try to establish an internally consistent framework, and reflect on others' systems (even those proposed 2,500 years ago). From weekly religious sermons to Emmanuel Kant's categorical imperative, we ponder right and wrong, and how to derive the principles that distinguish between them. Moral reasoning is not just a pastime of priests and philosophers. We frequently argue with family, friends, and colleagues about our own dilemmas, and we debate the choices of others.

Early research by Jean Piaget and later by Lawrence Kohlberg examined how children begin to defend their moral choices. Children were presented with a moral dilemma and then asked about the reasons for their judgment. Kohlberg found that young children focus on avoiding punishment, whereas older children, with more social experience, increasingly demonstrate an understanding that rules have to be followed for the greater good. Eventually some of the children justify their choices by reference to internally consistent theories about moral principles.

Given that we differ in our moral reasoning, we may expect to find many stark differences in our moral judgments. However, recent studies suggest that some assessments about fairness, harm, and cooperation are almost universal. For instance, imagine a situation in which you are the driver of a trolley that is about to run over five people, and you can flick a switch to turn the trolley onto a side track, where it will instead kill only one person. It is generally regarded as morally correct to save the five at the cost of one. Yet most people agree that it is not permissible to save five people in need of organ transplants by killing one person with healthy versions of all those organs. We instantly know which is right and which is wrong, even if we cannot actually articulate the rule that underlies this judgment. In fact, judgments about moral responsibility are complicated, usually involving distinctions between intended and unintended outcomes and between actions and omissions.

Research suggests that people's moral intuitions often precede their explicit moral reasoning. We tend to have instant affective reactions to scenes of moral violations.[7] The reliability of these responses have led some researchers to suggest that humans may possess a universal moral

---

7 These draw in part on ancient emotional assessment systems. Disgust, for instance, has been co-opted to drive us not only to avoid pathogens but also to avoid violation of sexual and other moral norms. Note that many of the more complex emotions, such as gratitude and guilt, may have evolved reasonably recently in the context of cooperation regulation. Anger toward cheaters, for instance, ensures that we punish or deter those who violate cooperative rules and moral conduct.

grammar that is partly innate (in a way that is analogous to Chomsky's notion of a universal grammar of language). But it is also possible that this is culturally transmitted. In any case, it is clear that we can override our moral intuitions: although we might be inclined to seek information that confirms our intuitions, we can also go against our gut reaction and revise first impressions. When people decide to become vegetarians for ethical reasons, their emotional reactions may change as a result. We even rationally employ intuition as a tool. For example, in school I sometimes found myself unable to decide which of two essay questions to pursue in an exam. I flipped a coin, only to monitor my gut reaction as to the outcome. If it was one of relief, I'd go with the coin; otherwise I'd override the toss.

Emotional responses to mental scenarios can be powerful and are essential to our conscience. We can experience shame and humiliation as a result of imagining how others see us, and we sometimes express this through blushing. Likewise, we can experience embarrassment or pride when imagining certain past and potential future events. Our current decisions can therefore be motivated by anticipated or "pre-experienced" emotional responses to hypothetical dilemmas, past misdemeanors, and foreseen events. For instance, anticipated regret stops us from pursuing many things we would thoroughly enjoy in the moment but we anticipate we will be embarrassed about after the fact. We generally do not go out to an expensive restaurant when we know we won't be able to pay the bill.

Our everyday actions are profoundly guided by our capacity for such self-reflective reasoning and planning. We can simulate both the affective and the practical consequences of our actions for our future selves (as well as for others and for the greater good). Yet recent research suggests that this reasoning is marred by certain biases. For example, we tend to systematically exaggerate our anticipated emotions. We tend to anticipate that we will feel happier achieving a goal than we actually will when we reach it. And when we fail, we tend to feel less unhappy than we anticipated. Dan Gilbert and colleagues suggest that part of the reason for these biases is that we anticipate the gist and ignore the details of future events. We might create a mental scenario simulating the pleasure of a vacation without imagining the nuisances of transportation and bad service. Another reason for such biases may be that exaggerating positive and negative outcomes helps us get motivated to choose future-directed actions in the first place. After all, the future is uncertain, and the present pressing. Consideration of the future and the moral consequences of one's choices needs to compete with more immediate, and more certain, pleasures. If the fear of future failure and the anticipation of a future reward are somehow magnified, it may make it easier to pursue prudent future-directed actions.

In general, for self-reflective moral reasoning to compete with more ancient, immediate urges, we needed to acquire a certain level of "executive functioning" (such as the executive power, discussed in Chapter 5, to decide which of several options to pursue). We need self-control: being able to inhibit one impulse in favor of another. For example, in reciprocal altruism we need to resist the temptation to cheat and secure short-term benefits because there is a greater future price to pay in form of a prison sentence, fractured trust, or a diminished reputation.

Children initially have great difficulty with such executive control. In a study known as the marshmallow test, the psychologist Walter Mischel examined under what circumstances young children became capable of controlling their impulses in simple situations. Children were given the option between having a small reward (such as a marshmallow) immediately and waiting for a larger reward later. He found that by age four many children demonstrate some capacity to delay gratification. Whether children delay depended on various factors, such as the nature of the reward and the length of the delay. Another important factor was whether the reward was present or not; delaying gratification is more difficult in the face of temptation. Merely thinking about the reward reduced the time children could delay. Looking at a picture of the reward enabled children to delay longer than when they had the real reward in front of them. Even imagining that a real reward was just a picture increased children's capacity to delay. When we become aware of such effects, we can deploy strategies to increase self-control. Differences in children's self-control predict outcomes dozens of years later, including numerous measures of health, wealth, and success.

Adults, as we saw in Chapter 5, can delay gratification for hours, years, and even lifetimes. Thus our self-reflective moral reasoning can gain control of our actions, desires, and thoughts.[8] We can override biological urges— even the will to live and reproduce—with our moral convictions. We can create moral philosophies, pursue noble causes, and follow high ideals. We can make deliberate choices and may be said to have free will. The price we pay for these powers is that others hold us responsible for our freely willed actions.

---

8 Darwin wrote: "The highest possible stage in moral culture is when we recognize that we ought to control our thoughts."

THOUGH THE WORD "PERSON" IN everyday language refers to any human being, in law and philosophy a "person" is usually a self-conscious entity that is able to choose its actions. Persons are recognized as having rights and duties. In this sense, an infant is not a person. If the entity cannot choose—for instance, because it does not have the executive control to inhibit certain action—it cannot be held morally responsible. If your actions are forced by someone else (e.g., you are pushed over a ledge), it is not an act of free will, and you are not personally responsible for the consequences that follow (e.g., for what you damage in your fall), as you would be if you had elected to jump. Similarly, if you cannot engage in self-reflective reasoning about your choices and their consequences, this lack of self-consciousness has critical implications for moral and legal responsibilities. If you were drugged, for example, your action might not be regarded as the product of your free will. However, if people think you had control and should have foreseen the bad consequences of an action, they tend to demand retribution or penance.

Consider the inventor Thomas Midgely, who introduced the idea of adding lead to gasoline to stop engines from knocking. He later helped develop commercial chlorofluorocarbon (CFC) for use in refrigeration. For many decades lead and CFC were used in cars and refrigerators. Both turned out to be among the worst pollutants the world has ever seen. But is Midgely himself guilty for having caused more pollution than anyone else in the twentieth century? I do not know if he could have foreseen the consequences of his inventions. The same action and outcome can lead to quite different assessment of moral responsibility, depending on your judgments about the person's foresight, control, and intentions.[9]

Even preschoolers distinguish intentional from unintentional acts. Yet, as you might recall, mind reading can become rather difficult. For instance, as Donald Rumsfeld observed in 2002, there exist "unknown unknowns, the [things] we don't know we don't know." Most people would grant that you cannot be morally responsible for unknown unknowns. But you may well be held responsible for things you know you do not know— you could have made a greater effort to find out.

Establishing moral responsibility can be a complicated matter, as can be observed in virtually any court. Naturally there is great incentive for

---

9   Some legal systems place more emphasis on outcome rather than intention. Scholars of jurisprudence, of course, continue to debate the philosophy of law, personhood, and responsibility.

people not to be found guilty, so everyone tries to construe situations to their advantage, sometimes through deception and lies. What is worse, the accused may not only deceive others but also deceive themselves. Deception is common in nature, but humans can go one step further and self-deceive. For instance, we avoid unwelcome information. People generally stop searching for new information quickly when they like what they have found so far, but they search significantly longer when they do not like it. In a medical saliva test, if changing color is said to indicate illness, people finish the test quickly, but when it is said to indicate health, people wait much longer. Robert Trivers and my colleague Bill von Hippel reviewed many studies that show that people search for, attend to, and remember information in apparently self-deceptive ways. Recall, for instance, that people remember their own good behavior better than their bad behavior, but show no such bias when recalling the behavior of others. Thus it's no surprise that perpetrators and victims tend to remember events in ways that are biased in favor of their situation. The perpetrator thinks he was well-intended, reasonable, and justified, whereas the victim recalls the perpetrator's action as malicious, irrational, and unwarranted.

In a sense, this is quite the opposite of self-consciousness. It looks like we are systematically misrepresenting our self to ourselves. How can we be both deceiver and deceived? Von Hippel and Trivers argue that self-deception evolved out of more ordinary deception between people. We have not eradicated the problems of free riding and cheating, but have come up with ways of coping with them. We look for signs of deception and enforce honest cooperation. To cheat, people in turn explore new avenues to exploit others' trust. These two pressures have created an arms race between deceiver and deceived. Self-deception is simply another layer of complexity, improving one's chances of deceiving others while maintaining a facade that enforces cooperation. Rumsfeld might have called this phenomenon "unknown misleading." That is to say, we sometimes do not appear to know that we are misleading others (even if deep down we may have an inkling of the truth[10]). And if you believe your own deception, you are not going to show the telltale signs of lying—making it harder for

---

10 To call this self-deception, rather than just a bias, we need to postulate that somewhere the person does know both the truth and the lie. One may call this an unknown known, to add another permutation, when one ignores or suppresses available information. There is some experimental evidence that supports this by showing that under certain circumstances—for example, through distraction—people do demonstrate access to the truth even if they seemed to have been convinced by their own lies.

others to detect your cheating. Furthermore, if you do get caught, punishment is likely to be reduced. Moral judgment and retribution for willful deception is harsher than for unintentional "mistakes."[11]

Thus, even concentration camp guards may have had a moral perspective that allowed them to believe in subhumans, the Aryan master race, and the moral leadership of the führer. Most evil in this world is perpetrated by people who think, at some level, that they are doing the right thing. The fight of good versus evil, when examined from both sides, is often the fight between two definitions of what is good. People can behave outrageously badly in the name of good. Just think of the Spanish Inquisition or modern suicide bombers. Whatever we do, we tend to find reasons for thinking we are right.

However, our self-reflective capacities allow us to recognize when we are wrong, as much as we do not like to admit it. Our nested thinking enables us to examine our choices, evaluate our evaluations, and question our morality. We can acknowledge our inner demons and develop strategies to keep them in check. Scientists conduct double-blind experiments to avoid tricking themselves. We can identify our own hypocrisies and strive toward ideals. We can change our mind based on a new analysis of the facts. We can feel remorse and try to make amends for our mistakes. We can beg for forgiveness. We can venture to be the change we want to see in this world.

> The fact that man knows right from wrong proves
> his intellectual superiority to the other creatures;
> but the fact that he can do wrong proves his moral
> inferiority to any creature that cannot.
>
> —MARK TWAIN

WHEN YOUR DOG DEFECATES ON the carpet, it may look guilty. You hope it understands that it has done wrong, as you do not want a repeat of the

---

11 A similar social explanation might also account for the systematic biases in our affective forecasting. In order to elicit cooperation on a future-directed project, one may benefit from exaggerating the positive consequences of success or the negative consequences of failure. You may be a lot better at recruiting help if you believe your own claims about future affect. In cases in which your prediction turns out to be wide of the mark, you will attract less punishment if you seem to have believed your own predictions than if you appear to have intentionally misled others.

mess. The dog may fear punishment, but does that mean it has a con-science or a moral sense? Do nonhuman animals show signs of any of the three levels of morality de Waal distinguished?

Our closest animal relatives certainly appear to have qualities reminis-cent of both our demonic and angelic sides. They can act in profoundly prosocial and antisocial ways, helping or hurting others. Following World War II it was widely held that only humans were capable of the atrocities of war. Animal conflicts with members of the same species are typically restrained and do not cause serious injury. However, as noted earlier, Jane Goodall found that our closest living relatives go on raids and brutally kill other chimpanzees. The strong bonds between adult males and the hostility toward individuals from other groups led Goodall to argue that chimpanzees are at the threshold of the human capacity for destruction and cruelty. What differentiated these chimpanzee raids from human war, Goodall submitted, was planning and language.

Like humans, chimpanzees sometimes violently take what they want. And they may violently defend what they have. Thus they have at least two of Hobbes's reasons for human quarrels—although I do not know of evidence that they plot preemptive strikes and deploy credible deterrence strategies.[12] Note that females can be as merciless as males, as the follow-ing observation by Goodall illustrates:

> At 1710 Melissa, with her three-week-old female infant, Genie, clinging in the ventral position, and followed by her six-year-old daughter, Grem-lin, climbed to a low branch of a tree. . . . Passion and her daughter [Pom] cooperated in the attack; as Passion held Melissa to the ground, biting at her face and hands, Pom tried to pull away the infant. . . . At one point Passion snatched the baby away but Melissa seized it back, biting Passion's hands. Passion, leaping around, seized Melissa from the rear and bit her deeply into her rump (the wound actually penetrated the rectum just above the anus). Melissa, ignoring this savaging, strug-gled with Pom. Passion then grabbed one of Melissa's hands and bit the fingers repeatedly, chewing on them. Simultaneously Pom, reaching into Melissa's lap, managed to bite the head of the baby. Melissa still held on, and Passion seemed to try to turn her over. Then, using one foot,

---

12 The human strategy, if based on reason (rather than somehow on an innate predisposi-tion), would seem to require some foresight and mind reading.

Passion pushed at Melissa's chest while Pom pulled at her hands. Melissa, still clinging to the baby, bit Passion's foot while Pom held and bit one of her hands. During the entire fight all the participants screamed loudly. Finally, Pom managed to run off with the infant. At this point Gremlin—who had been trying to help her mother throughout the fight, but had been repeatedly pushed out of the way—hurled herself at Pom, and Melissa managed to retrieve the infant; but almost at once Pom got it back and ran off. Pom climbed a tree with the corpse (for the baby is thought to have died when Pom first bit into its forehead) . . . Melissa tried to climb also, but fell back, seemingly exhausted, as a small dead branch broke. She watched from the ground as Passion took the body and began to feed.

Fifteen minutes after the loss of her infant, Melissa again approached Passion. The two mothers, in silence, stared at each other; then Melissa reached out and Passion touched her bleeding hand. As Passion . . . continued to feed on the infant, Melissa began to dab her wounds. Her face was badly swollen, her hands lacerated, her rump bleeding heavily. At 1830 Melissa again reached Passion, and the two females briefly held hands.

This passage brought tears to my eyes the first time I read it. Passion and Pom's behavior was absolutely appalling; moreover this was not an isolated incident. Over a four-year period the pair was responsible for killing at least two, possibly six, other infants. Such acts contradict all our notions about cooperative group living. Yet there exist countless examples of human murderers who, in spite of their gruesome disregard for others, did not destroy otherwise cooperative societies. Chimpanzee infanticide may happen to about 5 percent of infants. It has also been observed in other animals, such as lions and hyenas. Cannibalism is a behavior that stirs particularly strong moral disgust, but it has been reported in human tribes across the globe. Baboons and gorillas, like chimpanzees, have been known to feed on killed infants. However, Goodall's anecdote illustrates not only horrible aggression toward a group member but also the resumption of nonviolent relations. Chimpanzees tend to reconcile soon after conflicts end. Passion embraced Melissa at 1842.

Chimpanzees are capable of extensive prosocial behavior, even though they do not have the Hobbesian remedy of governments that monopolize violence to enforce laws and civilize its citizens. They have been observed acting in ways that look like attempts to comfort others who are suffering. Researchers analyzed spontaneous aggressive incidents and found that by-standers frequently kissed, hugged, and groomed victims of aggression. Victims were more likely to receive such attention than other animals, es-pecially immediately after serious (as opposed to mild) fights. Chimpan-zees, like humans, reassure and console the distressed. Macaques, on the other hand, showed no such signs of compassion.

When we mentally picture scenarios such as Melissa's struggle, we have emotional responses. Seeing another's emotion can also trigger a similar or related emotion. Chimpanzees' physiological responses to seeing negative emotions in video suggest that they experience negative emotional arousal as well. There is some evidence that chimpanzees read emotions in fa-cial expressions.[13] For example, they could spontaneously match pictures of positive and negative emotional expressions to relevant video scenes depicting a favorite food or a veterinary procedure. The consoling of dis-tressed chimpanzees, then, may be driven by a sympathetic concern for their emotional experience.

There is evidence that other species feel sympathy. Various classic ex-periments suggest that rodents are sensitive to pain of their cage mates. Rats have been found to press a lever to lower a distressed rat from a har-ness. In recent studies, rats have shown prosocial behavior that is difficult to explain without invoking empathy. For instance, rats freed a cage mate from a restrainer even when social contact was prevented and there was no direct reward for the liberator. They even did this when there was choco-late enticing them in another container.

Sympathetic concern may help explain some widely publicized feel-good stories of animals helping humans in distress. For example, when a three-year-old boy fell into the gorilla enclosure of Chicago's Brookfield Zoo, a female gorilla cradled the child and brought him to an entrance twenty meters away where keepers could retrieve him. However, often there are also lean alternative interpretations. In this example, the gorilla was hand-reared because her mother had neglected her, and the zookeepers had trained her to avoid neglecting her own offspring. During this process they had used a doll and rewarded the gorilla for bringing it to them for

---

13   Chimpanzees seem to use such expressions to signal to others, to beg from them, to show submission or dominance, or to request reconciliation. There are parallels in human and great ape emotional facial expressions, as is evident in numerous photographic comparisons between human and chimpanzee available on the internet. Note, however, we often mistake a chimpanzee fear response, showing both upper and lower teeth, with happiness.

inspection. Such training, rather than empathy, might plausibly explain why she brought the unconscious child to keepers, but we do not know for certain.

There are a few other stories of helping, such as Washoe, the sign-language–trained chimpanzee, acting as a lifesaver, which may not be that readily explained in lean ways. A female chimpanzee had been introduced to a small island enclosure but became distressed. In an attempt to jump the moat she landed right in it. Washoe, although hardly knowing the newcomer, is said to have instantly come to the rescue. She jumped an electric fence, held onto a post to step securely into the water, and extended her hand to pull the drowning chimpanzee to safety. This is one of the most extraordinary anecdotes of animal heroics.[14]

Frans de Waal has made the case that our primate relatives are fundamentally good-natured. There are signs that they have the basic moral sentiments, sympathy and reciprocity, which he associates with level 1 morality. Recall that many primates groom each other and so build alliances. They may subsequently help each other in different contexts. Chimpanzees form long-term affiliative relationships that one might call friendships. Recent laboratory experiments have shown that chimpanzees can be very helpful. For instance, in one study chimpanzees were found to unlock the cage door for another chimpanzee, especially when the caged chimpanzee had done the same for them in the past. Chimpanzees also help humans, for example, by picking up and returning things that were accidentally dropped, as human infants do. Chimpanzees can be kind.

In contrast, when it comes to food chimpanzees are often unhelpful, sharing only occasionally and haphazardly. Mothers rarely give their offspring food, and when they do they usually only offer leftovers such as fruit peels. Humans consider such behavior selfish. We feed our children for many years after they have been weaned. We are often polite and offer guests the best we have. Chimpanzees do not show any such hospitality. At times they collaboratively hunt for monkeys, bush pigs, and other prey and may share spoils. However, the sharing is usually not fair.[15] Generally chimpanzees are competitive over food and share because they are harassed by beggars—that is, they share to avoid having to defend the food.

---

14 Washoe also infamously bit neurosurgeon Karl Pribam's finger off (and then signed, "Sorry.")

15 One study suggests that a male that does share meat with a female improves the likelihood of later copulation.

In some studies, chimpanzees failed to help other chimpanzees obtain food even when it was at absolutely no cost to them. Presented with an apparatus with two choices—a food reward for itself and the same for a familiar other, or the food reward for self and no reward for the other—they showed no concern for their neighbor's welfare. However, a recent study reported some prosocial choices. Similar research with marmosets, tamarins, and capuchin monkeys demonstrated helping. Some primate species, including bonobos, also share food sources more readily in the wild than common chimpanzees.

Limits to sharing severely restricts what kind of cooperation chimpanzees can accomplish. Two chimpanzees may pull together on a device to obtain food that they could not have gotten by themselves, and then share the reward if it is available in two distinct places. However, when the reward is lumped together on one plate, the more dominant animal is likely to eat most, if not all, of the reward. Cooperation subsequently breaks down.[16] Interestingly, three-year-old human children were found to share more with others in a collaborative task than in other circumstances, whereas chimpanzees did not share preferentially with those with whom they collaborated.

Sharing of information is another form of helping. As we have seen repeatedly, humans are driven to link up their minds, whereas nonhuman primate communication does not seem to work this way. There are two common exceptions to the lack of informing: food and alarm calls. At first glance, in these circumstances the informing of group members appears to benefit the others and not so much the caller. The caller may lose food or attract the predator. However, when examined more closely, such calls may be beneficial to the caller because by attracting others to a food source, one may gain protection against predators while feeding. Similarly, alarm calls may recruit others for potential defense. Curiously, food and alarm calls tend to be given even if all the other group members are already present, somewhat undermining the otherwise tempting conclusion that they intend to actively inform the others.

Given the apparent limits in mind reading, language, mental time travel, and reasoning, we have already encountered various restrictions to the kind of help animals may be able to offer. It has become clear, however, that parallels to the human building blocks of morality (level 1) exist

---

16 Social carnivores, such as hyenas, rely more heavily on cooperative hunting than primates do. This may expose them to stronger selective pressure favoring cooperation and sharing.

in other primates. Chimpanzees can be brutal, but they can also be kind. They exhibit signs of sympathy, helping, and sensitivity to reciprocity.

Does this evidence not suffice to conclude that nonhuman animals can live moral lives? The psychologist and activist Marc Bekoff has argued that it does. When morality is defined as "a suite of interrelated other-regarding behaviors that cultivate and regulate complex interactions within social groups," various animals may indeed qualify as moral. It may be appealing to embrace nonhumans as moral equals, but this romantic perspective sets the bar very low. Morality, as we have seen, is more than a suite of behaviors. To humans, the same behavior can be right or wrong depending on intentions, control, and the relevant norms.

SOCIAL ANIMALS HAVE TO GET along with their group members. But does this mean that they demonstrate elements of level 2 morality? Recent work on rhesus macaques suggests they evaluate photos of in-group members more positively than photos of out-group members. So perhaps intergroup biases have ancient roots, and there may be diverse social pressures to facilitate group cohesion. Even without laws, courts, and police, animals may need to act in certain ways to reward conformity to norms and punish violations.

Yet can animals be said to have social norms at all? Some prominent researchers think not. Michael Tomasello argues that agreeing on norms requires "shared intentionality," as discussed in Chapter 6, and therefore is limited to humans. Without a drive to link minds and share goals, it is indeed unclear animals could establish social norms.[17] Children derive and enforce social norms not only because of others' authority or knowledge of their benefits, but because they have a sense of belonging to a group with certain norms: "We do it like this, and not like that." When children are shown one way to play a game and are later confronted with someone else doing it differently, they tend to protest.

However, other researchers believe social norms exist in some nonhuman animals. The anthropologist Shirley Strum argues that baboons have social norms that are enforced by the group. When an immigrant

---

17 In one study, for example, chimpanzees cooperated in tasks that involved concrete goals but not social goals. Even in the tasks with concrete goals they made no attempts at reengaging a human when she stopped cooperating. Children, however, try to communicate to get the adult to recommit to the shared concrete or social goal.

male frightens an infant, the troop will reliably mob him. They do this again and again until he stops. Strum believes that this is evidence for the norm that adults "ought" not to frighten infants. Given the serious risk of infanticide, there is a leaner interpretation available. The other baboons may simply be protecting the infant.

Another prominent case argues that primates operate with social norms of fairness. Capuchin monkeys were trained to exchange a pebble for a slice of cucumber. When another monkey received a grape—a much more desirable reward—in this exchange, the first monkey started to refuse to work for cucumber. Does this mean capuchins have a sense of fairness and refused to cooperate because they felt shortchanged? Killjoy skeptics suggested that the primates may refuse to work out of frustration from not getting a grape, rather than because of inequity. Subsequent experiments found that merely seeing a grape makes the cucumber less attractive, regardless of whether there is another animal around. Other studies have found some support for the fairness explanation in chimpanzees and dogs. The debate continues.

Chimpanzees do not show any overt signs of guilt and shame, such as facial blushing, that would signal that they had violated their own conscience.[18] There is also little evidence to suggest that animals police others' conformity to norms (if indeed they have them) and punish transgressions. There are some reports of high-ranking primates breaking up fights, but it is difficult to establish whether they did so because of "community concern" or because they simply wanted an end to the disturbing raucous. In one study rhesus monkeys that suppressed food calls were subject to increased aggression from group members. This apparent punishment could be interpreted as enforcement of a social norm. However, the punishing individuals may have been directly affected. There are few signs that any bystanders (those not directly affected) reward and punish obligatory and forbidden acts. Nor do I know of any evidence that animals like Washoe, after rescuing the young chimpanzee, reap status and respect from group members acknowledging their bravery. (However, this may reflect absence of evidence, rather than evidence of absence.) As we saw, in humans third-party reinforcement and promotion of moral norms are critical. Given the

---

18 Darwin already noted that only humans blush. Blushing may reflect awareness of how we appear to others and has been argued to be a remedial display for one's own inadequacies, reducing chances of social ostracism.

limits of communication and intentional teaching in the animal kingdom, it is difficult to see how animals could pontificate and moralize.

It remains possible that chimpanzees and other social animals have precursors of social norms, but there is little reason to believe that they have anything like human moral codes.

> A moral being is one who is capable of reflecting on his past actions and their motives—of approving of some and disapproving of others; and the fact that man is the one being who certainly deserves this designation, is the greatest of all distinctions between him and the lower animals.
>
> —CHARLES DARWIN

EVEN IF WE ACCEPT VARIOUS parallels in the building blocks of morality, de Waal argues that level 3 clearly sets humans apart from animals. Even Bekoff agrees. Humans engage in self-reflective moral reasoning and judgment. We assess the intentions and beliefs that underlie actions, and there appears to be no evidence that animals do anything remotely similar.

Moral self-reflection requires the capacity for flexible mental scenario building, which we have seen is essential for many of the traits discussed so far. We can reason about our past, present, and future motives, beliefs, and actions. This reflection allows us to direct our actions, and even our thoughts and desires, in deliberate ways. As the philosopher Christine Korsgaard noted, only humans have the capacity for normative self-government. And, I might add, only humans have created real governments with powers to pass new laws, judges that assess violations, and prison guards that execute the verdict.

Yet methods for testing human moral reasoning and judgment are not easily transferred to animal subjects. To probe the conscience of young men who claimed to be conscientious objectors to the military draft, German authorities used to ask them to imagine situations such as: What would you do if your girlfriend was raped in the woods and you were carrying a gun? These assessments require language as well as an explicit capacity to build and reflect on mental scenarios. There is no evidence that other animals ponder counterfactuals about what they would do if they were confronted with a certain dilemma. In courts, as in gossip, humans regularly reconstruct past events to determine guilt and responsibility. The limits we have discussed so far would appear to make self-reflective moral reasoning impossible for animals.

The only potential sign of moral value judgment comes from the ape language projects. A recent detailed analysis of a database from over eleven years documented the use of the lexigrams "good" and "bad" by two bonobos (Kanzi and Panbanisha) and one chimpanzee (Panpanzee). The symbol "good" was used a few hundred times over this period and "bad" was used even less (24 times by Kanzi, 174 times by Panbanisha, and 83 times by Panpanzee). For example, on one occasion when asked, "Do you know how you've been acting?" the chimpanzee replied, "Bad." Does this indicate self-reflective moral reasoning? Without further elaboration and careful testing it remains unclear what they are really saying. A lean interpretation could suggest the ape may have simply associated the sign with certain actions, as the human caregivers frequently use these expressions in specific circumstances (e.g., "Panzee's being bad"). On a different occasion one of the bonobos was eating a plum and then pressed the "good" lexigram, presumably commenting on the taste, rather than on a moral question. The occasional appropriate use of these symbols suggests that one could perhaps study judgment in these human-reared apes. The researchers interpret their finding as evidence for precursors of morality, not as a sign of self-reflective moral reasoning.

In sum, the evidence for morality in animals declines as we move from de Waal's level 1 to level 3. At level 1 there is reasonably good evidence that other animals may have something like compassion, and there are examples of reciprocal cooperation between unrelated individuals. At level 2, there are a few signs that our closest relatives exert pressures that support cooperative group living, but there is no compelling case for animals moralizing explicit norms and third parties punishing/rewarding moral violations/virtuous acts. At level 3, there is as yet no evidence that nonhuman animals engage in self-reflective moral reasoning.

THE ANIMAL RIGHTS LAWYER STEVEN Wise has made the case that chimpanzees and bonobos should be given legal personhood. Animal law is now taught at several esteemed universities. Personally, I think it is about

time that we explicitly extend our moral concern to include consideration
for the suffering of our closest surviving animal relatives. But should great
apes be given the rights of a person?

We saw that personhood usually entails notions of self-consciousness
and control. Great apes might recognize themselves in mirrors, but given
the reviewed evidence they do not seem to be (self-)conscious about their
own knowledge and intentions or about the long-term consequences of
their actions.

There is, however, some evidence that they have the executive control
to (at times) suppress immediate urges and delay gratification. We saw in
Chapter 5 that chimpanzees, unlike most other animals, can delay receiving a small reward in favor of a much greater reward for a few minutes. In
one study chimpanzees waited longer when they could distract themselves
by playing with toys, and they played more with the toys during periods
in which they could have taken the smaller immediate reward than during
periods in which they could not. These results suggest that chimpanzees
may be able to gain some control over their urges. However, does this
mean they can take responsibility for their choices?

A group called the Great Ape Project, headed by the philosopher Peter
Singer and others, has strongly argued for the inclusion of great apes into
our community of equals, in which we accept legally enforceable rights.
In particular, they argue for the right to life, the protection of individual
liberty, and the prohibition of torture. In 1999, New Zealand banned experimentation on great apes as a result, and other countries are now in
the process of following suit. Yet to give rights entails more, because with
rights come responsibilities—such as respecting others' rights. Although I
am sympathetic to anything that would improve the treatment of captive
apes and protect apes in the wild, given the evidence there is little hope
that they could become full members of our moral community.

Though we may be perfectly happy to extend the right to life, liberty,
and freedom from torture to apes (and so would be willing to prosecute
someone who kills an ape), would we be equally happy with the other
side of the coin? Would we be willing to put an ape on trial for murder? In 2002, Frodo, a twenty-seven-year-old chimpanzee studied by Jane
Goodall, snatched and killed a fourteen-month-old human toddler, Miasa
Sadiki, in Tanzania. I do not remember calls for a trial. Moreover, should
we police ape-ape rights violations? There would be little point in prose-
cuting male orangutans for rape or a chimpanzee for infanticide. During

the European Middle Ages, animals were in fact frequently put on trial for immoral acts such as murder or theft. They were given lawyers and penalties that matched those given to humans for similar crimes. For instance, in 1386 a court in Falaise, France, tried and convicted a sow for murdering an infant. The hangman subsequently hung the pig in the public square. Her piglets had also been charged but, upon deliberation, were acquitted because of their youth.

Without compelling evidence that animals can reflect on their choices and consider the moral consequences of their actions, we cannot seriously bind them into a social contract. They are not persons according to the law and should not be held accountable.[19] I would advocate new laws that charge us with clearer obligations to better treat, protect, and respect animals, because we can derive a moral principle according to which it is wrong to mistreat other animals. We can expand our circle of sympathy to include other creatures and take their needs and preferences into account.

The last few hundred years have seen dramatic changes as a result of people reflecting on rights, cruelty, empathy, and the greater good. Steven Pinker has documented how violence has drastically decreased from our everyday lives, even taking the extraordinary brutality of the world wars or the Rwandan genocide into account.[20] In recent times, people have placed greater value on compassion; they increasingly reflect on their choices and recognize the advantages of peace and civility. We now cooperate globally, and our moral assessments are communicated and enforced more swiftly and effectively than ever. Slavery, torture, rape, dueling, and executions are no longer common aspects of most people's lives. In spite of the many terrible things humans do, we seem to have become a lot nicer.

We care. We extend rights not only to all humans but increasingly to life more generally. Animal cruelty has only recently become widely frowned on and is, thankfully, on the decline. Animal husbandry and slaughter are increasingly regulated. Animal ethics committees check research proposals, including whether the research question and approach justify the means. Blood sports, from cock fights to fox hunting, are becoming extinct. Wild animals are increasingly attracting tourism rather

---

19 This is not to say that they should be considered legal "things." Laws may indeed require a radical overhaul to go beyond the dichotomy of persons and things. Should sentient beings be considered the same as a chair?

20 At least proportionally the number of violent deaths have fallen.

than hunters. Whale watching is bigger business than whaling. Many are increasingly feeling morally responsible for the fact that our actions, from pollution to deforestation, are causing the destruction of animal habitats and the extinctions of species. Once we are aware of the consequences of our actions, we become morally obliged to take them into account. Our self-awareness and attitudes to life on Earth have changed drastically in recent decades. It is quite likely that we have to change them a lot more if we want our closest wild animal cousins to be around in the future. I leave you to draw your own moral conclusions. I know you can.

TEN

Mind the Gap

[Man] owes his success to certain things which
distinguish him from other animals: speech,
fire, agriculture, writing, tools, and large-scale
cooperation.

—BERTRAND RUSSELL

LIKE MANY A SCHOLAR BEFORE and since, Bertrand Russell confidently asserts that certain traits set humans apart from animals. Although we appear to excel in many domains, such claims are not typically founded in any thorough comparison. In fact, if you set the bar low, you can conclude that parrots can speak, ants have agriculture, crows make tools, and bees cooperate on a large scale. We need to dig deeper to understand to what we owe our unique success. In the previous six chapters I have described what current evidence suggests separates us from other animals in the domains of language, mental time travel, theory of mind, intelligence, culture, and morality. In each domain, various nonhuman species have competences, but human ability is special in particular respects—and these have much in common.

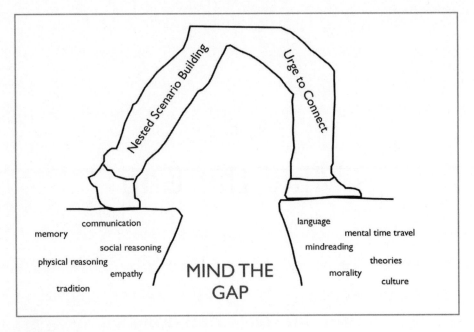

FIGURE 10.1.
Grossly simplified illustration of the conjunction of two foundational capacities
stepping across the gap from ape to human minds.

In all six domains we repeatedly find two major features that set
us apart: our open-ended ability to imagine and reflect on different situations, and our deep-seated drive to link our scenario-building minds
together. It seems to be primarily these two attributes that carried our ancestors across the gap, turning animal communication into open-ended
human language, memory into mental time travel, social cognition into
theory of mind, problem solving into abstract reasoning, social traditions
into cumulative culture, and empathy into morality.

Humans are avid scenario builders. We can tell stories, picture future
situations, imagine others' experiences, contemplate potential explanations, plan how to teach, and reflect on moral dilemmas. Nested scenario
building refers not to a single ability but to a complex faculty (recall my
theater metaphor), itself built on a variety of sophisticated components
that allow us to simulate and to reflect.

A basic capacity to simulate seems to exist in other animals. When
rats are in a well-known maze, the sequential firing of so-called place cells
in the hippocampus suggest that the rats can cognitively sweep ahead,

considering one path and then the other, before making a decision about where to go. Appropriate place-cell sequences have also been recorded during sleep and rest, suggesting a neural basis for the learning of the maze layout and its options. The challenges of navigation may well have selected for the fundamentals of mental scene construction. Moreover, we saw that great apes have demonstrated several other relevant capacities. They can think about hidden movements, learn and interpret human symbols, solve some problems through mental rather than physical computation, have complex sociality and some traditions, console each other, recognize themselves in mirrors, and show signs of pretense in play and deception. Great apes have a basic capacity to imagine alternative mental scenarios of the world. In certain contexts their abilities are comparable to those of eighteen- to twenty-four-month-old human children.

Human development of mental scenario building explodes after age two, however, while great apes' capacities do not. Children spend a considerable amount of their waking life in fantasy play. They conjure up and untiringly repeat scenarios with props such as dolls and toys. Thinking, in a fundamental way, is imagining actions and perceptions, and it has been argued that in play children test hypotheses, consider probabilities, and make causal inferences not entirely unlike (adult) scientists. Play certainly provides opportunity to practice, to build up expectations, and to test them. Children take on roles and act out narratives of what happens in certain situations. Gradually, they learn to deliberately imagine scenarios and their consequences without having to act them out. They learn to simulate mentally. They learn to think.

Eventually, children can imagine an almost limitless array of events. They begin to deploy counterfactual reasoning in which they contrast what did happen with scenarios of what did not happen. They increasingly consider what might happen in the future. A key to our open-ended, generative capacity is our ability to recursively embed one thing in another, as it enables us to combine and recombine basic elements such as people, objects, and actions into novel scenarios. Such nesting is also essential for reflection: our capacity to think about our own thinking. Nested thinking allows us to reason about the mental scenarios we entertain (just as we can draw pictures of ourselves drawing a picture).

We can connect diverse scenarios into larger plots. Narratives provide us with explanations for why things are the way they are and with opportunities for prediction or scheming how they will be. We can reflect on the relationships between past experiences and construct complex plans with embedded if-then steps. Even our closest animal relatives have not demonstrated this open-ended nesting ability. One explanation for this lack is that they do not have the necessary working memory capacity to embed information recursively. Michael Corballis has championed the idea that recursion is uniquely human. We came across evidence for this repeatedly, and it is easy to see its relevance in other contexts such as architecture, cooking, and music. You can apply your own recursive thought to come up with more domains.

Children also learn how thought can control action. For nested scenario building to provide adaptive advantages, one needs to gain some executive control, for instance, to inhibit present desires in the name of more prudent long-term goals. More generally, natural selection could not have gotten its teeth into private daydreaming; our scenario building must have had outward consequences for survival and reproduction. Our minds matter. Crucial here is the audience of our mental plays, to stick with the theater metaphor. The neurobiologist Bernard Baars suggests that consciousness is a broadcasting system: it disseminates single momentary messages throughout the brain to achieve coordination and control. The information we are aware of is made available to all subsystems.[1] Consciousness provides cohesion and allows a complex nervous system to pull in one direction. Conscious simulation of scenarios may therefore allow us to pursue complex plans. As long as we can manage our more immediate urges, our mental simulations and reflections can gain control over actions. To a significant extent, we can become masters of our own destiny.

Individual simulation is flexible and powerful but also a risky way of making decisions that can lead us fatally astray. In the heat of Australia's north a river may appear inviting for a swim—until you note the sign about the crocodiles. Individually, we often miscalculate, harbor false expectations, and become confused as to which option to pursue. Nested mental scenario building is not a crystal ball, nor is it a logical supercomputer. For flexible scenario building to really take off as the ultimate survival strategy, it required a second leg to stand on.

---

1 The theater metaphor does not suggest, of course, that there is some ghost in the machine watching the mental plays. That idea is a nonstarter because it merely shifts the problem to the question of who is watching the mental play in the ghost's head. Baars's notion of broadcasting across the system, on the other hand, is a possibility.

Our ancestors discovered that they could dramatically improve the accuracy of their mental scenarios by increasingly connecting their minds to others. We give each other advice—for instance, by posting signs about the possible presence of crocodiles. We can broadcast our imaginary play not only throughout our own system but to others around us. We exchange our ideas and give feedback. We ask others, and we inform them—for instance, by recounting what it was like when we were in a similar situation. We take an interest even without knowing whether anything important or useful comes of it. There are individual differences in how much an interest people display in what certain others have to say, but we are generally driven to wire our minds to those around us. Our expectations and plans are subsequently a lot better than they could have been if we didn't listen. It is generally good advice to consider advice—preferably from a variety of sources—before making up your own mind.

We saw that nested scenario builders can benefit from cooperating with other scenario builders in many other ways. For instance, our audience can be recruited for common goals. We can hatch complex plans, divide labor, and pledge cooperation. We can accumulate our achievements and pass them on to the next generation. To ensure all this happens, we appear to be hardwired with an insatiable urge to connect our minds.

Primates are social creatures, and evidence that social pressures have driven the evolution of primate intelligence is mounting. Humans have taken this sociality to another level. Unlike other primates, children sob to attract attention and sympathy. We ask what's wrong and try to make things better. We look each other in the eye, share what's on our minds, and absorb what is on the other's. This urge to connect must have been crucial to the establishment of signs and words that allow us to effectively read others' minds and express our own. We are driven to wire our mental scenario builders into large systems of scenario builders. We can learn from others' experience, even second- or thirdhand. Our drive ultimately led to today's networks of mobile phones and social media that let us exchange our minds across the globe.

As Michael Tomasello and colleagues have demonstrated, we make and pursue shared goals where our closest relatives do not. Even two-year-old children outperform great apes on tasks of social learning, communication, and intention reading. Other animals may give alarm calls and food calls but otherwise do not show many signs of a drive to share their experience and knowledge with others. Again, in all six domains this cooperative drive is evident and plays a significant role. Language is the primary means by which we exchange our minds. We talk to each other about the past and make plans about the future. We read and tell each other what is on our minds. We reason and solve problems collectively. We build social narratives that explain the world around us. We teach, and we learn from each other. And we argue about what is right and what is wrong. These examples serve to remind us how pervasive the urge to connect is. Those who lack this drive have severe social difficulties (and may be diagnosed as autistic). Our urge to connect was essential for the creation of cumulative cultures that shape our minds and endow us with our awesome powers.

Our capacity for nested scenario building even allows us, drawing on past experiences, to imagine others' advice internally.[2] So you might ask yourself what your mother would have said about the situation you find yourself in. We care about whether our parents, friends, heroes, or gods would be proud of what we do, even if they no longer exist (or never did). We can consider what others might remember us for. These thoughts can be important drivers motivating us to go beyond satisfying immediate personal self-interests in pursuit of "higher" notions of honor, valor, and glory.

We might aspire to nobility in character and virtue in action. We can invest heavily in unselfish actions, such as fighting oppression or pollution or helping a club, a person, or an animal. When we take on a cause, we seem to become part of something bigger and from such endeavors may derive some of the deepest feeling of meaning. One of the most remarkable things about humans is that we can strive to make some kind of difference. We may deliberately practice random acts of kindness, spread the word, fight injustice, teach the next generation, or start a revolution. Without the urge to connect our minds, such traits could not exist.

In sum, nested scenario building and the drive to link our scenario-building minds turned ape qualities into human qualities. They created powerful feedback loops that dynamically changed much of the human condition. They carried us where other animals could not go.

---

2 Hearing voices is quite normal. Relax. The trouble starts when you attribute these internal voices to external sources.

THE GAP, THEN, MAY IN one sense be much smaller than we might have anticipated. Only a couple of basic differences appear to have evolved between ape and human over the last six million years. In another sense, the gap clearly is vast. The two qualities represent profound differences that have had countless cognitive, emotional, and motivational consequences.

Children's play shows how the two legs work in lockstep to open dramatic new possibilities. Children often play together by acting out imaginary scenarios. They coordinate roles, such as doctor and patient, and play out how events might unfold. These social constructions teach children about particular situations and how to negotiate their scenario-building minds. Children learn from each other through imitation and instruction, adopting and enforcing socially inherited rules. They love playing games. They usually are desperate to win, even if no concrete rewards are involved.

We already saw how play gives children opportunities to practice thinking. They begin to select skills they want to get better at, typically with encouragement from their parents, and playfully rehearse. As we saw in Chapter 5, our scenario-building minds even allow us to practice just mentally rather than physically, and we can improve as a result. We provide our own feedback, or imagine the feedback we think others might give, without actual reward or punishment. Our capacity to create future scenarios and to consult others with expertise allows us to gradually become experts in our own right. Much of the human diversity of expertise is a function of the fact that different people expend effort in learning and perfecting particular skills. Just think of the Olympics. Even within each area of expertise we differ vastly from each other. Each of us is unique. Our extensive cooperation and division of labor means that groups, and the individuals within it, benefit less from everyone being good at the same thing than from being good at a host of complementary skills. Groups composed of individuals with high polyvalence may have had a particular advantage. Perhaps we have evolved to be so individually different because our species is extraordinarily social and cooperative.

In addition to play, children love stories. Stories contain lessons learned by others, and even entirely fictional narratives give the listener potentially useful information. Most stories involve a hero overcoming obstacles and enable listeners to learn how to solve problems—and what virtues are worth pursuing. In short, stories allow us to acquire experiences without having to leave the house or risking life and limb. The desire to

listen to stories, to be entertained by them, is therefore a powerful facilita-
tor of cultural transmission.

These days we can watch or listen to stories anytime, anywhere, at the press of a button. Virtual theaters beam other people's scenarios into our minds at extraordinary rates. Although we may enjoy such narratives alone, the sharing of stories used to be entirely social events. At a minimum one person shows or tells another. When stories are told to an audience of several people, everyone tunes in to one and the same scenario. Members of the audience have similar thoughts, emotions, and learning experiences. Stories, whether told or enacted, thereby help synchronize values, moral norms, and expectations in the group. Human cultures and belief systems are to a large extent defined by their stories. Stories of ancestors and origins shape our social identity: who we are and where we come from. Stories provide meaning and explanation.[3] Stories create bonds between people.

Whereas chimpanzees groom each other to bond and keep the peace, human cultural groups bond through sharing mental experiences. People gather at festivals to reaffirm their cultural identity through rituals and celebrations. They wear specific clothes, engage in traditional ceremonies, perform and act, tell stories of legends and ancestors. We sing and dance together, and we enjoy watching others perform. We hold concerts, parades, and shows. We celebrate birthdays and weddings, commemorations, and seasons. Animals do not appear to care about any of this. Contrary to how animals are depicted in children's cartoons, such as the film *Madagascar 2*, they do not seem to like to party.

Interestingly, humans who entertain others tend to have a sexual selection advantage. Artists, actors, and musicians typically have more partners than less entertaining types. Such a benefit is a powerful incentive for engaging in creative endeavors. As with other expertise, to improve their capacities entertainers practice their performances. And as we do with mental scenarios, tools, or sentences, we combine and recombine basic elements to create novel entertainment in acts, dance, and music.

---

3 A narrative can explain how we get from one point to another, and it can lead to identification of the causal links involved: the grass is wet because it rained this morning. Explanations in turn help us predict: it is raining now, so the river will rise. Crucially, they also provide opportunity for control. We often do not know the actual causal chains involved, but stories suggest what might have worked in the past. Many technologies and rituals have been invented to try to control things that matter to us, sometimes causally successful (e.g., making fire) and sometimes not (e.g., making rain).

Moreover, in many cultures, entertainment, parties, and ceremonies are fueled by mind-altering substances or practices, such as meditation or hypnotically repetitive dance. Psychoactive drugs, in particular alcohol (though the list spans all the way from coffee to cocaine), are powerful lubricants in many social exchanges. Animals have the capacity to form drug addictions—which is why they are often used in drug research—but drugs play a peculiar role in human societies. Sages, oracles, witches, and shamans have long used drugs for the express purpose of expanding their minds in quests for wisdom and access to worlds of spirits, gods, and fortunes. We know little about how far back such activities go, nor how much of a lasting influence they have had on the evolution of our minds. Obviously some drugs are dangerous and can lead to all sorts of problems. Nonetheless, many people go to great lengths and are willing to take great risks to manipulate their mental states through drugs. Whether socially sanctioned or not, our stories, rituals, and entertainers are full of them.

The difference between human and ape is clearly not only intellectual or rational in nature. Our scenario-building and -sharing capacities are also essential for our desire to let go of our worries and indulge in intoxication, celebration, passion, and excess. These facets of the human mind drive some of our most peculiar behavior. In ancient Greek mythology Zeus's son Dionysus represented these tendencies, whereas Apollo stood for harmony, order, and reason. The struggle between such opposites is a topic of much of our popular culture, from the fights between the Freudian id and the superego to those between Dr. McCoy and Spock. Trying to balance these forces is the decision-making ego, or Captain Kirk if you prefer. Our mind is a complex beast, and I won't pretend to be able to do all of it justice here. I have tried to focus on the basics that separate our minds from those of our closest animal relatives.

Our interconnected scenario-building minds have created virtual worlds of kaleidoscopic variety. Together we agree on shared immaterial ideas, ideals, and other make-believe. For instance, we invent social roles, institutions, and symbols, which we collectively imbue with specific authorities and powers. Referees, idols, CEOs, officers, priests, banks, and national symbols serve important functions in our communities and are critical for regulating our extremely complicated webs of cooperation. But their reputations, powers, and responsibilities only exist in our collective minds. Animals cannot perceive them. We merely imagine them together and act as if they are real. And so, for us, they are.

We have evolved a cultural world. Our collectively imagined ideas and concepts shape the very fabric of our reality. As children grow up, they face selection pressures within this artificial environment. Human cultural evolution and the evolution of human minds are inextricably intertwined. We shape the environment in which our children develop, what they learn, what they value and believe. Biologically oriented scientists sometimes underestimate the power of socialization and culture.[4] We should not. To a large extent we socially construct our minds. It is easy to see why one may be tempted to conclude that the gap, then, is the result of our cultural nurturing of growing minds. However, it is also the case that no amount of this same socialization can turn a goldfish, cat, or horse mind into a human mind, so far as we know. There is some evidence that great apes raised by humans, and so "enculturated," perform slightly better than their kin in the zoo on a few measures.[5] Still, we cannot turn them into anything like the reflective chimpanzee Kafka imagined. Humans are uniquely prepared to cross the gap and acquire our cultures. Therefore, I conclude with a discussion of how we appear to be biologically adapted to the cultural world our ancestors created.

> No other animal passes so large a portion of its
> existence in a state of absolute helplessness, or
> falls in old age into such protracted and lamentable
> imbecility.
>
> —JOHN HERSCHEL

OUR PREPAREDNESS FOR CULTURE COMES at a price. As Herschel observed, humans are extraordinarily weak and dependent in infancy and often also in old age. A newborn human must indeed be one of the most defenseless beings on the planet: it takes many weeks before it can even

---

4 And social scientists sometimes underestimate the power of biology. Both factors are typically important. When asked whether nature or nurture contributes more to personality, the neuropsychologist Donald Hebb is said to have astutely replied: "Which contributes more to the area of a rectangle, its length or its width?"

5 It is not entirely clear if this means that the enculturated minds are enriched or whether the captive zoo animal minds are deprived. The appropriate comparison point can only be wild apes raised in a normal social environment.

hold up its head and a year for it to walk. Yet brain growth outside the womb is crucial for our mental plasticity and hence our capacity to inherit culture from our social environment. Human brains grow more outside the womb than their closest relatives' brains do—both in absolute terms as well as proportionally (human newborns have only about 28 percent of their adult brain size, whereas chimpanzee newborns have an average of some 40 percent).

Stephen Jay Gould has argued that humans, in a sense, avoid growing up altogether. We continue to learn and remain playful and curious into old age—characteristics that for most mammals are found only in the young. The retention of juvenile traits into adulthood is called neoteny. And it is a great trick that with a few genetic changes can bring about vast consequences. Consider the axolotl, or Mexican walking fish. This common inhabitant of many aquaria has legs, can walk, and looks most peculiar for a fish. The reason for this is that it is not a fish at all but a neotenous amphibian. It is a salamander that has failed to grow up, remaining in the larvae stage without undergoing the metamorphosis into an adult. The flat face of a chimpanzee infant resembles that of a human. However, as they grow up, chimpanzees look ever less like us. Perhaps Gould was right when he argued we are a neotenous ape.

The anthropologist Barry Bogin has made the case that only humans have, in addition to infancy and juvenility, two new developmental stages before reaching full maturation. In mammals, infancy is defined as the period of maternal nursing. Juvenility is the stage after weaning and before sexual maturity. Primates in general have a relatively extended infancy period characterized by rapid growth, which stops with the eruption of the first permanent molars and the end of mother's lactation. In chimpanzees these events occur around age four. In humans, however, breastfeeding typically ends by about age two to three (or earlier), yet the first permanent teeth only emerge around age six. The period between weaning and first permanent teeth Bogin calls "childhood"—perhaps a less well-established term could have avoided confusion. In any case, Bogin's proposal is based on clear biological markers.

Human childhood in Bogin's sense of the word is characterized by a slowing of the rate of growth, immature dentition, immature motor control, and, importantly, dependence on food and care from others. Brain growth during childhood is severely advanced relative to skeletal and muscle growth, with the brain peaking in mass at around age seven. In other primates, after infancy, offspring can find food for themselves and have the teeth to show for it. When human mothers stop lactating, however, they still need to provide food for their child (or need others to do that job). Transition to this stage makes it possible for the mother to breed again. Chimpanzee birth intervals are nearly twice as long as those observed in typical human hunter-gatherers. Therefore, even though humans reach sexual maturity at a much later stage than apes, humans can reproduce much more frequently.

Juvenility is the next stage. It is characterized by the capacity of the youngsters to fend for themselves and yet be sexually immature. In most animals this period is short—in mice, for instance, it is a few days—but in primates it is quite long. In baboons, for example, there are three to four years between weaning and puberty. Juvenility in humans begins around age seven, with the increased secretion of androgens leading to the growth of pubic hair and a changed sweat composition. This phase is known as adrenarche and also occurs in gorillas and chimpanzees. Juvenility is characterized by a decline in growth rate and typically ends with the achievement of sexual maturity and adulthood. In humans, however, there is another stage that Bogin argues to be unique: adolescence.

The life cycle of animals is typically characterized by an initial acceleration and subsequent deceleration in growth rate. Human adolescence, however, is marked by a renewed skeletal growth spurt after years of decreasing growth rates. Adolescence starts typically around age eleven or twelve in females and a year or two later in males. Females may fully complete this stage as late as age nineteen and males a couple of years later. The period is characterized by exploration, sensation seeking, and various changes in social relationships, as well as by vulnerability to mental disorders such as schizophrenia and depression. As is widely known, in the process of achieving social and economic maturity, adolescents often challenge established cultural practices.

While brain size and basic structure are in place by human juvenility, throughout adolescence cortical grey matter becomes thinner, while white matter increases.[6] In line with these changes in the brain, the adolescent mind becomes increasingly capable of controlling behavior: adolescents improve in focusing attention, discipline, and resisting temptation. Exec-

---

6 This reflects that synapses are pruned and axons become increasingly mylenized.

utive self-control gradually improves. Even on very simple tasks such as "when a light appears on the left of the screen, look to the right; when it appears on the right, look left," inhibition errors only gradually decrease across adolescence. Adult levels of control are reached at the end of this period, at which point physical growth is complete.[7]

Whether or not one agrees with Bogin's definitions of developmental stages, it is clear that human development takes dramatically longer than that of any other primate and that it is characterized by unusual growth patterns. This developmental path provides ample opportunity for our second inheritance system to gain traction and transmit local culture.

A key to making these developmental changes possible must have been cooperative breeding. People other than the mother—including the father, uncles and aunts, grandparents, or even unrelated group members—help teach, protect, and provision offspring. Studies on hunter-gatherer groups show that families benefit from social norms of food sharing. Adults often intentionally acquire more food than they can consume to distribute among the group. The wider family typically helps raise the kids.

Human child-raising structures are not limited to Western nuclear families. The Musuo in Chinese Himalaya, for instance, have a system in which men do not invest in raising their own children but instead look after the children of their sisters and aunts. In this way they solve a fundamental problem of paternal parental investment. Before genetic testing, males could never be entirely certain that their purported children were in fact theirs. The Musuo avoid the problem of potential cuckolding altogether by supporting their less closely related, but more certain, kin—the children of their female relatives. There are diverse human child care arrangements, but what they all have in common is the support of nonmother members of the group.

Cooperative breeding is not a uniquely human trait per se.[8] However, humans provide unusual amounts of care to their children. In hunter-gatherer societies, prior to modern prophylaxis and medical care, about 50 percent of newborns survived into adulthood. Although this figure might strike us

---

7 However, some aspects of the brain, such as white-matter connectivity, continue to mature even after adolescence. In particular, a white-matter tract called the uncinate fasciculus, implicated in socio-emotional processing linking orbitofrontal cortex, amygdala, and temporal regions, matures surprisingly late. It is found to peak in the mid-thirties.

8 For example, marmosets breed in small groups in which all adults, males and females, contribute to carrying and supporting the group's offspring.

as horribly low, it is high compared to figures from other animals. Many animals do not invest at all in their offspring and have accordingly low survival rates. Fish, for instance, lay millions of eggs for only one to reach maturity. Lions, on the other hand, have few offspring, invest significantly in them, and are lucky to see 15 percent survive to adulthood. Primates with their extended development invest even more than most other mammals and tend to achieve a survival rate higher than lions, with chimpanzees reaching a figure of about 38 percent. Parental investment and survival rates appear to go hand in hand.

While some great apes outlive their fertile period, most stay reproductively active until they die. Chimpanzees start reproducing from about age thirteen onwards, and females, of whom fewer than 10 percent live to age forty, tend to reproduce until the end. In humans, on the other hand, females go through menopause and often live for several decades without reproducing. Human groups often contain a sizable cohort of postreproductive elders—posing an evolutionary mystery. Postreproductive individuals draw natural resources away from those of reproductive age. Such a phenomenon makes no evolutionary sense, unless these individuals contribute in some other way to the survival and reproduction of their genes.

The explanation is that postreproductive elders increase inclusive fitness. Grandmothers provide a range of support to grandchildren and are often crucial in raising them. Evidence from traditional societies suggests the presence of a grandmother reduces mortality rates of infants. In some populations increase in the postreproductive period was found to be associated with increased numbers of grandchildren. Clearly, this novel life stage helps propagate genes.

Humans generally live longer lives than apes, surviving to age seventy and beyond even in hunter-gatherer societies. The elderly can draw on a greater wealth of experience and knowledge—or memes, if you like—than younger individuals. They act as living links to previous generations and so may harbor crucial information about challenges the group may face, in areas such as foraging, predation, natural disasters, and enemies. In a sense they function as the libraries of nonliteral cultures: they are critical to retaining its "meme pool" (recall that crystalized intelligence does not typically decline with age). All human societies have notions of wisdom they respect. Those regarded as wise are typically thought to be compassionate, knowledgeable, experienced, and reflective. They have superior insight in significant life matters and are therefore consulted in times of trouble or when far-reaching decisions are to be made. This is another way through which the elders can contribute critically to the success of their children and children's children. Younger group members, in turn, revere and support the elderly, even when they eventually fall into the protracted "imbecility" Herschel lamented.

WE ARE AN EXTRAORDINARILY COOPERATIVE primate. Children learn that others around them are prosocial and helpful. Humans impart vital mental skills and knowledge across generations. And so we managed to accumulate wisdom and technologies, passing them on from one mind to another over immense spans of time. This practice allowed us to exploit what John Tooby and Irvine Devore called "the cognitive niche": through reasoning, planning, and cooperation we overcome the defenses of plants and prey, as well as the threats of predators and competitors. Our mental scenario building and rapid exchange of useful information have enabled us to flexibly deal with new challenges and outwit other creatures that could only adapt to us through the much slower traditional process of natural selection.[9] Wherever humans traveled, they could quickly accumulate critical information to compete with other large animals. Indeed, there is evidence that humans frequently caused mass extinctions of prey animals shortly after arriving on new shores.

Which leads me to our prehistory and the creation of the gap: What steps occurred on our path from the last common ancestor with chimpanzees to modern humans? And what forces brought these changes about?

---

9 Fast-reproducing smaller organisms, such as ants and bacteria, remain difficult to control and are arguably more successful than humans, at least in terms of number of individuals, diversity, and distribution.

ELEVEN

The Real
Middle Earth

WHEN MY MOTHER DIED, SHE was buried in the local cemetery in Vreden, the town in Germany where she was born and where she gave birth to my siblings and me. Around the time of her funeral, an archaeological dig found eighty graves nearby—dated at over three thousand years ago. Hearing this made me think of the many generations before me that eked out a living in our hometown. Did my ancestors live in the area for all this time? Perhaps. But it's likely that my lineage is more varied than that and possibly includes famous forebears such as Genghis Khan and Cleopatra.

Let me explain. In many cultures one's surname is passed on through the male line. My father's father's father was called Suddendorf, as is my son. Through church records we can trace back eight generations to an ancestor called Dirk. If I had a time machine and visited Dirk, would I recognize him as my forefather? Names can be misleading—you are, of course, just as closely related to your mother's family as to your father's. You have 2 parents, 4 grandparents, and 8 great-grandparents. Perhaps you know all their names. If you assume, for simplicity's sake, that people reproduce by around age twenty-five, then this means that your ancestry includes 16 great-great-grandparents one hundred years before your birth and 32 great-great-great-grandparents twenty-five years earlier. Do you know all their names? When you move back eight generations, you should find 256 direct ancestors. Besides Dirk there were 255 other ancestors to whom I am equally closely related. Shared name or not, if one of them had not had children, I would not be here today.

If I traveled back four more generations, I would have 4,096 direct ancestors to visit. The numbers increase rapidly as we move back further in time. Following the same logic, four hundred years before your birth there are 65,536 ancestors. Go back six hundred years, and the figure is a staggering 16,777,216. Genghis Khan died in 1227 CE, and the number of my direct ancestors in his generation would appear to be over two billion. You see, even if he hadn't spread his seed across Eurasia, I would have a good chance of being his descendant. So would you, if you have Eurasian ancestry. And if this calculation was correct.

There is a serious flaw in this number crunching, of course. These exponential increases eventually become nonsensical. If you go back to Cleopatra some two thousand years ago, such logic would mean you had more than a septillion ancestors (a thousand billion times a thousand billion or, more precisely $1.2 \times e^{24}$). The numbers exceed not only the number of people alive but even the number of people who ever lived. Why? The calculation wrongly assumes that ancestors are themselves not related to each other, when in reality they often are. Sometimes they are closely related, as is evident in the famous inbreeding of the royal families of Europe. Yet slightly more removed inbreeding happens regularly when, for example, people find lovers in their own village and its environs. Nonetheless, if Jesus had a surviving bloodline, many, rather than a select few, could claim direct descent. The oldest continuously recorded family tree is that of Confucius, who, over eighty generations after his death, has well over two million registered descendants.

Some human groups have been in relative isolation for long periods and would hence not have had much opportunity for mixing with other groups. Australian Aboriginals have probably been isolated for the longest. With perhaps the odd influx of Indonesian or Melanesian visitors, they have been separate from other humans for about two thousand generations (i.e., more than fifty thousand years). They can claim the purest bloodline, as it were. Most of the rest of us are mongrels.

IN ORDER TO UNDERSTAND THE origin of the gap, we must first understand where we come from. With modern genetics we are now able to examine the ancestry of people even without historical records. Most DNA gets mixed from one generation to the next, ensuring variation among offspring. However, there is some DNA that does not get recombined, which helps geneticists to reconstruct our ancestry. The Y chromosome (which females lack) is passed on exclusively from father to son. Mitochondrial DNA is in the cell body and is passed on via the female's egg, not the male's sperm. That is, you have the same mitochondrial DNA as your mother and her mother, and your siblings share it too. Occasionally, however, even these nonrecombining parts of DNA change through random mutations, becoming genetic markers. As a result of common descent, people of the same region often share the same markers. When some of them migrate, they take the markers with them, distinguishing themselves from populations already present in the new land. Thus are geneticists able to trace the history of human migration through DNA. By comparing DNA from people of different regions, they can calculate where their most recent common ancestor lived. Given that random mutations occur at a relatively constant rate (though some DNA regions do attract faster mutation than others), they can also estimate when common ancestors lived.

While some have raised legitimate concerns about the potential for abuse of this new genetic knowledge by racists and insurance companies, it opens up a wonderful new avenue for reconstructing our past. For example, the finding that there is more genetic variation within Africa than outside of it tells us about our common origins. In mitochondrial DNA, for instance, all non-Africans can be subdivided into two lineages, or so-called haplogroups (M and N). Other lineages exist only in Africa. In other words, Swedish, Japanese, Aboriginal, and Maya DNA are much more similar—that is, they are more closely related—than the DNA of two different groups of Africans. There is a simple explanation for this pattern: people evolved in Africa, and a sub-group of Africans moved out and populated the rest of the world.

Analyses of the Y chromosomes of men from around the world have yielded an estimate for the most recent common ancestor—Adam, if you will—of about 60,000 years ago.[1] All humans alive are patrilineally descended from him. Analyses of mitochondrial DNA have suggested that our most recent female common ancestor—we can call her Eve—lived between 150,000 and 200,000 years ago in East Africa. The first thing to note here, then, is that she lived a long time before "Adam." A reason for this time gap is that males and females are not equivalent in their reproductive potential. Women bear between none and about a dozen children during their lives, whereas men can sire from zero to hundreds of offspring. This means that a few men can be grandfather to a great many people. Though hundreds of people may share the same grandfather, they cannot all have the same grandmother. And so the last common grandmother is found much further back in the family tree than the last common grandfather. This is not to say that only an individual male or female was alive at these times. There would likely have been tens of thousands of individuals alive, many of them producing descendants. The bottleneck represents only the most recent common ancestor to all of us, even if most of us are also related to Adam's and Eve's companions.

DNA studies have an increasing impact on our understanding of our origins,[2] providing an independent source of information beyond what is known from fossils. Fortunately, many of the major findings do appear to agree with what is known through the archaeological record. The oldest anatomically modern human fossils are dated at almost two hundred thousand years old: the Omo fossils from Ethiopia, discovered by a team directed by Richard Leakey.

The earliest mitochondrial haplogroup (L0) includes the Khoisan of Southwest Africa, famous for click sounds in their languages. From this group three African lineages—L1, L2, and L3—split off. People belonging to the L3 group spread rapidly over the globe between sixty thousand and eighty thousand years ago and gave rise to the two haplogroups of all non-Africans (M and N) that are estimated to have migrated out of Africa

---

1 A recent study challenges this, putting this date as far back as 142,000 years ago. Incidentally, it has been suggested that Genghis Khan is responsible for one particular Y chromosome lineage that is present in about 8 percent of men in a large region of central Asia, but in only 0.5 percent of men elsewhere.

2 You can now get your own DNA tested to reveal your deep family tree (e.g., https://genographic.nationalgeographic.com).

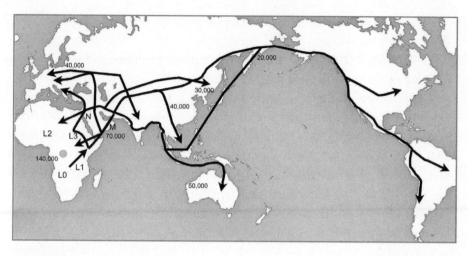

FIGURE 11.1.

Rough sketch of human migration as currently indicated by DNA and fossil evidence.

around that time. They dispersed into Asia, reaching Australia some ten thousand to twenty thousand years later. They settled in Europe around forty thousand years ago. From East Asia they moved into the Americas and quickly spread across the New World.

Over millennia our ancestors proceeded to settle most landmasses of the world and profoundly alter the landscapes they found. It was less than a thousand years ago that they finally settled the last major habitable lands they named Aotearoa, what the English-speaking world calls New Zealand. Aotearoa was devoid of humans, even of mammals, other than a couple of bat species, and Maoris arrived on canoes to find a land of plenty. Maori had brought chickens with them, which they called Moa, but stopped breeding them when they discovered the enormous birds available in the bush. The large flightless native Moa were not used to being predated by mammals, and as Maori flourished and hunted them, the birds rapidly went extinct, possibly within a century. As protein became increasingly scarce, Maori came to rely on shellfish and even, at times, on cannibalism. This was not an unusual turn of events. When humans first encountered lands where milk and honey flowed, they enjoyed and exploited them. Mass extinctions of local megafauna frequently followed. In some cases, the ecosystem collapsed, as happened on Easter Island. In other cases, severe periods of scarcity may have occurred before people managed to establish a new sustainable equilibrium.

Many of our migrating forebears, however, did not find virgin land. Instead, they encountered worlds already occupied by other hominins who had established themselves there much earlier. *Homo erectus* had lived in Asia for well over a million years before modern humans appeared. Neanderthals were living in Europe and West Asia. Denisovans were living in Siberia. And in Indonesia a small people reminiscent of Tolkien's hobbits lived on the island of Flores. We do not know how these encounters unfolded, but we do know that these ancient cousins disappeared and that, in the case of Denisovans and Neanderthals at least, some interbreeding with the newcomers occurred. In some circumstances, the new migrants probably established peaceful relations and learned important local knowledge from the experienced natives. In other circumstances, particularly in times of scarce resources, they would have had to compete. Ultimately, our ancestors prevailed. But who were these ancient folk, and where did they come from? What do we know about the evolution of hominins and their minds?

IN 1925 RAYMOND DART REPORTED the discovery in South Africa of the fossil of a three-year-old child (the Taung child) with a mixture of ape and human features. He called this species *Australopithecus africanus* (Southern ape); on eventual discovery of further specimens, it was hailed as "the missing link." Subsequent finds showed that a great range of upright-walking, big-brained hominins used to roam this planet, sometimes in overlapping territories. Some were our forebears; others were different branches of an increasingly bushy looking hominin family tree.

These links are "missing" only in the sense that these hominins are now extinct. This is not to say, however, that the fossil record is complete and no information is missing. In fact, new evidence is unearthed quite frequently, and the record misses all manner of things that do not have sufficient permanence. Whereas teeth and stone artifacts can exist for millions of years, soft tissue and plant material quickly decompose. Whole civilizations built on bamboo would leave little trace in the remote future. A lot of information is undoubtedly missing from the record, much of which will never be established, and scholars debate about what does exist. For example, disputes occur between paleoanthropologists who recognize a distinct set of fossils as a new species and those who argue that it should be considered part of normal variation of an already described species. Some paleoanthropologists are more inclined to "split," and others are more inclined to "lump."[3] Putting such arguments aside, the current picture of the hominin past is detailed and intricate, featuring many distinct hominin species, and there is no frantic search for a vital piece missing from the puzzle. Figure 11.2 shows the variety of hominins that are currently thought to have roamed the Earth over the last six million years. I will discuss them in turn further below.

How can we breathe life into fossils and begin to understand their minds and behavior? A typical starting point is to compare them to extant animals living in similar environments. Surviving apes have long served to inspire speculation about the life history, appearance, and behavior of our ancestors. It is tempting to simply assume the last common ancestor with animals, the point of departure on our way to evolve the current gap, was "chimpanzee-like." However, this assumption is problematic. For instance, should we assume the last common ancestor was more like a common chimpanzee or more like a bonobo? As we have seen these two species of *Pan* are quite different from each other, yet they are both equally closely related to us. How could we know whether our ancient ancestors were aggressive, male-dominated apes like common chimpanzees or oversexed egalitarian creatures like bonobos? In fact, it is quite possible they were neither. We have changed dramatically over the last six million years, and both chimpanzees and bonobos have had the same span of time to evolve their own idiosyncrasies. Simple analogies, then, do not suffice.

As we saw in Chapter 3, the distribution of traits among closely related species can be used to make inferences about the homological origin of these traits in shared ancestors, and the last common ancestors of great apes probably had a sophisticated mental repertoire. It is likely that they could think beyond what was immediately available to their senses;

---

3 Even in biology there is debate about what qualifies as a species in living organisms. One simple rule of thumb is that members of the same species should be able to produce viable offspring. Members of different species do not interbreed. Naturally, such a test cannot directly be conducted with fossils unless DNA can be extracted. Instead, arguments typically center on the comparison of small fragments of bone and teeth. Not surprisingly, there are disputes. The discoverers of a new set of fossils have an incentive to "split" rather than to "lump," because if their find is a new species they are its famous discoverers, and they can even give the species a name; whereas, if their find is yet another specimen of an already described species, there will be less excitement, and they may struggle to find funding for more digging. Even the most skeptical "lumpers," however, agree that there were a number of distinct hominin species.

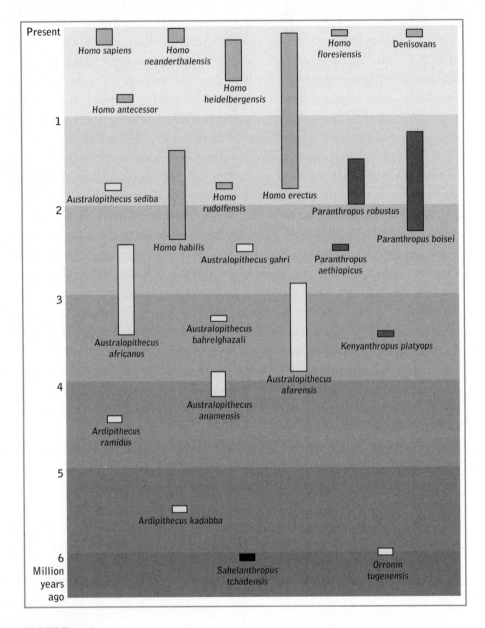

FIGURE 11.2.
Currently widely recognized hominins of the last six million years. "Splitters" would subdivide some species, such as *Homo erectus*, into different species such as *H. erectus*, *H. ergaster*, possibly *H. pekinensis*, and *H. georgicus*, whereas "lumpers" merge them. Some "lumpers" consider Neanderthals, for instance, to be a subspecies of *Homo sapiens* (i.e., *H. sapiens neanderthalensis*). Bars represent estimated age of fossils. Elongated bars indicate that multiple fossils of different ages have been found. Bar shades represent common assignment of distinct genera.

visually recognize themselves and reason about hidden displacement; make tools, copy others, and recognize being copied; maintain multiple social traditions; and reason by exclusion. With such mental attributes at its disposal these common ancestors embarked on a journey that eventually led to our human minds.

The study of what happened on that subsequent journey, of how we evolved the capacities that today set us apart from even our closest living relatives, has to rely on other kinds of clues, such as fossils and genetics. I am neither a paleoanthropologist nor a geneticist, so I can only sketch this path from my cautious reading of the currently available literature.

There are numerous theories about the subsequent anthropogenesis—about how, when, and why our ancestors became human. Some scholars highlight the role hominin bodies had, such as the capacity for aimed throwing, better cooling of the brain, and the ability to run long distances. Others have argued that changes in our sociality—collective punishment to enforce rules, the consequences of pair bonding and cooperative breeding—were the prime movers of our success. Yet others see specific innovations as the key, such as the control of fire and the development of cooking, or the invention of the baby sling or of symbols. Chances are, most of these factors played some part, but it is not clear how prominent a role each played in determining our ancestors' path. Below I sketch the major facts currently known about our journey[4] and evaluate their implications for the emergence of nested scenario building and the urge to connect our minds. Let's have a brief family reunion.

ACCORDING TO MOLECULAR EVIDENCE, THE last common ancestor of humans and any other living animal lived between 6 and 7 million years ago. Recent analyses suggest that the story may be more complicated than was long assumed. Hominin and chimpanzee lineages appear to have initially split 7 million years ago but then hybridized, finally splitting for good after 6.3 million years. Some such hybridization may also have

---

4 It may be reassuring that several more or less plausible accounts of anthropogenesis are available. Unfortunately, plausible ideas are easier to come by than decisive evidence. For instance, the aquatic ape theory proposes our ancestors returned to the sea (similar to dolphins or seals), or at least the shore, for a significant period of time. Adaptations to a watery environment might explain a range of peculiarities of the human condition, but alas there appears to be no direct evidence hominins spent any periods adapting to aquatic lives.

occurred with other species, such as the ancestors of gorillas. Chimpanzees and humans are more closely related to each other than to gorillas, but some sections of human DNA are more similar to that of the gorilla than to that of the chimpanzee. It is possible that some interbreeding occurred after the initial split from gorillas some 10 million years ago. The subsequent journey to humankind (just like these initial steps) was neither simple nor direct.

There are several relevant fossils from this time period. The oldest specimen is a skull from the Chad estimated to be over six million years old. This so-called *Sahelanthropus tchadensis* has a large brow ridge, small canines, and a cranial capacity of 365 cubic centimeters, equivalent to that of a modern chimpanzee. In 2001 researchers reported six-million-year-old fossils of a different probably upright-walking hominin, *Orronin tugenensis*, dubbed "millennium man." It is unclear whether these fossils are of an ancestor of humans, of chimpanzees, or of both. A fossil of a potential early hominin from a period following the split from the last common ancestor was described in 2004: *Ardipithecus kadabba* lived over five and a half million years ago. It appears to be the precursor of a species of which we now know a lot more.

In 2009 researchers described in detail a near complete fossil skeleton of an over 4.4 million year old hominin species called *Ardipithecus ramidus*. Its brain capacity was still equivalent to that of a modern chimpanzee (300–350 cubic centimeters), but it already had reduced canines and walked on two feet when on the ground. Yet *Ardi* still had a grasping toe for climbing trees and had flexible wrists unlike those of chimpanzees. It had been widely assumed that when our common ancestor climbed down from the trees, it walked on its knuckles just as modern chimpanzees and gorillas do. The flexible wrist of Ardi, however, suggests that this was not a shared ancestral trait. Furthermore, there are signs chimpanzee and gorilla knuckle walking differ from each other and perhaps evolved independently. It may be that bipedalism did not evolve from quadruped ground locomotion but rather derived from bipedal clambering in trees, as is frequently observed in orangutans.

The most common account for the rise of bipedalism and the split between the lines that led to modern apes and us is called East Side Story. Coined by French anthropologist Yves Coppens, the theory points to a massive geological event and its climatic consequences. Some eight million years ago tectonic plate movements resulted in the Great Rift Valley and its peaks on the western rim that now separate East Africa from the rest of Africa. Whereas the west of the continent continued to enjoy the precipitation required for rainforests, the climate and vegetation of the east changed dramatically. As a result, while apes in the west continued to live in the same habitat, their cousins in the east had to adapt to an increasingly different environment, as forests turned into open savannah.Our ancestors had to find ways to make a living away from the trees. In line with this scenario, virtually all early hominin fossils come from this part of Africa.[5] It is likely that climate change and the rise and spread of grasslands have repeatedly played major roles in our evolutionary history.

Bipedalism is not necessarily the obvious solution for an ape adapting to the savannah. For one, it comes with some serious side effects, including back problems and hemorrhoids. Bipedalism also required a realignment of the spine and a narrowing of the pelvis, restricting the size of the birth canal and resulting in painful and dangerous birthing. This, in turn, appears to have forced a gradual shifting of infant skull and brain growth into the postnatal period. As previously discussed, the upside of this development is that social and cultural input could more effectively shape the infant's maturing brain. Thus this apparent design flaw may have been a crucial step in the evolution of our minds.

Bipedalism did not make us fast sprinters—our ancestors' speed would have been no match for savannah predators such as lions and hyenas. In fact, there is increasing evidence that early hominins were seriously preyed on. How did they manage? One advantage is that bipedalism frees the hands to carry things, potentially opening up new opportunities for defense. A human hand is capable of power and precision grips that allow effective clubbing and throwing. Early signs of these adaptations appear in *Ardipithecus* (and are well developed in early *Homo* some two to three million years later). Striking with rocks and sticks may have allowed otherwise defenseless early hominins to fend off carnivores, such as the many species of saber-toothed cats that roamed the savannah. At times, throwing rocks may have been sufficient to repel predators even before they attacked. Rather than the image of "man the hunter," some

---

5 Some recent evidence challenges this widely assumed scenario. *Sahelanthropus tchadensis* was found to the west of the Great Rift Valley, and *Orronin tugenensis* appeared to have lived in forested environments.

researchers have therefore argued that "man the hunted" may have driven initial steps in our evolution.

It is tempting to use my scenario-building mind to expand on this story and highlight how selection would have led to increased mental scenario building and foresight. It is easy to imagine that natural selection had particularly long teeth back then, giving those who were prepared to defend themselves a distinct advantage over those who were not. Having appropriate defenses ready when it mattered most would have been crucial, possibly leading to the carrying of objects such as stones and clubs for defensive use. Alas, we do not know whether they actually used such defenses, and signs that hominins carried arms only appear millions of years later.

THE ARDIPITHECUS FOSSILS COME FROM an East African region known as the Afar Triangle in Ethiopia, a source of many extraordinary fossil finds. In 1974 Donald Johanson and colleagues discovered a near-complete skeleton of a hominin now known as *Australopithecus afarensis*, famously called Lucy. The following year the team unearthed a group of thirteen individuals of the same species. They were a successful genus that survived for over two million years.

FIGURE 11.3.
Reconstruction of "Lucy," *Australopihtecus afarensis*, 3.2 million years old. *(All skull reconstructions by Bone Clones (www. boneclones.com), photographed by Sally Clark.)*

When Raymond Dart discovered the first Australopithecine, he appeared to have found the missing link not only between ape and human bodies but also between their minds. The primary visual cortex of humans is relatively smaller than in other primates, and so the boundary of this area—the lunate sulcus[6]—is relatively forward in apes and more backward in modern humans. Dart reported that the primary visual cortex of *Australopithecus* resembled a human's more than an ape's; he concluded that these hominins had devoted more brain resources to storage and association of information than to vision. Unfortunately, inferences about ancient brains are based only on endocasts—casts of the internal bone of the cranium—and there have been some persistent disagreements about what might be gleaned from them.[7]

The oldest known Australopithecines, *A. anamensis*, lived some 4.2 million years ago. *A. afarensis* roamed Africa from 3.9 to 2.9 million years, weighed between thirty and forty kilograms, and stood a little over a meter tall. Their cranial capacity was similar to or slightly larger than that of modern apes, with a mean of 458 cubic centimeters (their South African relatives, *A. africanus*, had a mean of 464 cubic centimeters). Their pelvis and legs are appropriate for bipedal locomotion, and the question of their bipedalism was apparently settled once and for all by a most extraordinary find: Mary Leakey, while playing Frisbee, stumbled across their fossilized footprints. When a volcano erupted in northern Tanzania 3.7 million years ago, covering the ground with ash, subsequent rain turned it into something akin to plaster. Numerous animals walked across this surface and left permanent tracks. Among them were the footprints of an upright-walking hominin family.[8]

---

6 A bit of the brain that folds outwards is called a gyrus, whereas the crease in between two gyri is called a sulcus. The primary visual cortex is divided from the parietal lobes further forward by a deep crease called the lunate sulcus.

7 The Taung child fossil includes a unique natural endocast, but nonetheless there remained debate about the location of the lunate sulcus. Dart's conclusion was challenged and defended by two of the most influential scientists in this field, Dean Falk and Ralph Holloway. In a personal history Holloway recounts this and many other disagreements with Falk, including a confrontation at a conference in which both proponents were armed with conflicting endocasts of the same Australopithecine skull (one showing an occipital-marginal sinus, the other not)—leading him to speculate about the damage a thrown plaster endocast might do.

8 Curiously, there is also evidence that while *Australopithecus* feet were functionally like those of modern humans, some contemporary hominin species continued to be able to comfortably climb trees. A recent find of a 3.4-million-year-old foot still had an opposable big toe like that characteristic of *Ardipithecus*.

Australopithecines were not only walking upright but also beginning to lose their fur. Evidence for this derives from a surprising source: lice. These notorious parasites tend to be specialized to particular hosts and cannot survive long without them. Most primates host only one type of louse. Humans, however, host three. One is the head louse. With humans' loss of fur,[9] it seems to have evolved to survive on the head, leaving the crotch available for another species of louse. The pubic louse, DNA comparison suggests, migrated to our ancestors some 3.3 million years ago. The closest relative to the human pubic louse is the louse specialized on gorillas, so we apparently got crabs from the ancestors of gorillas. More importantly, what this finding means is that 3.3 million years ago our head and pubic hair were already sufficiently separated by hairless body regions to enable two types of lice to live in distinct habitats on the same host. In other words, Australopithecines were not covered entirely with fur. (Incidentally, the third human louse is the body louse that lives in clothes. This species diverged from head lice between 170,000 and 83,000 years ago, suggesting that by that time humans had proverbially left the Garden of Eden and were regularly wearing clothes.)

Australopithecus afarensis may have used stone tools. In 2010 researchers reported 3.39-million-year-old bones with cut marks for the removal of flesh and percussion marks for access to marrow. Their conclusion has been disputed. However, some hominins certainly discovered the power of rocks a few hundred thousand years later. The oldest stone tools are currently about 2.5 million years old and are associated with fossils of yet another species, *Australopithecus garhi*. This Australopithecine had a cranial capacity of 450 cubic centimeters and was found in the rich Afar Triangle. Cut marks on bones of various ungulates indicate the tools were used to butcher carcasses.

A variety of other hominins appear to have walked the Earth during the same period. Fossils from the Chad, west of the Great Rift Valley, although resembling *afarensis*, have been proposed to belong to a different species (*Australopithecus bahrelghazali*). Meave Leakey and colleagues reported a 3.5-million-year-old fossil with such an unusually flat face

---

9 The claim of hairlessness is somewhat misleading, as it is not the loss of hair follicles but the replacement of thicker terminal fibers by thinner, shorter, and more transparent vellus fibers that creates our lack of apparent fur cover.

that they proposed a new genus: *Kenyanthropus platyops* (although lump-ers reckon it may merely be a squashed Australopithecine). In 2010 two partial skeletons of yet another distinct Australopithecine were found in South Africa. This species has been called *Australopithecus sediba* and lived as recently as 1.8 million years ago. While there is debate about the exact number of Australopithecine species, it is clear that quite a few existed, all with diverse features. Given the frequency of recent discoveries, sev-eral more species likely remain to be found. With overlapping ecological niches, there would have been some competition for resources. Closely related species would have evolved along different lines to exploit different resources (like chimpanzees and gorillas do today).

Australopithecines eventually diverged along a minimum of two dis-tinct evolutionary paths. Some became increasingly stocky and developed enormous jaws with which they could chew hard foods such as nuts and fibrous vegetation. Once known as "robust" Australopithecines, they are today more commonly classified as the distinct genus *Paranthropus*. They had large chewing muscles, attached to the top of the skull on a distinctive sagittal crest. Such a crest is absent in humans but prominent in exten-sive chewers such as gorillas.[10] The teeth are large and ideal for grinding.

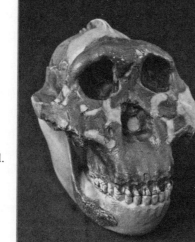

FIGURE 11.4.
Paranthropus boisei
(KNM OH 5),
"Nutcracker Man,"
1.8 million years old.

---

10 Given the potential interbreeding between hominins and ancestral gorillas noted earlier, *Paranthropus* may be the offshoot of such hybridization. This in turn could also mean that our ancestors may have gotten crabs from mating with *Paranthropus*. But I am merely speculating.

During times of plenty, they probably ate a variety of foods, including animal protein. During times of scarcity they could fall back on hard foods others could not eat.

At least three different species of *Paranthropus* are widely recognized. The earliest one is *P. aethiopicus*, followed by *P. robustus* and *P. boisei*. *Paranthropus* cranial capacities were similar or slightly larger than those of earlier Australopithecines: *P. boisei* had an average of 481 cubic centimeters and *P. robustus* of 563 cubic centimeters. They used their smarts to employ stone and bone tools to dig up tubers. Wear patterns on these tools indicate that they also used them to access foods such as termites. Unlike chimpanzees, who probe for termites with twigs, *Paranthropus* used bones to break open the mounds. Although they used tools to get food, there is so far no evidence that they changed the shape of these implements.

P. boisei is one of the greatest success stories of all hominins that ever graced the Earth, surviving for over a million years. It is unclear why they eventually went extinct. One possibility is that their feeding adaptation became increasingly specialized, such that when a radical environmental change affected their main food sources, they were unable to survive. Cyclic glaciations that began some three million years ago increasingly intensified. In such times of instability, rapid adaptability would have been crucial. It is also possible that other hominins contributed to the demise of their cousins. For hundreds of thousands of years *Paranthropus* lived side-by-side with the earliest creatures science calls humans.

SOME AUSTRALOPITHECINES BECAME MORE GRACILE and evolved into *Homo*. The genus *Homo* has a smaller, less projecting face; a smaller gut; more efficient bipedalism; smaller teeth; and a bigger brain than *Australopithecus* and *Paranthropus*. A recent study suggests that a mutation in a gene crucial for masticatory muscles appeared around 2.4 million years ago, leading to a reduction of chewing strength and removing, according to the authors, an evolutionary constraint on the expansion of the cranium and brain.

The earliest member of the genus *Homo* is a species Louis Leakey and his colleagues called "handy man" or *Homo habilis*, because of their association with what were at the time the oldest known stone tools. The oldest fossils of this species are 2.4 million years old; their cranial capacity averaged around 600 cubic centimeters. The stone tools *Homo habilis* manufactured, called Oldowan tools because they were initially found in Tanzania's Olduvai Gorge, consisted of modified river cobbles. By removing flakes of a cobble, *H. habilis* produced sharp-edged tools, effective for cutting hides and removing meat from bone. Bigger stones were used as hammers to crack bones. In this way *Homo habilis* were able to consume large animals such as antelopes, rhinoceroses, and hippopotamuses.

Although meat was part of the diet of earlier hominins, just as it is part of the diet of modern chimpanzees, these stone tools would have significantly increased access to this high-energy food source. This is not to say that *Homo habilis* were efficient hunters. A large part of their diet included plant materials, and much of the meat was probably scavenged. On at least some of the fossil bones other carnivores' bite marks precede the marks of stone tools. *Homo habilis* might have competed with scavengers for what predators left behind. They repeatedly moved carcasses to special sites, where they used their stone tools to butcher them. A recent find from a 1.95-million-year-old butchering site from Turkana includes bones from aquatic animals such as turtles, fish, and crocodiles. These animals are rich in nutrients critical for brain growth, leading to speculation that a changing diet played a key role in allowing early *Homo* to grow bigger brains. Increased cognitive powers, in turn, may have been used to safely and effectively secure high-quality foods.

Butchering sites show that *Homo habilis* transported objects sometimes several kilometers from their source; it is possible they also carried tools on their person to be ready for future use. Clubs and stone tools could have been used to strike approaching predators or poachers. It has been argued that early stone tools were of good throwing size and that the hand of early *Homo* was able to perform the power and precision grips necessary for clubbing and aimed throwing. In other words, these hominins may have been armed.

Eventually, *Homo* used weapons to hunt, but it remains unclear when the practice started. Weapons are likely to have played a key role in hominins' evolution from prey and scavenger into a top predator, and they were undoubtedly used in conflicts within and between hominin groups. Yet while a capacity for precise striking and clubbing is self-evident in *Homo habilis*—after all, that is how the stone tools were made—it is not clear when hominins began to use aimed throwing in any systematic way. (Incontrovertible evidence for projectile weapons appears much later in the archaeological record.)

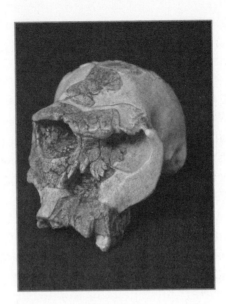

FIGURE 11.5.
Homo habilis
(OH-24),
1.8 million
years old.

One particular fossil dated 1.8 million years ago already had a much larger cranial capacity of 775 cubic centimeters. Some scientists have hence argued that this find deserves a separate species name, *Homo rudolfensis*, but it remains possible that this is part of normal variation or even sexual dimorphism.

Homo habilis existed from South Africa to Ethiopia and survived until 1.4 million years ago. By the eastern shores of Lake Turkana they lived side-by-side for several hundred thousand years with another hominin species, one that was ultimately even more successful: *Homo erectus*.

SOME SCHOLARS ARGUE THAT THE genus *Homo* really began with *Homo erectus* rather than *habilis*. As adults they stood up to 1.8 meters tall and had large cranial capacities with a mean of 1,000 cubic centimeters—more than twice that of Australopithecines and modern apes. In contrast to Australopithecines, *erectus* females were not much smaller than their male counterparts. Although they sported a low forehead with massive brows and lacked a chin, *Homo erectus* looked and walked a lot like modern humans. Their footprints are essentially indistinguishable from ours.

In 1984 Richard Leakey and colleagues discovered a nearly complete 1.6 million-year-old skeleton of a *Homo erectus* near Lake Turkana, known as Nariokotome Boy. He stood 1.6 meters tall and had a cranial capacity

 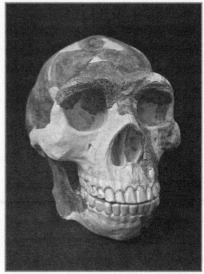

FIGURE 11.6.A–B

Two examples of *Homo erectus*: on the left is a 1.6-million-year-old skull
without jaw from Africa; on the right is a 450,000-year-old skull from China
(also called "Peking Man").[11]

of some 880 cubic centimeters and body proportions much like those of a
modern human. It was initially thought that the boy died around the age
of twelve, based on the denture and the overall size of the body. However,
detailed study of the growth patterns point to an age of only eight years, implying *Homo erectus* grew up a lot more quickly than modern humans do.

Whereas the earlier hominin cranial capacities do not differ statisti-
cally from those of modern apes,[12] those of early *Homo* demonstrate a
significant increase. A larger brain requires more nutrients to run. *Homo*

---

11   There is debate about whether erectine fossils should be split into several species. The
most common split is made, based in part on somewhat thinner cranial bones and less pro-
nounced brow ridges, between African and Asian populations, with the former labeled *Homo
ergaster* and the latter *Homo erectus*. Some splitters even reserve the name *Homo erectus* only for
Indonesian specimens and refer to the Chinese fossils as *Homo pekinensis* and Georgian fossils as
Homo georgicus. In spite of the anatomical and geographical differences, these fossils have much
in common, and I adopt a lumper position here and refer to them collectively as *Homo erectus*.

12   The numbers of fossil specimen on which these estimates are based are often small,
leaving statistical comparisons with limited power. For example, in a review by Robson and
Wood, estimates are based on one *Sahelanthropus tschadensis*, eight *Australopithecus africanus*,
ten *Paranthropus boisei*, six *Homo habilis*, thirty-six *Homo erectus*, seventeen *Homo heidelbergensis*,
and twenty-three Neanderthals.

erectus, like modern humans, had relatively large bodies, small guts, and small teeth—features that suggest they underwent a fundamental change to a higher-quality diet. The most common explanation for this is that they consumed more meat. The increase in protein-rich food could explain the reduction in gut and might have enabled the increase in brain size.

The reason for better access to meat may have to do with advances in locomotion. Tendons, ligaments, and bones changed in ways that favor stable and efficient long-distance running. Modern humans are by no means great sprinters compared to other mammals, but we do have extraordinary endurance. Like hyenas and migratory ungulates such as wildebeest (and unlike most other savannah animals, let alone primates), humans can run for extended periods of time, potentially giving them a distinct advantage when it came to hunting and scavenging. Running to wherever the vultures gathered may have improved the chances for *Homo erectus* to beat other scavengers to the feast. A key constraint for long-distance running is overheating.[13] Reduced fur and whole-body sweating allow humans to dissipate heat rapidly. Instead of nasal breathing typical of apes, we increase ventilation during strenuous activity by mouth breathing. It is possible that *Homo erectus* was already so prepared and may hence have become capable of pursuing prey until they got within close enough range to strike. It may have even allowed them to run down prey to exhaustion. Such hunting requires persistence and perhaps an increased degree of foresight, and it may have selected for better mental scenario building and exchange to support cooperative hunting. Modern human hunters read tracks to determine what prey to pursue. While we do not know when track reading first emerged, it may have been a crucial step toward recognizing that one thing can symbolically stand for another.

It is likely that *Homo erectus* made significant strides toward becoming a top predator. It has been suggested that they outhunted other large carnivores, many of which may have gone extinct as a result. Much of this is speculation, and there are alternatives to the idea that improved hunting and scavenging drove the evolution of *Homo erectus*. For instance, Kristen Hawkes and colleagues argue that the grandmother effect began to take

---

13 Overheating has also been argued to be a key problem for large brains. It has been proposed that elaborate cranial venous circulation evolved to act as a radiator for the increasingly large brain of hominins.

hold and was the crucial advantage. Increase in brain size and adult body weight suggest that *Homo erectus* lived longer lives (though estimates of the proportion of old to young fossils are still low). Climate-driven changes in female foraging—perhaps the gathering of tubers—and food-sharing practices may have led to changes in life history and ecology. The key foods may have been gathered rather than hunted.

Richard Wrangham has championed the case that the cooking of food was the crucial step. Perhaps the smartest thing any of our ancestors ever did was to figure out how to control fire.[14] Cooking enables us to break down foods that are otherwise not digestible, rids them of common parasites, and stops them from spoiling quickly. Burned earth associated with *Homo* and animal fossils comes from as early as 1.6 million years ago. However, it is unclear whether the fires were started deliberately or whether they were wild fires that burned through the area. Fire, once brought under control, enabled not only cooking but many other advances in terms of vision and heat, attack and defense. Spending nights around the campfire might have, over many millennia, been a catalyst for a number of milestones in human evolution, fostering traditions, communication, and innovation. Fire gave us immense new powers (remember Prometheus), enabling us to drive away predators and drive out prey from their hideouts. It allowed us to change entire landscapes and forge countless tools.[15] However, there is little evidence to support the idea that control of fire was present, let alone widespread enough, to account for the rise of *Homo erectus*. The most convincing evidence for early fire use is 790,000 years old (though a recent report suggests fire use by 1 million years ago).

Homo erectus was the first hominin species certain to have dispersed across the old world. The oldest *Homo erectus* fossil currently comes from East Turkana and is 1.8 million years old. Almost equally old fossils were found in the country of Georgia; by 1.2 million years ago they are

---

14 The importance of fire for Paleolithic hominins, especially the discovery of how to make fire, is powerfully dramatized in Jean-Jacques Annaud's wonderful 1981 movie, *Quest for Fire.*

15 From 164,000 years ago there is evidence of heat treatment of stone to improve flaking properties. Wood tips could be hardened, and adhesives softened; eventually new powerful materials could be formed, such as ceramics and metals. But these events only occurred thousands of years ago, not many hundreds of thousands of years ago.

evident in Spain. The migration, however, appears to have first gone rapidly eastward.[16] By 1.6 million years ago *Homo erectus* were living in China ("Peking Man") and Indonesia ("Java Man"). They made these new territories their own and persisted for a very long time. It used to be thought that they died out a few hundred thousand years ago, but new evidence suggests that in pockets, at least, they survived until quite recently. New dating of three Indonesian skulls has yielded estimates of between 40,000 and 70,000 years ago. Another study suggests that the descendants of "Java Man" may have survived until as recently as 27,000 years ago. If these dates are correct, then *Homo erectus* lived on Java for 1.6 million years.

Homo erectus must have had some distinct adaptive advantages. They did not sport obvious new biological weapons such as claws, venom, or incisors, but they had a bigger brain. Recent analysis of teeth found that *Homo erectus* had much more variation in complexity of microwear compared to earlier hominins. This finding suggests not specialization but a new degree of behavioral flexibility. Rather than purely focusing on one type of food, be that meat, cooked tubers, or something else, they appear to have relied on a great diversity of foods. This, in turn, suggests changes in their mental capacities were the primary cause of their success, rather than any specific diet or behavior. They may have made some serious headway in both key aspects of the gap: open-ended, reflective mental scenario building and connecting of minds.

Some capacity for mental scenario building is evident in the tools they made. The Acheulean stone-tool industry is the most successful tool set ever created. It includes symmetrical hand axes and cleavers, flaked on both sides of the stone rather than just one. Bifacial hand axes are teardrop shaped and can be very flexibly deployed. They were useful for butchering and for working plant materials. They sit comfortably in one's hand (see Figure 11.7) and allow one to cut effectively for long periods. The creation of these tools implies a capacity to plan, to create a mental scenario of the

---

16 This is not to say that there was a deliberate exodus. It is likely they simply kept spreading into habitable areas. For instance, climate change turned much of Eurasia periodically into grasslands, and predators may have simply followed their prey species. With subsequent cooling, *Homo erectus* groups would have had to retreat to accommodating refugia that provided protection in these new areas. Temporary isolations, in turn, lead to local changes. The role of refugia and climate change in human evolution is currently hotly debated, especially in light of the apparent diversity of hominin forms (including diverse *Homo erectus*) that used to walk this planet.

way to the final product. The stones were carefully selected for suitability in weight and size, and precise blows were used to change them gradually into the desired form. The stone was first worked roughly to create the shape and then with more subtle flaking to produce the sharp, symmetrical tool. The toolmakers needed to understand a thing or two about the properties of the stone to predict how it would break and splinter. Making a bifacial hand ax is a difficult craft—as my fingers had to discover the hard way. Even an experienced knapper takes significant time to make such a tool. This effort was not expended for one-time use. These tools were carried over distances and used repeatedly, implying some foresight of their future utility.

Perhaps the most curious aspect of these versatile tools is that they were made in a standardized fashion. The design stayed much the same for well over a million years. The earliest example appears some 1.76 million years ago in Africa. The biface I hold in Figure 11.7 was made almost 1.5 million years later. It is the most lasting piece of technology our lineage has ever produced. Although tools became thinner and more trimming flakes were removed, there was little variation over time. These multipurpose tools gave our ancestors a distinct edge—in the literal and metaphorical sense of

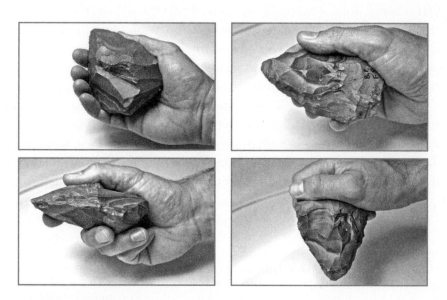

FIGURE 11.7.A–D
Acheulean bifacial hand ax, approximately two hundred thousand years old, from modern Israel *(photo Sally Clark, stone tool courtesy of Ceri Shipton).*

the word.[17] They were often made out of flint; as with modern computer chips, some of our most amazing powers derive from silicon.

The fact that these tools were made the same way for so long shows that *Homo erectus* were capable and motivated to learn from each other the skill of shaping them. They were maintaining social traditions with high fidelity. On the other hand, the lack of variation shows little of the cumulative character that marks our modern culture. Why did they not regularly improve on the design? Perhaps their minds were not ready for it. The ratchet effect that Tomasello highlighted may not have been operating yet. Nevertheless, there was some increase in complexity, and other inventions of a more perishable nature may well have been accumulated. Tools made out of wood or hide are less likely to leave traces, and important inventions, such as new means of communication and cooperation, may leave none.

Our drive to connect our minds is primarily expressed through language. In line with the social learning the Acheulean tool tradition implies, several theorists have suggested that *Homo erectus* developed a more sophisticated communication system. William Calvin, for instance, made a case based on the claim that *Homo erectus* increasingly employed aimed throwing to bring down prey, in particular with bifacial hand axes. He has argued that reliance on one-armed, aimed throwing selected for a refinement of motor sequence capacities in the left hemisphere and that these were later adapted for speech. Holloway has claimed endocasts of *Homo rudolfensis* and *Homo erectus* already indicate changes in the brain that look like a Broca's area—an area in the left hemisphere implicated in speech production.

Others point to the sizes of endocasts to infer that *Homo erectus* were already linking their minds differently from the way their ancestors did. As noted earlier, whereas primates groom each other to form group cohesion, human group members typically bond by talking to each other and sharing experiences. Robin Dunbar highlights the role of gossip as a means to bond and has come up with a way to estimate when a switch from grooming to gossip might have occurred. He has shown that the larger the

---

17 Note that the first successful migrants to Asia did not carry the Acheulean tool technology. Bifacial hand axes appear much later in the East Asian record, and it is unclear whether they represent convergent invention or whether they were carried across by new migrants. *Homo erectus*'s early migration was done with the more primitive Oldowan tools that had long been used by *Homo habilis*. This suggests it is not the tools per se that made a difference but the minds that employed them.

typical social group of a primate species, the larger their neocortex ratio. By plugging values for extinct species into the formula that describes this relationship, he estimated the different group size various hominins were likely to have lived in.[18] If correct, then the last three million years saw a steady increase in group size from about 60 in Australopithecines to over a hundred in *Homo erectus* and up to 150 in modern humans.[19] Once you have an estimate of typical group size, you can predict (at the risk of introducing another estimation error) the time that each species is likely to spend on grooming. While Australopithecines may still have been able to maintain group cohesion through grooming some 20 percent of their waking time, *Homo erectus* would have had to spend a third of their days grooming each other. Such a requirement would seriously cut into the time available for other essential behaviors. Perhaps our capacity to connect our minds emerged around the time *Homo erectus* could no longer effectively connect through physical touch alone.

We lack hard evidence for an increased urge and capacity to link minds.[20] Yet one recent find may indicate a marked shift in the social mind of early *Homo*. A 1.77-million-year-old skull from Dmanisi in Georgia is of an elderly man without teeth and with advanced subsequent bone loss—implying he survived for a significant time without being able to chew. The other group members may have done the chewing for him.

There is much we do not know, but the minds of *Homo erectus* were clever enough to allow them to spread across the old world and survive in diverse habitats for hundreds of thousands of years.

---

18   When models reliably describe the relationship between two factors in the present, one can use one factor to make inferences about the other in the past. For instance, there is a relation between the thickness of primate thigh bones and overall body weight, so the thickness of a fossilized bone can be used to estimate the weight of the individual. However, Dunbar's estimates are problematic. Neocortex ratio is the volume of neocortex divided by the volume of the whole brain. Although we can measure the cranial capacity of hominin skulls to estimate the volume of the whole brain, how much of this is neocortex remains unclear. Furthermore, group size may be influenced by other variables, such as climate and food availability. Neanderthals had very large brains, but most other evidence suggests they lived in rather small groups in the cold European climate.

19   According to Dunbar, this number is the size we can keep track of, the number of friends we can name, the number of people we invite to funerals and weddings, and the number that, when exceeded, leads to the break-up of a typical hunter-gather group. Whether this is our natural group-size limit is difficult to assess.

20   In some cases heavy objects were apparently moved over kilometers, suggesting cooperation and possibly shared intentionality. However, none of these are conclusive.

SOME DESCENDANTS OF HOMO ERECTUS became robust and acquired modern human characteristics. These fossils are sometimes lumped together under the umbrella label "archaic" *Homo sapiens* or premodern humans—another set of not-missing links. A variety of potentially separate species coexisted from 800,000 years ago onwards in Africa, Asia, and Europe. The oldest member of this group is a hominin from Spain called *Homo antecessor*. It had a relatively flat face, a high domed forehead, and a cranial capacity of just over 1,000 cubic centimeters. The first archaic fossil to be described was a robust and chinless jawbone from near Heidelberg, Germany, leading to the name *Homo heidelbergensis*. Many archaic fossils have since been found in central Europe, with a few in Africa and Asia. *Homo heidelbergensis* lived between 600,000 and 150,000 years ago. They stood up to 1.8 meters tall and were robust, with immensely thick brow ridges and a large brain with an average cranial capacity of 1,200 cubic centimeters.

Homo heidelbergensis probably cooperated in various ways. A spine of a five-hundred-thousand-year-old specimen from Spain shows a hunched back. Such a condition would have meant significant incapacitation, suggesting that this forty-five-year-old received help and support from others. They seemed to have cooperated to bring down big game; numerous remains were found in association with stone tools. Unlike with *Homo habilis* butchering sites, the bones now show initial stone-tool cut marks and only later carnivore teeth marks. Remarkably, four-hundred-thousand-year-old spears were discovered near Bielefeld, Germany, together with ten butchered horses and flake stone tools, strongly suggesting cooperative big game hunting. It is possible that these spears were used primarily to thrust, but they may well have been thrown, and replicas suggest an effective range of some fifteen meters.

Making these spears would have required stone tools. Making a tool to make other tools suggests some capacity for nested scenario building. Archaic *Homo* may have made other significant strides toward modern minds. Ash deposits and charred bones suggest the capacity to control fire. In 2010 evidence for stone blades from Kenya were dated at over five hundred thousand years old, suggesting some planning to repeatedly produce blades from a single core.

From around three hundred thousand years ago onwards, Acheulean bifaces were increasingly replaced by smaller cores and flakes (known as the Levallois technology). Producing these tools requires complex steps

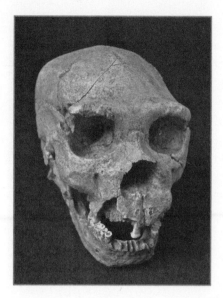

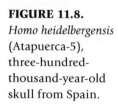

FIGURE 11.8.
Homo heidelbergensis
(Atapuerca-5),
three-hundred-
thousand-year-old
skull from Spain.

and the coordination of subgoals. A nucleus is shaped and a striking plat-form prepared to reach the overarching goal of producing one specific smaller flake (or several). Carrying a core, rather than a bifacial hand ax, made it possible to produce specialized tools on the spot as required. New tools were invented. The hafting of stone tips, for instance, appears to have begun around that time (although some recent evidence suggests it may have emerged as early as five hundred thousand years ago). That is, tools were produced not only out of an existing object but by adding parts. Composite tools strongly suggest a capacity to imagine hierarchical, *nested*, mental scenarios. To make a spear with a stone point, for instance, one has to make the shaft, the point, and the binding in separate steps. Archaeologists, such as Stanley Ambrose and Ceri Shipton, suggest that the emergence of composite tools, such as stone-tipped spears and knives, indicates a dramatic change in mental capacity. Assembling units in dif-ferent configurations produces new tools, much like assembling words in different configurations produces new sentences.

ARCHAIC SAPIENS LIKELY GAVE RISE to both our own species and our most famous prehistoric cousins, the Neanderthals. The first recognized Nean-derthal fossils were unearthed in 1856 in the Neander Valley in Germany. Today over two hundred fossils have been described. Neanderthals lived

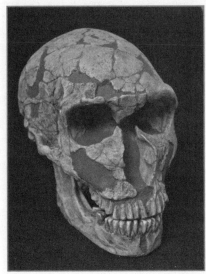

FIGURE 11.9.
Homo neanderthalensis
(La Ferassie 1), fifty-
thousand-year-old
skull from France.

from about 160,000 years ago to as recently as 27,000 years ago, during a period of extreme climate changes that would have posed major survival challenges. They were a stocky people adapted to the cold, with large noses and protruding brows. The size and shape of their inner ear canals were distinctly different from that of modern humans. Some of them had pale skin and red hair.

Neanderthals lived in small groups, at times occupying caves, and ranged across ice-age Europe to the Middle East. Recent evidence suggests they may have moved much farther east than previously thought, up to western Siberia. They were game hunters and largely depended on meat for sustenance. Their front teeth show unusual wear, potentially because they used them to hold items. Evidence from Gibraltar indicates that they exploited a variety of food sources, including fish and dolphins (which they may have scavenged from the beach). There is some indication, such as bones opened to extract marrow, that some Neanderthals practiced cannibalism.

In 2010 DNA from three 38,000-year-old female Neanderthals from Croatia was extracted to construct a draft sequence of their genome. Evidence thus far suggests the last common ancestor of humans and Neanderthals lived between 440,000 and 270,000 years ago. Mitochondrial DNA extracted from thirteen Neanderthals subsequently indicated there was great genetic variation before 50,000 years ago, but that younger spec-

FIGURE 11.10.
Mount Carmel Caves on the Mediterranean coast close to
Haifa. The caves were used for over half a million years, with
traces from the Acheulean to modern humans. For thousands
of years Neanderthals and modern humans seem to have
lived here side by side, or at least repeatedly visited the region
(with warmer weather modern humans moved north, and
with colder weather Neanderthals spread south). This is
where they probably interbred.

imens from Western Europe showed much less variation. This result may
be due to a mass extinction around that time, during which cold periods
engulfed Europe, and the surviving few recolonized as the weather improved. As noted in Chapter 1, the results of ancient DNA comparisons
show that modern non-Africans have inherited some Neanderthal genes.
So in a sense, Neanderthals are not entirely extinct after all.

The current evidence suggests, however, that the degree of intermixing
was small, so the general idea that modern humans replaced earlier populations is still mostly true. Neanderthals probably interbred with modern
humans in the Middle East around eighty thousand years ago, when they
cohabited the region for thousands of years (see Figure 11.10). Modern
humans eventually migrated into a Neanderthal-inhabited Europe some
forty thousand years ago. As they moved farther west, Neanderthals began

to disappear, and by thirty thousand years ago only a few remained on the western coasts of Spain and Portugal.

The traditionally brutish image of Neanderthals has been revised in recent years. Their brains were as large as—and some even larger than—ours, with a mean cranial capacity of 1,426 cubic centimeters. Neanderthals survived into relatively old age in spite of severe disabilities caused by wounds, arthritis, and broken bones, suggesting that they cared for the sick and had some capacity for empathy, cooperation, and social norms. Furthermore, evidence of Neanderthal burials suggests that they may have had some sense of the future beyond an individual's lifetime and possibly held spiritual beliefs. They evidently used fire and wore clothes, and they hunted not just old, injured, or weak animals but prey in their prime. Their minds were undoubtedly much more like ours than that of an ape.

Neanderthals show clear evidence of having made composite tools and therefore of possessing some capacities for nested thought. Did they have our urge to connect their minds and our means to do it? Dental growth patterns suggest that Neanderthals may have grown up more quickly than modern humans (and even faster than earlier archaic *Homo*), which would indicate less time to educate the young. However, recent findings raise the possibility that Neanderthals engaged in some human social signaling: they may occasionally have worn shells as jewelry and used ochre as paint. It has even been posited that some of them produced cave paintings. Archaeologists are hotly debating whether modern humans left these signs of symbolic thought, whether Neanderthals copied these behaviors from modern humans, or if the behaviors independently emerged in our cousins. It is certainly possible that they had symbolic capacities to exchange their minds.

Yet did they have the physical capacity to speak? The cognitive scientist Philip Lieberman has long argued, based on reconstructions of their vocal tracts, that Neanderthals would not have been able to make the phonetic elements necessary for speech. The discovery of a 60,000-year-old Neanderthal hyoid bone, a free-floating bone located in front of the voice box (larynx), challenged this conclusion because it was essentially modern in size and shape, suggesting that the Neanderthal voice box was much like that of modern humans. A recent find of a hyoid bone from an earlier archaic *Homo sapiens* suggests this modern form may already have emerged some 500,000 years ago. By contrast, a 3.3-million-year-old hyoid bone of an *Australopithecus afarensis* is more similar to that of modern great apes.

Further clues come from recent DNA analyses. Neanderthals were found to have the modern version of a gene strongly implicated in human speech.[21] This gene appears to have emerged before the split between the lines leading to Neanderthals and modern humans, and it therefore points to an ancient origin. There are many pieces to the speech puzzle, however, and it remains uncertain when hominins started to speak.

To make matters worse, language may not have started with speech at all—it is possible that gesturing preceded talking. Although attempts to teach other primates to speak have been failures, some great apes have successfully learned to use gestures to communicate. One stumbling block is that primate articulations, unlike their hand movements, are primarily emotive and not under voluntary control. Michael Corballis has championed the idea that gestural communication became more sophisticated in early hominins, perhaps initially based around declarative pointing and iconic mime. Language may have gradually evolved in the gestural domain. Only in modern humans, Corballis argues, was gestural language supplanted with speech—which you could argue are gestures produced by the tongue. Sign languages are languages that simply use a different modality (and we still tend to gesticulate when we talk). If this account is correct, then even if Neanderthals and earlier hominins could not speak, they might have had sign language.

ANATOMICALLY MODERN HUMANS GO BACK some 200,000 years. Analyses of teeth suggest that by 160,000 years ago our ancestors had a modern human life history. Yet compelling evidence for a fully modern human mind emerges significantly later. The oldest clear evidence for a human burial, for example, is a 119,000-year-old site at Skuhl Cave, Israel (see Figure 11.10). The earliest evidence for ornamentation and potential ethnic signaling are 82,000-year-old shell beads from Morocco with signs of wear and traces of ochre. Slightly younger beads (75,000 years old) were found at Blombos Cave in South Africa. In this same cave were found the earliest known examples of abstract-looking designs: geometric patterns engraved in ochre and in bone. Evidence for long-range projectile weapons in the

---

21 A mutation of the FOXP2 gene in a British family is associated with profound difficulties in articulation. The gene is implicated in Broca's area gaining vocal control.

form of potential arrowheads also appears around this time, as does evidence for compound adhesives. Recent finds from South Africa show that
by 40,000 years ago artifacts such as bone tools and poisoned arrowheads
are nearly identical to those still in use today by San hunter-gatherers. This
is evidence of continuity in material culture and lifestyle.

The oldest known representational paintings, dated at thirty-two
thousand years ago, are located in the Chauvet Cave in France, portrayed in
Werner Herzog's film *Cave of Forgotten Dreams*. From this period onwards
symbolic capacities are expressed in a range of other domains including
sculptures, engravings, and even musical instruments. Controversial new
dates from the prehistoric cave of Geissenklösterle in Germany suggest that
figurines and flutes were produced even earlier, around forty thousand
years ago. Among the early works of art is the famous Hohlenstein-Stadel
figurine, which comprises lion and human parts and so demonstrates a
creative capacity to combine basic elements into novel constellations.

Yet most early cave art lacks any obvious narrative. The oldest apparent scenario with a story line I could find comes from one of the most
wonderful caves ever discovered. In 1940, eighteen-year-old Marcel Ravidat
looked for his dog, Robot, on a hill near the river Vezère when he stumbled on a hole in the ground. He and three friends returned four days later
and forced their way into the tunnel armed with knife and torch. They
discovered Lascaux Cave, its walls covered in colorful paintings of some
nine hundred animals, such as bulls, horses, and stags. About seventeen
thousand years earlier human painters went into the cave with reindeer-fat
torches, made scaffolds, and then painted the walls with extraordinary realism. In the lowest part of the cave, known as the Pit, is the only painting of
a human and what appears to be a depiction of an event (see Figure 11.11).

One might interpret the scene as a hunt gone wrong. The human figure
is lying on his back with a bison charging toward him. The bison appears
to have been hit by a spear and looks as if its entrails are falling out. Next
to the fallen man is a stick with a bird on it—possibly a spear thrower. On
the other side of the man are two parallel series of three dots and a woolly
rhinoceros. Perhaps the rhino disemboweled the bison and caught the
man. Perhaps not. Diverse interpretations have been advanced, including
some esoteric proposals involving stars, spirits, dreams, or trance. Most
commentators agree that the painting tells a story, although it is far from
obvious what that story is.

FIGURE 11.11.
Reproduction of the Pit scene at Lascaux. The first
illustrated story?

LESS THAN A KILOMETER FROM Lascaux, Roger Constant dug on his farm
in search of an entrance to the famous cave. He was unsuccessful but instead found a seventy-thousand-year-old burial site: a Neanderthal in fetal
position surrounded by bear bones.

Neanderthals were no longer around when Lascaux's famous paintings were made, but until their eventual demise, they seem to have been
evolving culturally. Michelle Langley and colleagues have shown that their
behavioral complexity increased over time, just as it did in anatomically
modern humans. In some instances the technologies of modern humans
were advanced, but in others they were not and were even inferior. There
are signs that earlier inventions were lost for thousands of years before
they reemerged. The archaeological record complicates the simplistic view
that modern humans gradually ratcheted up sophisticated technology and
displaced populations of brutish Neanderthals. It increasingly looks like
local factors, such as climate, population densities, and conflicts, played
a major role in the diffusion, accumulation, or loss of particular cultural
inventions.

One potentially telling difference between post-Neanderthal humans
and earlier hominins is that the ratio of younger to older adults shifts in the
fossil record. An analysis of fossil teeth from over 700 individuals found
numbers of old to young hominins gradually increased from late Australopithecines (37 old, 316 young) to early *Homo* (42 old, 166 young) to
Neanderthals (37 old, 96 young). Post-Neanderthal Stone Age Europeans,
by contrast, are disproportionally represented by older individuals (50 old,

24 young).[22] This points to a better survival rate and hence opportunity to accumulate cultural information. An individual was considered old when it was twice the age of expected reproductive maturity. In other words, the individual could conceivably have been a grandparent. Only when older age becomes reasonably common might we expect the grandmother effect to have had significant traction.

In some localities, new evidence suggests, hominins other than Neanderthals and anatomically modern humans persisted successfully until very recently. As noted in Chapter 1 recent genetic analysis of thirty-thousand-year-old fossils found in the Denisova cave in southern Siberia revealed that the DNA belonged neither to a modern human nor to a Neanderthal. Instead, the data suggest that these Denisovans are the Asian descendants of an earlier migration out of Africa some five hundred thousand years ago that split into Neanderthals in Europe and Denisovans in Asia. Moreover, like the Neanderthals, Denisovans appear to have interbred with modern humans. They contributed some genetic material to present-day Melanesians. The picture of hominin ancestry is increasingly becoming more colorful and diverse.[23]

Another major find was *Homo floresiensis*, tiny hominins from the Indonesian island of Flores. They stood one meter tall and were discovered in 2004, at a time when the world was watching Peter Jackson's film adaptation of Tolkien's *The Lord of the Rings*. The label "hobbit" immediately stuck. Indeed, people began to wake up to the notion that Middle Earth is not that outlandish an idea. Until recently in prehistory, a range of distinct hominins have been roaming this Earth, not entirely unlike hobbits, orcs, elves, and dwarfs did in Tolkien's novel. The hobbits on Flores had small brains (around four hundred cubic centimeters) and yet appeared to have made sophisticated stone tools. The discrepancy led to some skeptical debates about whether these specimens are deformed or diseased modern humans, a debate that also occurred when the first Neanderthals were unearthed. Current evidence suggests these were hominins with attributes akin to *Homo habilis* or *Homo*

---

22 There are some problems with such ratios, given that the populations compared are not from the same time period. Consider, for instance, that bones of older individuals are more susceptible to decay because of bone mineral depletion; thus the more ancient the sample, the fewer bones of older individuals may survive.

23 In 2012, strange eleven-thousand-year-old to fourteen-thousand-year-old fossils from a Chinese site were described. The skull looked very distinct from that of modern humans, featuring thick brow ridges and lacking a modern chin. They cooked large deer and have hence been dubbed Red Deer Cave People. It is as yet unclear whether they represent any new species or are part of human diversity.

erectus (who, as you may recall, survived on neighboring Java until perhaps as recently as twenty-seven thousand years ago). The isolation on the small island of Flores may have resulted in so-called island dwarfism, a phenomenon observed in many animals. Elephants on Flores were tiny too (yet the local rats and lizards are unusually large). It's a mysterious isle of flowers, indeed.

Even if a group of fossils is eventually identified as a human pathology rather than the remains of a distinct species,[24] new finds appear with such regularity that we most likely underestimate the great diversity of hominins that used to populate this planet. Throughout most of our prehistory several species of upright-walking, intelligent hominins lived side-by-side with our own ancestors. We are merely the last one standing.

DARWIN APPEALED TO GROUP SELECTION to explain the rise of human faculties such as morality[25] and noted that "at all times throughout the world tribes have supplanted other tribes." Competition between and within groups might have increasingly selected for enhanced mental scenario building and better linking between minds, including advances in the domains of language, foresight, theory of mind, reason, culture, and morality. When a group cooperates to attack you, the most effective response usually is to cooperate in defense. The burgs and fortresses of old Europe and the Pa sites of Maori illustrate how frequently humans had to do this in recent times. Battle plans and strategies, weapons technology, organization and enforcement, bluffs and deceits, valor and heroism are only some of the qualities for which such persistent threat of conflict might select. It pains me as a pacifist to give so much credit to war and conflict, but it appears quite plausible (given our history of conflict and our current obsessions with less violent intergroup competitions, such as soccer[26]). However, as so often is the case, plausibility should not be mistaken for proof.

---

24 To complicate matters, strange fossils may also represent a pathology that is not in existence today.

25 Darwin wrote: "A tribe including many members who, from possessing in a high degree the spirit of patriotism, fidelity, obedience, courage, and sympathy were always ready to aid one another, and to sacrifice themselves for the common good, would be victorious over most other tribes; and this would be natural selection." However, group selection continues to be a hotly debated topic.

26 We have radically reduced the negative effects of group competition, war, and aggression through sports. They allow us to compete, hone our skills, increase cooperation, and focus our attention; they have taken the place of warfare in many aspects of our lives. Of course, some violence is also associated with sports—think of soccer hooligans—but this is a far cry from actual war.

Such intergroup competition may have had a fundamental role in the creation of the current gap between us and our closest relatives. Conflict and competition may have led not only to the supplanting of tribes but to the extinction of closely related hominin species and subspecies (just as human arrival may have led to the extinction of other megafauna). Indeed, such competition with other hominins may have selected for better scenario building and communication between group members.

Since there is currently no direct evidence for genocides of hominin species, the effect of such conflict is speculation. The oldest indication of violence between people currently comes from archaic *sapiens* in Spain some 250,000 ago. Several skulls demonstrate healed impact fractures, and several individuals show signs of having been butchered (and presumably consumed). Healed skull fractures have been interpreted as signs of attacks with weapons and have also been reported from Neanderthal and Paleolithic human sites. A Neanderthal in northern Iraq (Shanidar 3), dated at over 50,000 years old, was found with a penetrating lesion to the rib cage. Experiments suggest that the injury was likely caused by a long-range projectile, and the researchers have raised the possibility that modern humans might have been responsible. The jaw of a 30,000-year-old Neanderthal child was found among modern human remains at a cave in southwestern France. It bears cut marks that are distinctive indicators of slaughter. In spite of these few suggestive finds, it is not clear how common and significant violent encounters have been.

Compelling evidence of violence between humans comes from a twelve-thousand-year-old cemetery in the Sudan that contains a couple dozen individuals with chert projectile points embedded either in or next to their bones. From the subsequent millennia, there are various mass graves and numerous examples of wounds caused by weapons. Recall, for instance, the violent death of Ötzi the Iceman. Despite such finds, it is not possible to properly determine the scale of prehistoric conflicts.

Over the last decade the evidence had strongly favored an out-of-Africa account, in which modern humans displaced archaic forms wherever they met, over a multiregional account, in which ancient hominins such as *Homo erectus* evolved into humans in different regions of the world. But there has been a paradigm shift in recent years. As the new Neanderthal and Denisovan evidence shows, some gene flow with existing populations is evident, and there is a good chance more will be unearthed. African data also suggest interbreeding with archaic people.

The competing multiregional account is unlikely, since the genetic data point to common descent from East Africa. Compared to chimpanzees, human DNA is homogeneous and not as diverse as a simultaneous evolution in different parts of the planet would suggest. Given that some interbreeding clearly took place, for now, a combination of out-of-Africa and multiregional accounts probably best approximates the complex path to modern humanity.

Whether through competition, absorption, or other factors, our hominin cousins are no longer discernible, even if many of us carry some of their genes. Our closest remaining relatives now are apes in the equatorial jungles of Africa and Asia. We have come out of the last ice age as clear winners, spreading across the globe. We have subdued much of the natural world and domesticated plants and animals to serve our needs. We wield unheralded powers to create new worlds and have often been unimaginably ruthless in our conquests.

In this chapter I have tried to outline our current knowledge about our ancestors' extraordinary journey to becoming human as well as offer plausible scenarios about the forces that may have created the gap, without obscuring the difference between established facts and informed conjecture. In so doing I have exercised the two legs—the ability to generate and reflect on scenarios and the drive to communicate them—that, whenever they first came into being, amplified our skills above those of other animals.

Our capacities to generate accurate mental scenarios and exchange them widely and efficiently have improved dramatically in recent generations. My mother's mother came of age in a world without electricity, computers, or cars and with her education limited to Catholic doctrine conveyed by parents, nuns who taught her at school, and priests. She had little opportunity to make herself heard beyond her hometown. By contrast, my children are growing up in a world in which they can access ideas from virtually anyone anywhere, and they can in turn broadcast their own musings and discoveries across the globe. We have come a long way—and there are more changes ahead.

T W E L V E

Quo Vadis?

We are made wise not by the recollection of our
past, but by the responsibility for our future.

—GEORGE BERNARD SHAW

The greatest adventure is what lies ahead.
Today and tomorrow are yet to be said.
The chances, the changes are all yours to make.
The mold of your life is in your hands to break.

—J. R. R. TOLKIEN

HUMANS HAVE LONG THOUGHT THEMSELVES special—distinguished
from other creatures on this planet. In some sense we certainly are. We
have been extraordinarily successful by sheer numbers: there are more
than seven billion of us, and we constitute about eight times the biomass
of all other wild terrestrial vertebrates combined. After millions of years of
being one of several upright-walking hominins living in small clans armed
with stone tools, we now stand alone, wielding unrivalled powers. In a
mere five hundred generations we progressed from the Stone Age to smart
phones and space exploration. Advances in bio-, nano-, and computer
technology are rapidly opening up countless new frontiers.

A key to these advances is an invention that transformed our capacity
to both accurately create and effectively share mental scenarios: writing.

Current evidence suggests that the first script was not devised, as one might have expected, by poets, philosophers, or historians but by accountants in Near Eastern farming communities.[1] At the end of the last ice age some tribes began to abandon a hunter-gatherer lifestyle in favor of a sedentary agricultural existence, which enabled rapid population growth and was a catalyst for development.[2] Selectively planting the seeds of wild plants with desirable characteristics led to the domestication of high carbohydrate cereals such as wheat and barley, which are also easy to store. Wild herding animals, such as sheep and goats, were penned, and selective culling and breeding drove their domestication. Farming required planning, cooperation, and work, but it also produced surplus that allowed some people to increasingly focus on activities other than obtaining food. Through the distribution of grains, meat, and other goods, people could make a living in trade, building, prostitution, security, or administration. The first cities and temples were built; increasingly complex civilizations, including their languages, animals, and crops, started to spread. Economic activity expanded, and as early as 9,500 years ago, clay tokens were employed to bolster human memory. Different shapes, such as a cone or a cylinder, represented different units of merchandise, such as a measure of grain or an animal. A thousand years later, seals emerged and were used to identify the people responsible for the goods represented by tokens, opening up a brand-new world of accounting. Farmers, temple administrators, and traders could keep track of their transactions, including debts and pledges.

About five and a half thousand years ago Sumerian accountants used sealed, hollow clay balls to envelop tokens for longer periods. Once sealed, the content of the container could only be checked by breaking it, so someone clever came up with the idea of impressing the tokens on the outside of the wet clay container before putting them inside. Six impressions on the outside meant six corresponding tokens on the inside. It did not take long for people to realize there was no longer any point in having

---

1   Writing appears to have also been invented in other regions: in China, in Central America, and on Easter Island. It is possible that these are entirely independent inventions of script, though any sort of contact with people who write could have given someone an idea.

2   Those who continued to pursue a hunter-gatherer lifestyle have increasingly been marginalized. Few cultures have been able to maintain this ancient way of life and only at those locations that were least attractive for farmers.

the tokens on the inside. Subsequently, clay containers were replaced by impressed clay tablets, and the technique became widespread in Syria and Mesopotamia. Accountants then began to add pictographic symbols that were traced rather than impressed, such as small marks for counting or a simple picture of an ear of barley to represent barley. Signs were increasingly simplified, and the first cuneiform script, involving various arrangements of a basic wedge (*cuneus* in Latin), emerged over five thousand years ago. Eventually, scribes began to use writing to record things other than accounting. The rest is history—literally.[3]

Writing has allowed us to connect our minds and accumulate knowledge across time and space like never before. Early inscribed statues from the city Ur, for instance, record the name, title, and ancestry of a king, as well as what palaces and temples he built and what lands he acquired. Writing was used to record creation myths, laws, prayers, calendars, teachings, and funerary texts. Pharaohs and kings began to send couriers to disseminate written decrees, letters, and warrants. Postal services eventually emerged in Persia, Rome, and China, allowing ordinary individuals to exchange their thoughts and experiences. Ancient Greek historians began to systematically document wars and narrate significant events. Once fixed in writing, thoughts became lasting objects.

Some written teachings came to be revered as originating from divine sources, and sacred scriptures have had an incredibly powerful and enduring influence. Most of today's major moral traditions have roots going back to important thinkers some 2,500 years ago. During this period the Buddha offered his philosophy in India, Confucius in China, Socrates in Greece, and, perhaps a bit earlier, Zoroaster (Zarathustra) in Persia. It may even have been possible for a single person, such as the fictional grandson of Zoroaster in Gore Vidal's historical novel *Creation*, to have lived long enough to speak to all of these great men. Why did these influential moral

---

3 Writing systems in other parts of the world may have been invented for other reasons. In the Americas, for instance, concern for time and calendar may have been critical to the early symbols of Olmec and the later full-fledged script of the Mayans. In China the earliest undisputed scripts are divinations on animal bones, though there is some debate about potentially older origins. Little is known about the Easter Island script, Rongorongo, which has not yet been deciphered. Though humans have created many diverse scripts, the neuroscientist Stanislas Dehaene has made the case that they have much in common because of specific constraints on how visual information is encoded in our brains.

traditions emerge almost simultaneously? The reason may be less a coincidence of moral insight (or divine communication) than a function of the spreading influence of writing. Written down, moral teachings became standardized and so could proliferate like no oral traditions had done. Social norms became written laws. Even if interpretations of, say, the Old Testament differ widely, people can always return to the source.

Written words invite critical reflection, focused debate, and commentary. Readers at any time and place can assess and build on others' ideas. Aristotle's writings, for instance, influenced much of subsequent Western philosophy. If we depended on word of mouth alone, we might only have heard fragments of his thoughts, with no way of distinguishing the original ideas from what retelling has added or subtracted. With writing, we can still "hear" the voices of the writers centuries after they have died. Their scenario-building minds can still be wired to ours (even if the information flow is unidirectional). We can learn from the dead and, in a sense, collaborate with them across time, confirming one idea, debunking another, and qualifying a third. As Carl Sagan put it: "The library connects us with the insights and knowledge, painfully extracted from Nature, of the greatest minds that ever were, with the best teachers, drawn from the entire planet and from all of our history, to instruct us without tiring, and to inspire us to make our own contribution to the collective knowledge of the human species."

For much of history, texts were hand-copied, painted on silk, bamboo, or paper, making distribution limited and vulnerable to loss. Books were precious treasures. The great library of Alexandria, first organized by one of Aristotle's students some 2,300 years ago, systematically attempted to gather the world's knowledge, collecting hundreds of thousands of scrolls. It took accumulation of culture to another level. With its eventual demise much of the recorded thought of the ancient world was lost. Many documents were destroyed in a fire during Caesar's attack in 48 BCE, but the library continued to be a hub of science for several more centuries. Its last famous librarian was the female mathematician and astronomer Hypatia, who was murdered by followers of Alexandria's Archbishop Cyril in 415 CE. The great library waned as the dark ages beckoned.

Our drive to exchange our minds has fuelled a search for ever-more effective media. The invention of woodblock printing in China in the third century enabled more rapid copying. Europeans caught on later, with Gutenberg's printing press appearing around 1440 and revolutionizing mass production. Within a century the number of books in Europe is thought to have multiplied from tens of thousands to tens of millions. Written accounts and ideas became accessible to ever-larger audiences around the world. The printing of newspapers from the seventeenth century onwards attuned many people's minds to the same scenarios we call "current affairs."

Printing helped people focus their collective efforts on understanding nature. Books and journals created the opportunity for an intellectual interchange that cumulated in the Enlightenment: the age of reason. The first journal fully devoted to science, *Philosophical Transactions of the Royal Society*, was founded in 1665 and still appears regularly to this day. Researchers such as Isaac Newton reported their discoveries through the journal, and a hundred years later, Wilhelm Herschel would use it to tell the world about his profound astronomical observations. Without printing, the inductive scientific approach his son John Herschel advocated would not have gathered the momentum it did. With printing, scientists could efficiently share data, compare hypotheses, and systematically put them to the test. Findings could be rapidly and reliably disseminated, ratcheting up human knowledge dramatically. Numerous technological breakthroughs followed, as scientists and engineers solved problems and informed each other about exciting new opportunities—some, in turn, with far-reaching consequences on how we exchange our minds.

Postal services had long enabled remote communication, but the invention of the telegraph in the nineteenth century made long-distance mental connections instant. At the same time, motorized transportation opened up new opportunities for visiting family and friends who lived far away and so enabled us to more frequently indulge our urge to link up. We came to increasingly rely on various media to satisfy this drive. Telephone, radio, television, fax and email are more recent technological advances that have enhanced our capacity to connect our minds across time and space.

The rise of the internet and satellite networks make it possible to hook up virtually any human mind with any other anywhere (feel free to insert sarcastic remark about your local service provider here, if you like). The web offers access to others' writing from across the globe. Social media such as Facebook and Twitter occupy an important place in many people's daily lives, and by the time you read this there will be yet other ways of communicating. It may seem a mystery to older generations that these computer-mediated interactions are so extraordinarily popular, but they are logical extensions of a long historical trend. These media enable us to satisfy our urge to connect and inform each other in an instant, wherever we are, about any idea or whim we care to share.

We increasingly collaborate economically, politically, and intellectually within a range of global networks. More and more people use these technologies to discuss, complain, or gossip and to instigate and coordinate cooperative projects. Many a quirky idea or hobby that may have wilted in solitude in the past can flourish with input of like-minded people from elsewhere. Progress in science and technology has been extraordinary as a result of the rapid exchange of research reports through electronic journals and, increasingly, through open-access outlets that enable anyone with an internet connection to search and read the latest findings. Within a few generations our understanding of the nature of our world, and our place within it, has changed dramatically.

THE TEMPLE OF APOLLO AT Delphi is said to have been inscribed, "Know thyself." When Linnaeus published his famous classification of all living things, *Systema Naturae*, he included humans but did not provide a taxonomic description. Instead he simply wrote, "*Nosce te ipsum*," the Latin version of that same ancient phrase.[4]

It was long thought that our apparently unique position on Earth is miraculous. The knowledge accrued over the last couple of hundred years of systematic scientific inquiry has provided a different perspective on humanity. Only recently did we discover that other hominins used to share the planet with our ancestors. Now we know that the vast gap between humans and other animals is in part due to the disappearance of our most closely related species. By developing a perspective that encompasses hundreds, thousands, even millions of years, we are gaining a very different view on who we are. Our ancestors shaped much of this world, burning forests, draining swamps, and domesticating some species while wiping out others. With industrialization our forces grew monumentally, and it may only be our generation that is waking up to the fact we are rapidly

---

4　This means that his readers should not need a description, given that they are human themselves, but it also, perhaps inadvertently, points to self-knowledge as something that may set us apart from other animals.

altering this planet—potentially undermining our own future prosperity. Only with a clearer understanding of where we come from and who we are can we gain a better view of where we are heading.

Oracles, prophets, and diviners have long been in the business of telling the future. Like inventions in science fiction that inspire engineers, prophecies can guide people's actions in foreseeable ways. Indeed, in extreme cases, prophets of doom have led to mass suicide and utopian visions to revolution. Science has not only brought us a more systematic route to explanations but given us new tools for reliably forecasting the future—and shaping it.

We have begun to record changes systematically, giving us databases from which to make models and predictions. Accounting of past crop yields and related variables such as rainfall or temperature allows us to extrapolate future yields. Unique events, such as new inventions or the consequences of the introduction of a new species to an ecosystem, remain difficult to predict, but more continuous change is easily plotted. Statistical models link numerous variables and enable complex extrapolations in which we can even quantify the errors we are likely to make. We are increasingly able to predict what we care about, such as how long we are likely to live, when we will run out of certain resources, and what consequences our activities have on fauna and flora. Environmental impact studies are now commonplace. Computer simulations can be used to create scenarios of what will happen if we continue down one path and what is likely to happen if we change one parameter or another. We can compare alternative future situations in terms of possibility, probability, and desirability.

On an ecological macro level, current models predict profound changes in climate, atmosphere, and the oceans. The rapid decline of forest habitats, biodiversity, and resources such as oil or fish, as well as accumulating waste and pollution, are now widely recognized as global problems with significant future consequences. What can we do to avert disasters and create a sustainable future? It may turn out to be most fortunate that as we seem to be reaching the breaking point on many fronts, we are beginning to exchange our minds globally and are becoming self-aware of our interconnected fate.

Our future depends on how accurately we can build scenarios of the future and how well we can link our minds to cooperatively tackle global problems—abilities that set us apart from other animals (and that, arguably, got us into this mess). We face colossal challenges. The buck stops with us. There is no sign one of the other creatures on Earth will jump into the fray and sort things out. We are the only species capable of launching strategic cooperative efforts designed to address these challenges.

I HAVE REVIEWED HERE CURRENT evidence on the nature and origin of what makes us human. The data led me to propose that the peculiarity of the human mind primarily stands on two legs: our open-ended capacity to create nested mental scenarios and our deep-seated drive to connect to other scenario-building minds. These traits have had dramatic consequences for the way we communicate, our access to past and future, our understanding of and cooperation with others, and our intelligence, culture, and morality. We have managed to create a fast and efficient cultural inheritance system through which human groups have accumulated novel powers that ultimately allowed us to dominate much of the planet.

Of course this analysis is far from the last word on the gap. The key to Herschel's scientific method is to continue to test hypotheses by further observation and experimentation. There are at least two sides to this in the present case. On one side, the claim that nonhuman animals do not have these capacities requires further scrutiny. We need to establish more precisely the competences and limits of animals on some of the key components, such as working memory. If future research demonstrates, say, that an orangutan can compute certain problems recursively, then this will falsify part of the hypothesis.[5]

On the other side, we need to further examine if there are other factors that I overlooked that are critical to human uniqueness. Though we have seen the central relevance of the two legs (and the interplay of their consequences) in the six domains reviewed, more systematic work is required to examine if this is also the case in other domains. Have a fresh look at your

---

5 This need not lead to a wholesale rejection but rather to a refinement of the analysis. Ideally, researchers who argue that their findings show that a trait is not unique should also suggest where the difference lies instead (e.g., the recursive capacities might be limited to a particular domain). Ideally, the new hypothesis explains all that the earlier explained plus the new facts.

own intuitions about what sets human minds apart. Do the two factors ultimately play a key role in what is unique about the traits you consider?[6]

Some of my deductions will be proven wrong. New evidence will challenge results that inform the current investigation. Romantics and killjoys will continue to quibble, and new middle ground will be charted. As scientific progress accelerates, our understanding of the gap will become increasingly refined. I have no doubt that genetics, neuroscience, comparative psychology, and paleoanthropology have more surprises in store.[7] Having said that, the current picture of the gap is clearer than we could reasonably have expected even a few years ago. In this book I have offered a snapshot of this picture. I hope that you have enjoyed the show and that your own scenario-building mind has been inspired to ponder this mental fodder some more. Whether much of my interpretation turns out to be correct or not, I hope to have convinced you that there is a need to systematically research these questions in a science of the gap.

The book's focus on what sets us apart from other animals should not distract from the long list of traits we share with other animals. Our minds are wired in ancient ways. Many of our fundamental cognitive processes, emotions, and desires are not unique. For instance, frustrated humans can work themselves into a rage, just as chimpanzees do. People can become agitated, "flip," or go "ape shit"—although we generally try to rein in such outbursts. We punish violations of our social norms as we try to uphold a polite and civil society; our culture and morality help cultivate less aggressive and more socially compliant behavior. Still, our primate heritage cannot be denied:

---

6   I regularly ask students about their intuitions, and their answers often include the domains discussed here as well as a host of other possibilities. Recently they wrote essays on topics such as adornment, aesthetics, arts, celebrations, complex emotions, dance, democracy, engineering, games, greed, hospitality, humor, law enforcement, mathematics, medicine, music, religion, rituals, schizophrenia, sexual modesty, spirituality, sports, suicide, thirst for knowledge, and warfare. In most of these, to the extent that they are uniquely human at all, the case that the two legs play a major role in carrying our uniqueness can be made. However, in some, such as aesthetics, this is not so clear. Aesthetic projects, of course, may depend on the artist envisioning scenarios and on a drive to get a message across. But the basic notion that some things are more pleasingly aesthetic than others does not really depend on these traits. So there might be something extra here, though it is also possible that other animals have such preferences too. Clearly, a lot more comparative work needs to be done.

7   Perhaps this is the place to make one concrete futuristic prediction myself. I suspect that scientists will one day clone Neanderthals and other earlier hominins. I am not sure if I am optimistic or pessimistic about the consequences of this one, but it would be quite interesting.

Darwinian man, though well behaved, at best is only
a monkey shaved.

                            —GILBERT AND SULLIVAN

Reminders of our animal nature are a counterweight to the common view that humans are separate from the natural world, but they should not obscure the fact that we are peculiar indeed. There is no point belittling the extraordinary powers that separate us from other animals nor denying that we are a primate. It is time we established a more balanced view that acknowledges both the similarities and the differences between animals and humans. This may require letting go of some long-cherished notions of self-importance, but it should in no way diminish our sense of wonder about our peculiar existence. Know thyself.

A MORE PRECISE ANALYSIS OF the gap has some applied benefits we can throw into the bargain. Researchers often study mice, rats, and other animals in an effort to understand the genetic or neurological bases of human mental functions and associated disorders, but this only makes sense when these animals have some rudiments of the human trait. For characteristics that turn out to be entirely unique to humans, the use of animal models is, for the most part, profoundly misguided. A clearer understanding of the gap will give us a better framework for deciding when such research is not promising. It may save laboratory animals' lives and help researchers avoid many a garden path.

The mapping of the genomes of our closest living animal relatives can help unravel the mysteries of the human genome. It is critical to complement the genetic data with information about cognitive and other traits of nonhuman primates. Traits that turn out to be unique to humans are likely dependent on those aspects of the genome—and the nervous system—that are unique. Knowledge of the gap helps narrow down the search space for identifying the neurological and genetic bases of these traits.

By the same token, we can take advantage of knowledge about what we share with some of our closest relatives. For instance, consider the finding that humans and great apes share a capacity for mirror self-recognition and stage 6b object permanence, whereas small apes do not (see Chapter

3). Given the close relation between these species, the trait in great apes is most likely a homology—that is, we inherited it from a common ancestor that evolved it between eighteen and fourteen million years ago. A homology entails that the trait is driven by a similar inherited neuro-cognitive and genetic foundation. In other words, it relies on a basis that humans and great apes share, and that we do not share with gibbons. To identify what critical changes make the trait possible, then, one can focus research on those aspects of the brain or genome that great apes and humans have in common but that they do not share with small apes.

Evolutionary perspectives are increasingly proving useful in diverse areas of psychological inquiry. Though evolutionary psychology textbooks still often feature little discussion of our closest living animal relatives or even of our extinct hominin relatives, the material covered in this book illustrates how critical they are. A careful appreciation of the making of the gap, I would argue, is essential for a truly evolutionary perspective on the human mind. Perhaps one day Darwin's prediction that psychology will be based on a new foundation may yet be fulfilled.

WHAT ABOUT THE FUTURE OF the gap itself? There are three obvious possibilities: the gap could decrease, increase, or remain the same. The view that the gap is stable may derive from a belief that humans are no longer evolving—a perspective quite widely held. Could it be that cultural and technological advances mean biological evolution no longer matters to us? We have become so powerful in creating artificial worlds that we primarily seem to adapt the environment to us rather than the other way around. Modern medicine increasingly circumvents natural selection, and our global interconnectedness means there are no longer many isolated pockets in which human evolution can diverge. Has the evolution of our minds stopped?

On a moment's reflection this scenario seems rather unlikely, and even reeks of arrogance, as it seems to imply that we are the final product—the height and endpoint of evolutionary achievement. I find it difficult to believe that after four billion years of life forms changing on the planet, it all comes down to the perfection that is you and me. Given our past, it seems more likely that we are another segment in the long chain of evolutionary change. Tens of thousands of generations from now, if we manage to not go extinct, our descendants will look back at us as early humans. In fact, there is evidence that even over short time frames natural selection is effective in bringing about genetic changes in human populations. Furthermore, natural or human-made disasters can create isolation rapidly—as could human success: just think of the possibility of humans eventually populating other planets. Those who do go extraterrestrial may quickly find themselves isolated and available for separate evolutionary trajectories. In sum, it is highly unlikely that evolution will stop with us.

What, then, is the trajectory of the evolution of the human mind? Some data suggest that over the last ten thousand to fifteen thousand years, as population density increased, brain sizes *decreased*. Given associations between brain size and IQ, this may reflect a decline in mental ability over the time we gained most of our amazing technological powers. Potential reasons for this decrease are changes in nutrition and climate, as well as the possibility that our societies, with extensive division of labor and social safety nets, enable the less mentally endowed to survive where in the past they would not have reproduced. Many of us get by without having the basic skills of hunting and gathering that had been essential for our forebears. Perhaps as technology does more and more of the hard work for us, our artificial world will put ever less demand on our minds. It is imaginable that in the future we will all sit in our lounge chairs and play in virtual reality. Is it possible that our mental capacities are dropping and the gap will become smaller?

It seems unlikely that our minds will dumb down dramatically as long as humans are needed to design and maintain these artificial systems. However, the forces of natural selection on humans today are puzzling. The rich, successful, powerful, beautiful, and well-educated people seem to breed less, not more, than most of the rest of us. In other words, they appear to leave fewer copies of their genes in the next generation than those not blessed with these seemingly highly advantageous attributes. One may therefore worry humanity will slowly lose its edge and the gap eventually gets smaller as a result.

It is also possible, of course, that we cut short our own success story more dramatically. In addition to radically changing the environment, our arms races have resulted in weapons that enable us to annihilate each other many times over. War, terrorism, or mishaps could quickly result in a dramatic unraveling of our civilizations. If we somehow mess it up, our minds may struggle to rebuild, especially as we become ever more dependent on technology. As Einstein warned, "I do not know with what weapons World War III will be fought, but World War IV will be fought with sticks and stones." Countless civilizations have collapsed. In addition to violent conflicts, common causes include habitat destruction, soil and water management problems, overhunting and overfishing, introduction of new species, and overpopulation. As we are increasingly linked into one system, and we face many of those problems on a global scale, it is possible our modern civilization too will collapse one day for similar reasons. A potentially bleak future awaits in which few survive and other creatures are given a chance to close the gap.

A scenario like the one in the 2011 film *Rise of the Planet of the Apes*, in which we unleash a deadly pandemic while simultaneously enhancing the capacities of apes through biotechnology, is highly unlikely but not entirely out of the question. Genetic engineering has given us radically new powers to influence evolutionary pathways. Advances in biotechnology, such as the capacity to turn any cell into a stem cell and use it to grow body parts or entire organisms, will open up incredible opportunities. It is not far-fetched to assume that we may one day be able to alter brain development and enhance the minds of our closest remaining relatives. Humans are increasingly guiding evolution itself. Some call this "playing God."

Humans have played God for a long time. At least since the beginnings of agriculture humans have practiced what Darwin called "artificial selection." We encourage versions of plant and animal life that are useful to us and discourage those that are not. Artificial selection may also have been important in the shaping of our own species. Hitler's genocides and attempts at breeding a superior race may be the first things to spring to mind, but we socially guided our evolution long before any notions of eugenics. Capital punishment and banishment from social groups not only enforce social norms but select against certain undesirable attributes, such as tendencies to rage violently. Richard Wrangham and Brian Hare have argued that we have domesticated ourselves much as we have domesticated dogs and horses. Domesticated animals are not only less aggressive and more cooperative than their wild counterparts but also typically sport smaller brains. So this proposal is in line with recent human brain size reduction and the overall decline in violence and increase in cooperation that Stephen Pinker argued has characterized recent history.

We have gained some significant new capacities for what we might call "auto-artificial selection." Contraception is the most obvious, letting us curtail reproduction. Conversely, we can make sperm fertilize eggs in ways other than through sex. We will increasingly have the opportunity to deliberately determine not only the number of offspring but also their characteristics, from sex to disease resistance. Many people have understandable reservations about such interference. But imagine if you could make the genetic changes to stop your child from getting cancer, Alzheimer's, or whatever else has plagued your family tree. It is a small step from preventing disease to influencing the intellectual capacities of offspring or altering the shape of the nose. This direct interference in the genetic makeup of the next generation—"artificial mutation" rather than just artificial selection—may lead to drastic changes as we fast forward tens, hundreds, or thousands of generations into the future. We are increasingly acquiring the power to shape our own evolution, and we may well end up using it to acquire greater mind power.

I predict that the gap will widen. In fact, there are signs it already is widening. Over the past century humans have improved in their average performance on intelligence tests by about 3 percent every decade. Some evidence suggests that brain size, contrary to the trend of the last ten thousand years, may have slightly increased over the last 150 years. We have more nutritious foods and more stimulating education. We bolster our scenario-building minds with ever more refined machines and technologies that allow us to measure, model, and control the world in increasingly powerful ways. Through the internet and other electronic networks we are connecting millions of minds and bringing about an explosive growth in cultural accumulation. Answers to most questions are only a few clicks away. Science is accelerating, and greater knowledge in turn will open doors for the already foreshadowed biological, as well as electronic or chemical, enhancement of human mental powers. We are getting ever smarter—and, one can only hope, wiser.

There is a second way through which we may increase the gap. We could make ourselves appear more special on this planet by reducing the capacities of our closest animal relatives—moving the other side of the chasm. I do not mean we somehow dumb down the apes; I am referring to driving them to extinction. Their demise would turn other species into our closest living relatives, thereby widening the gap. Let's face it: we are in the process of doing just that. As we have seen, all the great ape species are endangered, and their numbers are primarily declining for one reason: human activity. Whether through habitat destruction, bush-meat consumption, or the pet trade, we are causing the demise of our closest animal relatives, perhaps not entirely unlike what we might have done to our upright-walking hominin relatives in the past.

There are, of course, humans who are desperately trying to stop the extinction of apes, and I encourage you to join them, but the current projections are bleak. In a couple of generations, our descendants might wonder at just how different they are from their closest remaining animal relatives: the monkeys. Apes may join Neanderthals and *Paranthropus* as half-forgotten creatures of the past. So our descendants may be even more baffled by their own apparent uniqueness (and possibly be distracted by questions about the importance of the fact that monkeys typically have tails whereas humans do not). Let's make sure they are more enlightened about the nature and origin of the gap. We better protect our tailless ape relatives carefully—for their own sake, as well as for the sake of our children.

We can consider the long-term consequences of our actions. We are the only species on this planet with the foresight capable of plotting a path toward a desirable future. Plan it for the apes. We are beginning to appreciate the drastic consequences our activities have had on Earth, and we can increasingly predict what repercussions our actions will have. So we are burdened with the responsibility of making the right decisions in the present. Humanity has a wondrous potential to cooperatively address impending disasters and to protect our own future as well as the future of our cousins on the other side of the gap. There are reasons to be hopeful. History is not just full of violence and cruelty but also full of heroism, kindness, and prudence. We have overcome many obstacles in the past, and we are better equipped than ever to look ahead and to collectively steer the ship out of troubled waters toward new frontiers.

ACKNOWLEDGMENTS

THIS BOOK HAS BEEN A long time coming. It is the accumulation of much recent work from across a wide range of disciplines. It is also the accumulation of my personal journey of discovery. Since I was a teenager I have wondered why we are the peculiar creatures we are, and much of my research has been ultimately driven by related questions. I kept on collecting pieces of information I saw as part of the larger puzzle until I had enough, I thought, to attempt constructing a coherent picture for more public display. It took more years than anticipated to complete this book, and telling the story ended up involving more self-disclosure than I had anticipated. Please forgive the indulgences.

The efforts of many people went into the writing of this book. First and foremost I want to thank my sounding board and proofreader, the gutsy Christine Dudgeon, who, in the name of science, dives with sharks, takes tissue samples to study their genetics, and makes them vomit to study what they eat. Thank you for your encouragement, support, enthusiasm, and love. Chris is my best friend and mother of our wonderful children, Timo and Nina. Our kids have been untiring participants in many an experiment and are the source of tremendous inspiration and joy.

My brilliant mentor Michael C. Corballis guided me through my master's degree, my PhD, and several subsequent collaborations. He is an academic gentleman, intellectual treasure, and exemplary scientist I cannot thank enough. From the early days of this project I would also like to mention the support of my late parents, Heinz and Hanni Suddendorf, as well as of Barbara Gerding, Pam Oliver, Richard Aukett, Shayne Carter, Owen Sweetman, Paula Nightingale, Matt Donaldson, Tina Forster, and Dave Rickard. I am grateful to many others with whom I discussed my ideas in New Zealand, Germany, and Australia. Thanks to Angela Dean and Darryl Eyles for the title suggestion "Mind the Gap" and for giving me Richard Holmes's exquisite book *The Age of Wonder*. The books that have inspired me, such as Jared Diamond's *Guns, Germs, and Steel* and Tim Flannery's *The Future Eaters*, cannot all be listed here. Instead I want to thank all those authors who contributed to the works listed in the full version of the references. It is only through having access to what oozed out of their minds that I could construct this view of the gap.

My first attempt at turning this project into a popular science book began in 2003 at the University of Queensland, and I would like to thank the support of my colleagues Ottmar Lipp, John McClean, Mark Nielsen, Virginia Slaughter, and Valerie Stone. Special thanks goes to Valerie for her contributions to the 2004 attempt to start telling this story. Thanks to my many students in the courses *Learning and Cognition*, *Evolutionary Approaches to Human Behavior*, and *Evolutionary and Comparative Perspectives on Cognition* who, over the years, have helped me clarify my thoughts and develop my passion for teaching. I have had the pleasure of working with many excellent postgraduate students at honors, master's, and PhD level, and I would like to especially note the contributions of my doctoral students Janie Busby, David Butler, Emma Collier-Baker, Jo Davis, Janine Oostenbroek, and Jonathan Redshaw.

I thank the Australian Research Council for supporting many of the research projects I reported on in this book. The UQ Early Cognitive Development Center has enabled my colleagues and me to study the minds of infants and children, and a special thanks here goes to Sally Clark as administrator and to the many parents and children who have donated their time to our research over the years. Numerous Australian and international zoological institutions have helped in allowing us to test nonhuman primates over the years. I need to single out especially the support of Alma Park Zoo and the zoos of Rockhampton, Perth, and Adelaide. The chimpanzees Cassie and Ockie have been particularly cooperative over the last decade. Thanks to Graeme Strachan and all supporters of Rockhampton Zoo for making their new enclosure, as well as female companions, a reality.

In the second half of 2010 I finally started writing *The Gap* in earnest on a sabbatical at the University of Auckland. Special thanks goes to Michael Corballis, Russell Gray, and Niki Harre for their hospitality and support. I made some serious headway during this time. Unfortunately, upon return to Brisbane in January 2011, we had to deal with the aftermath of our house having been flooded. Everything was put on hold as we recovered and rebuilt. I am deeply grateful to all the neighbors, friends, and countless strangers who helped us get back on our feet, and allowed me to refocus on the book writing within a few months.

By that point Peter Tallack's Science Factory literary agency had signed me on, and with his excellent support I swiftly obtained a contract and, at last, a hard deadline. My deep gratitude goes to him and to my editors at Basic Books, T. J. Kelleher, Tisse Takagi, and Melissa Veronesi, and Beth Wright at Trio Bookworks for their outstanding work and dedication to bringing this project to fruition.

I am grateful for useful comments from friends and experts who took the time to read my half-baked attempts. Note that none of them are responsible for (or guilty of) my opinions, however. Thanks to Emma Collier-Baker, Michael Corballis, Chris Dudgeon, Philip Gerrans, Colin Groves, Niki Harre, Bill von Hippel, Rachel Mackenzie, John McClean, Virginia Slaughter, Peter Tallack, and Jason Tangen for commenting on several chapters. Thanks to Michael Balter, Matt Donaldson, Andy Dong, Claire Harvey, Marc Hauser, Andrew Hill, Simon Lake, Michelle Langley, Chris Moore, Mark Nielsen, Mike Noad, Candi Peterson, Ceri Shipton, and Alex Taylor for thoughtful feedback on individual chapters.

Last but not least, I would like to thank supporters of another of my passions as they helped me stay sane and halfway balanced throughout all this—my football clubs, which have given me so much joy and community: FC Vreden, Borussia Mönchengladbach, Brisbane Roar, Brisbane Olympic FC (Sparta & Sharks), and the West End Partisans. We bond through a passion for the beauty of the game and are willing to sacrifice blood, sweat, and tears for something so inconsequential as the kicking of a ball. We humans sure have peculiar minds. Enjoy yours.

BIBLIOGRAPHICAL NOTES

(A more comprehensive version of the notes and bibliography can be downloaded from http://psy.uq.edu.au/gap)

Chapter 1: The Last Humans

3   **Wilhelm Herschel:** Holmes, 2008.
4   **John Herschel:** Herschel, 1830.
5   **Descended from the apes?:** For background on the murky history behind this quotation, see the Quote Investigator (February 9, 2011): http://quoteinvestigator.com/2011/02/09/darwinism-hope-pray.
5   *On the Origin of Species:* Darwin, 1859.
6   *The Descent of Man:* Darwin, 1871.
6   **(Footnote 3) 99.4 percent:** Wildman et al., 2003.
6   **the split occurred some six million years ago:** For genetic evidence, see Patterson et al., 2006; for relevant fossil evidence, see Brunet, et al., 2002; Haile-Selassie, 2001; and Senut et al., 2001.
8   **embodied cognition:** e.g., Isanski & West, 2010.
8   **judge a hill to be steeper:** Proffitt, 2006.
9   **Milan Kundera's astute reply:** Kundera, 1992.
9   **William James:** James, 1890.
9   **relief from pain:** Bateson, 1991.
9   **"In the distant future:** Darwin, 1859, p. 335.
9   **Even evolutionary psychology:** e.g., Barkow et al., 1992.
10   **textbooks on evolutionary psychology:** e.g., Buss, 1999.
10   **pioneers such as Wolfgang Köhler:** e.g., Köhler, 1917/1925.
10   **arguments for rapid transitions:** Gould & Eldredge, 1977.
10   *Homo floresiensis:* Brown et al., 2004.
11   *Gigantopithecus:* Ciochon, 1996.
11   **(Footnote 6) Traditionally these are:** Larick & Ciochon, 1996.
11   **(Footnote 6)** *Australopithecus sediba:* Berger et al., 2010.
12   **primary adverse force of nature:** Alexander, 1989.
12   *Guns, Germs, and Steel:* Diamond, 1997; see also Flannery, 1994.
12   **blankets infested with smallpox:** Ranlet, 2000.
12   *The Better Angels of Our Nature:* Pinker, 2011a.

12   goes back to prehistoric hunter-gatherers: Bowles, 2009; Keeley, 1996.
12–13 kill members of their own species: Goodall 1986.
13   Neanderthal inheritance: Green et al., 2010.
13   Denisovans: Krause et al., 2010.

Chapter 2: Remaining Relatives

15   our primate heritage: Groves, 1989.
16   it is social problems: Humphrey, 1976.
16   It is hardly an exaggeration: Köhler, 1917/1925, p. 293.
16   Primates are fond of grooming: e.g., Dunbar, 2010.
16   a vervet monkey mother: Cheney & Seyfarth, 1980.
17   Achieving high rank: de Waal, 1982; Goodall, 1986.
17   Dunbar established that the greater: Dunbar, 1992.
17   Taxonomists subdivide primates: e.g., Groves, 1989; Stanford et al., 2013.
18   apes grow up slowly: Bogin, 1999.
18   *Homo sylvestris*: Corbey, 2005.
18   Carl Linnaeus: Linnaeus, 1758.
19   widely used classification: e.g., Stanford et al., 2013.
20   They comprise four distinct genera: Geissmann, 2002. For more on
       gibbons, see http://www.gibbons.de.
21   a better model of what our hominin: Fitch, 2000.
21   critically endangered: All population estimates for the apes are based
       on the International Union for the Conservation of Nature (IUCN)
       Red list of Threatened Species 2012 (http://www.iucnredlist.org). See
       also reviews for the United Nations Environment Program, Great Ape
       Survival Partnership: http://www.unep.org/grasp.
22   Hainan black crested gibbon: Stone, 2011.
22   "Leakey's Angels": Fossey, 1983; Galdikas, 1980; Goodall, 1986.
24   stay in the subadult (unflanged) stage: Utami et al., 2002.
25   Carel van Schaik: van Schaik et al., 2003.
25   Anne Russon and Birute Galdikas: Russon & Galdikas, 1993.
26   first draft of the gorilla genome: Scally et al., 2012.
27   first initiated by Dian Fossey: Fossey, 1983.
27   Recent fecal analyses: Hofreiter et al., 2010.
27   Dick Byrne: e.g., Byrne & Russon, 1998.
28   a gorilla was observed using a stick: Breuer et al., 2005.
29   a previously unknown large population: "Wildlife Conservation
       Society Discovers 'Planet of the Apes,'" Wildlife Conservation Society,
       August 5, 2008, http://archive.wcs.org/gorilladiscovery/press-release
       .html.
30   the social lives of chimpanzees: e.g., Goodall, 1986.
30   "chimpanzee politics": de Waal, 1982.
30   boundaries that male groups patrol: Goodall, 1986.
31   Hunting primates: Boesch, 1994.
31   spear bushbabies: Pruetz & Bertolani, 2007.
31   seek out medicinal plants: Huffman, 1997.
31   ways of fishing them out: e.g., Whiten et al., 1999.

31   (Footnote 9) When Jane Goodall first: Goodall, 1964.
31   (Footnote 9) Liberian stamp: Whiten & McGrew, 2001.
31   stone hammers and anvils: Boesch, 1990.
31   stone tools 4,300 years ago: Mercader et al., 2007.
32   only described in 1929: Schwarz, 1929.
32   collaboratively hunt monkeys: Surbeck & Hohmann, 2008; Hofreiter et
        al., 2010.
33   a lot more sex: de Waal, 1996.
34   larger brains are more intelligent: McDaniel, 2005.
34   Do humans, then, simply have the largest brains?: For absolute and
        relative brain weights of various species, see Jerison, 1973; Roth &
        Dicke, 2005.
34   170 billion cells: Azevedo et al., 2009.
35   Douglas Adams: Adams, 1979.
35   Table 2.1: Jerison, 1973; Roth & Dicke, 2005.
36   (Footnote 12) absolute size is the better predictor: Deaner et al., 2007.
36   Andrew Whiten and I: Whiten & Suddendorf, 2007.
37   differences in internal organization: Preuss, 2000.
37   linearly scaled-up primate brains: Herculano-Houzel, 2009.
38   relatively smaller in humans: Holloway, 2008.
38   (Footnote 13) information flow has reversed: Noack, 2012.
38   first documented microscopic distinction: Preuss et al., 1999.
38   density is much higher in humans: Elston et al., 2006.
38   (Footnote 14) Von Economo neurons: Nimchinsky et al., 1999.

Chapter 3: Minds Comparing Minds

40   Daniel Dennett notes: Balter, 2012d.
40   for animal welfare: e.g., Lea, 2001; Wise, 2000.
40   "the senses and the intuitions: Darwin, 1871, p. 126.
41   trying to conceal the evidence: Lindsay, 1880.
41   Clever Hans: e.g., Wynne, 2001.
43   (Footnote 2) This case has been argued: Shettleworth, 2010.
43   (Footnote 3) mentally travel in time: Suddendorf & Corballis, 1997.
43   (Footnote 3) some killjoy explanations: Collier-Baker et al., 2004; Sud-
        dendorf & Corballis, 2008b.
43   (Footnote 3) chimpanzees can notice: Nielsen et al., 2005.
44   William James: James, 1890.
44   (Footnote 4) evidence that babies assess: Hamlin et al., 2007.
44   (Footnote 4) challenged by simpler explanations: Scarf et al., 2012.
45   various animals dream: Darwin, 1871.
45   pretend play in the second year: e.g., Leslie, 1987.
46   Kakama carried a log: Wrangham & Peterson, 1996.
46   Andrew Whiten and Dick Byrne: Whiten & Byrne, 1988.
47   Sue Savage-Rumbaugh: Savage-Rumbaugh, 1986.
47   The gorilla Koko: Patterson & Linden, 1981.
47   the chimpanzee Viki: Hayes, 1951.
47   stronger behavioral evidence: Leslie, 1987.

48 **neither extreme view should overlook inconvenient facts:** Whiten & Suddendorf, 2007.

49 **Jean Piaget:** Flavell, 1963.

49 **(Footnote 7): earlier than originally proposed:** Baillargeon, 1987.

50 **great ape genera have passed:** Call, 2001b; Collier-Baker et al., 2005.

50 **domestic dogs were one of:** e.g., Gagnon & Doré, 1994.

51 **dogs had "cheated":** Collier-Baker et al., 2004.

51 **(Footnote 8) marmosets and gibbons:** Mendes & Huber, 2004; Fedor et al., 2008.

51 **passing all of Piaget's object permanence tasks:** Call, 2001b; Collier-Baker & Suddendorf, 2006.

51 **Josep Call and others:** e.g., Call, 2004, 2006; Hill et al., 2011.

52 **research on mirror self-recognition:** e.g., Suddendorf & Butler, 2013.

52 **Darwin briefly described:** Darwin, 1877.

52 **Gordon Gallup developed:** Gallup, 1970.

52 **This experiment has been replicated:** For reviews, see Swartz et al., 1999; Tomasello & Call, 1997.

52 **(Footnote 9) gorillas pass:** e.g., Posada & Colell, 2007.

53 **By twenty-four months close to all:** e.g., Nielsen & Dissanayake, 2004.

53 **(Footnote 10) There is some variation:** Kaertner et al., 2012.

53 **(Footnote 10) Bedouin children:** Priel & Deschonen, 1986.

53 **human and chimpanzee infants:** Bard et al., 2006.

53 **baboons, capuchins, and macaques all fail:** Anderson & Gallup, 2011.

53 **conditioned pigeons:** Epstein et al., 1981.

54 **dolphins demonstrated mirror self-recognition:** Reiss & Marino, 2001; http://www.pnas.org/content/suppl/2001/05/02/101086398 .DC1/0863Movie2.mov.

54 **Two magpies and one elephant:** for magpies, see Prior et al. 2008 and http://www.youtube.com/watch?v=4mD8velB83w; for the elephant, see Plotnik et al., 2006 and http://www.pnas.org/content /suppl/2006/10/26/0608062103.DC1/08062Movie3.mov.

54 **we only have strong evidence:** Anderson & Gallup, 2011; Suddendorf & Butler, 2013.

54 **(Footnote 11): failed to make capuchin monkeys "pass":** Roma et al., 2007.

55 **(Footnote 12) not all passed the task:** Swartz et al., 1999.

55 **Bobtail squid:** Jones & Nishiguchi, 2004.

55 **On the richer side:** e.g. Gallup, 1998.

55 **associated with the emergence of self-conscious emotions:** Lewis et al., 1989.

55 **with the use of personal pronouns:** Lewis & Ramsay, 2004.

55 **Celia Heyes:** Heyes, 1994.

55 **Ulric Neisser:** Neisser, 1997.

55 **Josef Perner:** Perner, 1991.

56 **marked our participants' legs:** Nielsen et al., 2006.

56 **(Footnote 14) Visual self-recognition in live video:** Suddendorf et al., 2007.

56   (Footnote 14) Self-recognition in delayed videos: Povinelli et al.,
      1996; Suddendorf, 1999a.
56   (Footnote 14) mirrors and photos involves different: Butler et al., 2012.
57   in mirrors around the same time as: e.g., Lewis & Ramsay, 2004; Niel-
      sen & Dissanayake, 2004.
57   (Footnote 15): In a review of the research literature: Suddendorf &
      Whiten, 2001.
58   studies on gibbon self-recognition: Hyatt, 1998; Lethmate & Dücker,
      1973; Ujhelyi et al., 2000.
58   Over the course of a two-year: Suddendorf & Collier-Baker, 2009;
      http://rspb.royalsocietypublishing.org/content/suppl/2009/02/24
      /rspb.2008.1754.DC1/rspb20081754supp04.mpg.
62   mirror self-recognition evolved between: Suddendorf & Butler, 2013.

Chapter 4: Talking Apes

63   Thanks to words: Huxley, 1956, p. 83.
63   the bishop of Polignac: Corbey, 2005.
63   language is distinctly human: Corballis, 2003; Deacon, 1997; Hauser et
      al., 2002; Pinker, 1994.
64   animals also have: Hauser, 1996.
65   Symbols are *about* something: e.g., Perner, 1991.
66   develop this representational insight: e.g., DeLoache & Burns, 1994.
66   even twenty-four-month-olds can: Suddendorf, 2003.
67   known as forming a "meta-representation": e.g., Perner, 1991.
68   (Footnote 5) similar to the game Pictionary: Garrod et al., 2007.
69   (Footnote 6) less likely to suffer: Diamond, 2010.
71   Human languages are *generative*: e.g., Corballis, 2003.
72   Recursion is considered a key: e.g., Corballis, 2011; Hauser et al., 2002.
73   this generative grammar: For Chomsky's recent theorizing and the
      basic operation he now calls "merge," see Berwick et al., 2013.
73   defines the language faculty: Hauser et al., 2002; for a critique, see
      Jackendoff & Pinker, 2005; for the debate about recursion, see Corbal-
      lis, 2011.
73   (Footnote 10) Skinner had been enticed: Skinner, 1957.
73   predisposed to develop language: Pinker, 1994.
75   one hundred thousand years ago: Berwick et al., 2013.
75   not present in all human languages: Evans & Levinson, 2009; Everett,
      2005.
75   Part of the problem may be: Corballis, 2011.
75   computational models from evolutionary biology: Levinson & Gray,
      2012.
75   one study compared word order: Dunn et al., 2011.
76   Language is the source: de Saint-Exupéry, 1943.
76   Paul Grice: Grice, 1989.
77   (Footnote 11) I actually like the word: Suddendorf, 2008.
78   Michael Corballis and I have argued: Suddendorf et al., 2009b.

78 Friedrich Max Müller: Corballis, 2011; Radick, 2007.
78 "I wish someone would keep: Cited by Radick, 2007, p. 31.
78 Enter Richard Garner: Radick, 2007; Suddendorf et al., 2012.
79 evolution proceeding in leaps: Gould & Eldredge, 1977.
80 language first evolved in gestural form: e.g., Corballis, 2003.
80 Dorothy Cheney and Robert Seyfarth: Cheney & Seyfarth, 1990.
81 close examinations of communication systems: Hauser, 1996.
81 they learn the songs: Garland et al., 2011; Noad et al., 2000.
81 just enough to say: Smith et al., 2008.
81 alarm calls of prairie dogs: Slobodchikoff et al., 2009.
81 a hidden communication system: Mäthger et al., 2009.
81 (Footnote 12) Great apes have much more voluntary control: Premack, 2007.
82 I cried out a short and good "Hello!": Kafka, 1917.
82 control of the face and voice: Premack, 2007.
82 the African grey parrot Alex: Pepperberg, 1987.
83 Louis Herman: Herman et al., 1993.
83 Seals have also been trained: Schusterman & Gisiner, 1988.
83 A border collie, Rico: Kaminski et al., 2004.
83 Famous examples include: Washoe (Gardner & Gardner, 1969); Koko (Patterson & Linden, 1981); Chantek (Miles, 1994); Sarah (Premack & Premack, 1983); Kanzi (Savage-Rumbaugh et al., 1993).
83 Nim Chimpsky: Terrace, 1979.
83 dispute ensued: e.g., Savage-Rumbaugh et al., 1980; Fouts, 1997.
84 Valerie Kuhlmeier and Sally Boysen: Kuhlmeier & Boysen, 2002.
84 recent analysis of decades of data: Lyn et al., 2011.
84 (Footnote 14) Chimpanzees tend to have severe problems: Matsuzawa, 2009.
85 "Who are you?": Patterson, 1991.
85 The bonobo Kanzi: Savage-Rumbaugh et al., 1993.
85 (Footnote 15) Tamarins have been found: Fitch & Hauser, 2004.
85 (Footnote 15) starlings can learn recursive rules: Gentner et al. 2006; but see Corballis, 2007a.
86 Steven Pinker, for example, insists: Pinker, 1994.
86 Sally Boysen, for example, taught: Boysen & Hallberg, 2000; see also Matsuzawa, 2009.

Chapter 5: Time Travelers

89 Forethought is the most important: Russell, 1954, p. 179.
89 time travel will never become a reality: Holden, 2005.
90 Norbert Bischof: Bischof, 1985.
90 thesis into a monograph: Suddendorf & Corballis, 1997.
91 "It's a poor sort of memory: Carroll, 1871.
91 Clive Wearing: Wearing, 2005.
91 Endel Tulving: Tulving, 1985, 2005.
92 the ultimate function of this capacity: Suddendorf & Busby, 2005; Tulving, 2005.

92   Jennifer Thompson thought: Thompson-Cannino et al., 2009.
92   the reliability of eyewitness testimony: Loftus, 1992; Schacter, 1999.
93   Remembering episodes is a reconstructive process: Bartlett, 1932.
93   recall your own good behavior better: D'Argembeau & Van der Linden, 2008.
93   memory systems are inherently future-directed: Bar, 2011; Suddendorf & Corballis, 2007.
94   two sides of the same coin?: Schacter et al., 2007; Suddendorf & Corballis, 1997, 2007.
94   similar problems imagining future events: Klein et al., 2002; Tulving, 1985.
94   children's capacity to answer such questions: Busby & Suddendorf, 2005; Suddendorf, 2010b.
94   Introspectively, there are some: D'Argembeau & Van der Linden, 2004.
94   In old age we tend to report: Addis et al., 2008.
94   depressed and schizophrenic patients: Williams et al., 1996; D'Argembeau et al., 2008.
94   Brain imaging studies have found: Addis et al., 2007; Okuda et al., 2003.
94   there are some important differences: Suddendorf, 2010a.
94   imagine situations you have never experienced: Gilbert & Wilson, 2007; Suddendorf & Corballis, 2007.
95   Just as a theater production: Suddendorf & Corballis, 2007.
96   Take Ötzi: Suddendorf, 2006.
97   We differ also in how much we worry: Zimbardo & Boyd, 1999.
97   John Lennon sang: "Beautiful Boy (Darling Boy)."
99   shortcomings in any one of the components: Suddendorf & Corballis, 2007.
99   We can learn from others' memory: Social remembering can have negative and positive effects on memory accuracy, though both may be beneficial: Roediger & McDermott, 2011.
99   much of human conversation is: Szagun, 1978.
99   Stumbling on Happiness: Gilbert, 2006.
100   How parents talk to their children: Parent-child conversation and children's memory (e.g., Fivush et al., 2006) and future time concepts (Hudson 2006).
100   children begin to talk about past: e.g., Busby Grant & Suddendorf, 2011; Nelson & Fivush, 2004.
100   infantile amnesia: e.g., Bauer, 2007; Nelson & Fivush, 2004.
101   insist that they have always known it: Taylor et al., 1994.
101   In one study we told children stories: Busby Grant & Suddendorf, 2010.
101   We presented children with a curious puzzle: Suddendorf et al., 2011.
102   William Friedman: e.g., Friedman, 2005; see also Busby Grant & Suddendorf, 2009.
103   What's time?: Browning, 1896, p. 425.
104   psychologist William Roberts: Roberts, 2002.
104   The chimpanzee Panzee: Menzel, 2005.
104   Rats appear to use: e.g., Foster & Wilson, 2006; Tolman, 1948.

104   **Nicola Clayton, Anthony Dickinson:** e.g., Clayton & Dickinson, 1998; Clayton et al., 2001.
105   **such studies on other animals:** For reviews, see Dere et al., 2008; Suddendorf & Corballis, 2008a.
105   **if it walks like a duck:** Eichenbaum et al., 2005.
105   **episodic-like memory is neither necessary nor sufficient:** Suddendorf & Busby, 2003.
106   **should be able to control their future prudently:** Suddendorf & Corballis, 2010.
106   **Long-term planning:** Dawkins, 2000.
106   **bacteria demonstrate future-directed capacities:** Mitchell et al., 2009.
107   **The wasp always inspects the nest:** Fabre, 1915.
107   **taste predicts later sickness:** Garcia et al., 1966.
107   **cannot learn that a sound or a sight:** Garcia & Koelling, 1966.
108   **select a stick of the appropriate length:** Mulcahy et al., 2005.
108   **sometimes carry stones:** Boesch & Boesch, 1984.
108   **Norbert Bischof and Doris Bischof-Köhler:** Bischof, 1985; Bischof-Köhler, 1985.
108   **laboratory monkeys that were fed biscuits:** Roberts, 2002.
109   **delay gratification for several minutes:** Dufour et al., 2007.
109   **Perhaps the most prominent case:** Mulcahy & Call, 2006a. For critique of this evidence, see Suddendorf, 2006.
109   **(Footnote 6) One high-profile study:** Correia et al., 2007; see also Cheke & Clayton, 2012.
109   **(Footnote 6) In another study, two squirrel monkeys:** Naqshbandi & Roberts, 2006. For a killjoy critique, see Suddendorf & Corballis, 2008b.
110   **subsequent studies:** Osvath & Osvath, 2008. For a killjoy critique, see Suddendorf et al., 2009a.
110   **In another study ten chimpanzees:** Dufour & Sterck, 2008.
110   **An unusual report:** Osvath, 2009.

Chapter 6: Mind Readers

113   **Of all the species on Earth:** Zimmer, 2003, p. 1079.
114   **much as we do science:** Gopnik, 1993.
114   **simulating their experiences:** Gordon, 1996.
114   **requires mental scenario building:** Suddendorf & Corballis, 1997.
114   **(Footnote 1) "intentional stance":** Dennett, 1987.
115   **special affinity for social stimuli:** e.g., Moore, 2006.
115   **they prefer to look at open eyes:** Batki et al., 2000.
115   **The developmental psychologist Chris Moore:** Moore, 2013.
115   **infants start to point:** Liszkowski et al., 2004.
115   **motivated to keep making links:** e.g., Tomasello et al., 2005.
115   **(Footnote 2) blind children are typically delayed:** Peterson et al., 2000.
115   **(Footnote 3) in one study Israeli parents:** Feldman et al., 2006.
117   **Nested processes are also involved:** e.g., Dennett, 1987; Perner, 1991.

117   original paper on mental time: Suddendorf & Corballis, 1997.
117   William Hazlitt: Hazlitt, 1805, p. 1.
117   mental disorders are extreme versions: Baron-Cohen, 2002; Crespi &
         Badcock, 2008.
118   disorders of theory of mind: e.g., Baron-Cohen, 1995; Brüne & Brüne-
         Cohrs, 2006.
118   "On the lack of evidence": Penn & Povinelli, 2007.
118   David Premack and Guy Woodruff: Premack & Woodruff, 1978.
119   Heinz Wimmer and Josef Perner: Wimmer & Perner, 1983.
120   Extensive research on false-belief tasks: Wellman et al., 2001.
120   earlier in children who have older siblings: Ruffman et al., 1998.
120   better on language tasks: Astington & Jenkins, 1999.
120   deaf children: Peterson & Siegal, 2000.
120   attribute them to themselves: Gopnik & Astington, 1988.
120   how they come to know: O'Neill et al., 1992.
121   John Flavell: Flavell et al., 1983.
121   to lie is to knowingly implant a false belief: e.g., Suddendorf, 2011.
122   Robin Dunbar: Dunbar, 2007.
122   faux pas: Baron-Cohen et al., 1999.
123   manage the impressions we give: Tedeschi, 1981.
123   They form expectations: Csibra et al., 1999.
123   They copy what someone else intends: Meltzoff, 1995.
123   In fact, even toddlers: Clements & Perner, 1994; Onishi & Baillargeon,
         2005.
124   Ian Apperly and Stephen Butterfill: Apperly & Butterfill, 2009.
124   Henry Wellman, Candi Peterson, and colleagues: Shahaeian et al.,
         2011; Wellman & Liu, 2004.
124   shed tears to express: Humans are the only primates to shed tears:
         Bard, 2003.
124   (Footnote 5) Chimpanzee mothers and infants: Bard, 1994.
125   The work of primatologists like: e.g., Goodall, 1986.
125   tactical deceptions in primate societies: Whiten & Byrne, 1988.
126   Daniel Povinelli reported studies: Povinelli et al., 1990, 1992.
126   lean interpretations of their behavior: e.g., Povinelli & Eddy, 1996.
126   (Footnote 6) brain sizes have increased in tandem: Jerison, 1973.
127   capacity to reinterpret behavior in mental terms: Povinelli et al.,
         2000.
127   alternative explanation to this proposal: Suddendorf & Whiten, 2003.
127   (Footnote 7) chimpanzee aggressively chasing a female: de Waal,
         1986.
128   Michael Tomasello and Josep Call: e.g., Tomasello et al., 1999; Call et
         al., 1998.
128   Even dogs and monkeys: Emery, 2000.
128   In collaboration with Brian Hare: Hare et al., 2000.
128   Rhesus monkeys similarly: Flombaum & Santos, 2005.
129   In an extension of Hare: Hare et al., 2001.
129   some great apes appear to recognize: Tomasello & Carpenter, 2005.
129   distinguish accidental from purposeful: Call & Tomasello, 1998.

129   **discriminate between someone who is unwilling:** Call et al., 2004.
129   **distinguish appearance from reality:** Krachun et al., 2009a.
129   **when a competitor cannot see them:** Hare et al., 2006.
129   **Grey squirrels, for instance:** Leaver et al., 2007.
129   **Similarly, scrub jays:** Clayton et al., 2007. For a killjoy critique, see van der Vaart et al., 2012.
130   **no nonhuman animal has passed false-belief tasks:** Kaminski et al., 2008; Krachun et al., 2009b.
130   **no other animal has anything like a theory of mind:** Heyes, 1998; Penn & Povinelli, 2007.
130   **They may have a limited:** Call, 2001a.
131   **what they call "shared intentionality":** Herrmann et al., 2007; Tomasello et al., 2005.
131   **a collaborative task with an adult:** Warneken et al., 2006.
131   **poor at using and providing social cues:** Hare & Tomasello, 2004; Liszkowski et al., 2009.
131   **when the options are far apart:** Mulcahy & Call, 2009; Mulcahy & Suddendorf, 2011.
131   **virtually only to request:** Povinelli et al., 1997.
131   **only some 5 percent:** Lyn et al., 2011.
132   **A recent large-scale examination:** Herrmann et al., 2007.
132   **may not be comparable:** de Waal et al., 2008.
132   **they only reason about observables:** Penn & Povinelli, 2007.

Chapter 7: Smarter Apes

133   **Man is most uniquely:** Hoffer, 1973, p. 19.
134   **Bees use optic flow:** Chahl et al., 2004.
135   **Research on intelligence:** Neisser et al., 1996.
136   **"intelligence is what the tests test":** Boring, 1923.
136   **various indicators of "success":** Gottfredson, 1997.
136   **test scores have been increasing:** Flynn, 2000.
137   **resulting theories of intelligence:** Deary et al., 2010; Neisser et al., 1996.
137   **practical intelligence is quite distinct:** Sternberg, 1999.
137   **multiple intelligences:** Gardner, 1993.
138   **emotional intelligence:** Salovey & Mayer, 1990.
138   **Pinker offers the following definition:** Pinker, 1997, p. 62.
138   **(Footnote 4) William James called "having interest":** James, 1890.
139   **Man is a rational animal:** Russell, 2009, p. 45.
139   **numerous biases and heuristics:** Tversky & Kahneman, 1974.
140   **seven (plus or minus two) chunks:** Miller, 2003.
140   **a mere three to five chunks:** Cowan, 2001.
140   **working memory is the stage:** Suddendorf & Corballis, 2007.
140   **embedded processes are only possible:** Read, 2008.
140   **(Footnote 5) Alan Baddeley:** Baddeley, 1992, 2000.
141   **half of the variability in IQ:** Oberauer et al., 2005; Oberauer et al., 2008.
141   **steadily between ages four and eleven:** Alloway et al., 2006.

141   Graeme Halford: Halford et al., 2007; Halford et al., 1998.
141   and theory of mind: Gordon & Olsen, 1998.
141   crucial factor in human cognitive evolution: Balter, 2010.
141   (Footnote 6): Recent research suggests: Oberauer et al., 2008.
142   to *decontextualize*: Gerrans, 2007.
142   Robert Sternberg suggests: Sternberg, 1999.
143   The imagination is one: Darwin, 1871, p. 45.
143   recursion is a key mechanism: Corballis, 2011.
143   In so-called divergent-thinking: Suddendorf & Fletcher-Flinn, 1999.
144   Designing is the capacity: e.g., Suddendorf & Dong, 2013.
144   Some animals use tools: Bentley-Condit & Smith, 2010.
144   Queensland jumping spider: Wilcox & Jackson, 2002.
145   Embedded thinking: Suddendorf, 1999b.
145   J. David Smith and colleagues: Smith et al., 1995.
145   claim the middle ground: Smith et al., 2012.
146   Wolfgang Köhler's classic experiments: Köhler, 1917/1925.
146   In one study gorillas and orangutans: Mulcahy et al., 2005.
147   other species manufacture tools: Bentley-Condit & Smith, 2010.
147   use a tool to obtain another tool: Taylor et al., 2007.
147   Nathan Emery and Nicola Clayton: Emery & Clayton, 2004.
147   Ravens, for example, are capable: Heinrich, 1995.
147   Taylor and colleagues recently: Taylor et al., 2010.
148   They made the most elementary mistakes: Povinelli, 2000.
148   natural stick tools: Mulcahy et al., 2013.
148   Several other results suggest: e.g., Yocom & Boysen, 2011.
148   In one ingenious study: Mendes et al., 2007.
148   Rooks, corvids that are not known: Bird & Emery, 2009.
148   New Caledonian crows can learn: Taylor et al., 2011.
149   After ninety trials only one: Visalberghi & Limongelli, 1994.
149   Chimpanzees fared slightly better: Povinelli, 2000.
149   Povinelli and his colleagues Penn and Holyoak: Penn et al., 2008.
149   some chimpanzees can avoid the trap: Seed et al., 2009.
149   New Caledonian crows were recently shown: Taylor et al., 2009.
149   (Footnote 10) hidden humans as causal agents: Taylor et al., 2012.
150   One study also found that a chimpanzee: Gillan et al., 1981.
150   the argument by Povinelli and colleagues: Penn et al., 2008.
150   stark individual differences: Flemming et al., 2008.
150   such spontaneous inferences: Call, 2006.
150   Call placed food in one of two cups: Call, 2004.
152   Andrew Hill followed up: Hill et al., 2011.
153   Tetsuro Matsuzawa and colleagues: e.g., Kawai & Matsuzawa, 2000.
154   Ayumu could even beat humans: Inoue & Matsuzawa, 2007.
154   In follow-up research humans: Silberberg & Kearns, 2009.
154   Dwight Read: Read, 2008.
154   working-memory capacity in human evolution: Balter, 2010.
155   "laser-beam intelligence.": Cheney & Seyfarth, 1990; Premack, 2007.
155   "response breadth": Sterelny, 2003.
155   A classic study on zoo animals: Glickman & Sroges, 1966.

155 thirty-eight different ways: Whiten & Suddendorf, 2007.
156 (Footnote 14) Rates of behavioral innovation: Reader & Laland, 2003.
156 One study recorded the diversity: Parker, 1974a, 1974b.

Chapter 8: A New Heritage

157 The primary difference: Dennett, 1995, p. 331.
158 (Footnote 1) "linguistic relativity": e.g., Evans & Levinson, 2009.
159 Other animals cooperate: e.g., de Waal, 2005.
159 bacteria outnumber human cells: Sleator, 2010.
160 William D. Hamilton: Hamilton, 1964.
161 researchers used this exercise: Madsen et al., 2007.
161 Robert Trivers: Trivers, 1971.
161 (Footnote 4) the Cinderella effect: Daly & Wilson, 1988.
162 whether true altruism exists: e.g., Harman, 2010; Ridley, 1997.
162 (Footnote 5) Niko Tinbergen: Tinbergen, 1963.
163 may be called sociopaths: Mealey, 1995.
163 As Dawkins argued so persuasively: Dawkins, 1976.
163 (Footnote 6) cheater-detection mechanisms: e.g., Cosmides et al., 2005.
164 "indirect reciprocity": Haidt, 2007.
165 New forms of cultural learning: Tomasello, 1999, p. 526.
165 Cumulative culture has a role: e.g., Boyd et al., 2011; Dennett, 1995;
        Sterelny, 2003.
165 a second inheritance system: Whiten, 2005.
165 Richard Dawkins suggests: Dawkins, 1976.
166 debates about the precise similarities: Mesoudi et al., 2006; Sterelny,
        2003.
166 evolves in response to local demands: Sterelny, 2003.
166 (Footnote 9) Once the island of Tasmania: Flannery, 1994; Taylor, 2010.
167 a capacity for imitation from birth: Meltzoff & Moore, 1977; but see
        Suddendorf et al., 2013.
167 By nine months infants can: Meltzoff, 1988.
167 infants begin to imitate rationally: Gergely et al., 2002.
167 Mark Nielsen showed: Nielsen, 2006.
168 The psychiatrist Justin Williams: Williams et al., 2001.
168 (Footnote 10) Our proposal has garnered: e.g., Kana et al., 2011.
169 the chameleon effect: Chartrand & Bargh, 1999.
169 Research suggests that when: van Baaren et al., 2004.
169 (Footnote 11) behavioral synchrony is associated: Feldman, 2012.
169 a study on Mayan children: Maynard, 2002.
170 teaching appears to be a cross-cultural: Tomasello et al., 1993a.
171 We can pass on questions: Corballis & Suddendorf, 2010.
172 (Footnote 13) Elinor Ostrom: Ostrom, 2009.
172 reciprocal altruism requires: Stevens & Hauser, 2004.
172 Vampire bats: Denault & McFarlane, 1995.
172 Cuddlier primates depend: de Waal, 1989.
172 In one study chimpanzees: Melis et al., 2006a.
173 preferentially from high-status individuals: Horner et al., 2010.

173   In 1953 a Japanese macaque: Kawai, 1965.
173   psychic connections: Shermer, 1997.
174   diffusion experiments: Whiten & Mesoudi, 2008.
174   In one study researchers trained: Whiten et al., 2005.
174   In a recent study on orangutans: Dindo et al., 2011.
175   The systematic comparison has yielded: Whiten et al., 1999.
175   work on Sumatran orangutans: van Schaik et al., 2003.
175   Cetaceans, as well: Rendell & Whitehead, 2001.
175   New Caledonian crows: Holzhaider et al., 2010; Hunt & Gray, 2003.
176   different types of social learning: Whiten & Ham, 1992.
176   Even an octopus: Fiorito & Scotto, 1992.
176   Learning by copying others: Byrne & Russon, 1998.
176   (Footnote 14) often dubbed "emulation": e.g., Tennie et al., 2004.
177   chimpanzee and macaque infants: Ferrari et al., 2006; Myowa-
          Yamakoshi et al., 2004.
177   mirror neuron system was first: Rizzolatti et al., 1996.
177   Mike Noad observed: Noad et al., 2000.
177   in other humpback populations: Garland et al., 2011.
178   Off the coast of Brazil: Morete et al., 2003.
178   Louis Herman has shown: Herman, 2002.
178   notable exceptions are the great apes: Byrne & Russon, 1998; Russon &
          Galdikas, 1993.
178   mirror everything the chimpanzee: Nielsen et al., 2005.
178   documented in other great apes: Haun & Call, 2008.
178   chimpanzee Viki to copy: Hayes & Hayes, 1952.
178   "do as I do" paradigm: Byrne & Tanner, 2006; Custance et al., 1995;
          Miles et al., 1996.
178   Similar attempts with monkeys: Mitchell & Anderson, 1993.
178   (Footnote 15) One study did find: Paukner et al., 2009.
178   presented chimpanzees with a puzzle box: Whiten et al., 1996.
179   in other experiments chimpanzees: Whiten, 2005.
179   In a seminal study Victoria Horner: Horner & Whiten, 2005.
180   Figure 8.2: Photo reprinted from Nielsen & Widjojo, 2011, with permis-
          sion of Nova Science Publishers, Inc.
180   (Footnote 16) "transmission biases": e.g., Boyd et al., 2011.
181   Mark Nielsen recently examined: Nielsen & Tomaselli, 2010.
181   teaching in the animal kingdom: Hoppitt et al., 2008.
182   Adult whales have been observed: Guinet & Bouvier, 1995.
182   Christophe Boesch: Boesch, 1991.
183   In a recent study chimpanzees: Dean et al., 2012.

Chapter 9: Right and Wrong

185   Of all the differences between: Darwin, 1871, p. 97.
186   Johann Kremer: *Johann Kremers Tagebuch in Auszügen*, http://
          auschwitz-ag.org/unternehmen_auschwitz/6.2.7B.htm.
186   (Footnote 2) Alfred North Whitehead: Whitehead, 1956, p. 145.
187   According to Frans de Waal: de Waal, 2006.

187   "cold" and "hot" processes: e.g., McIlwain, 2003.
187   Michael Tomasello and colleagues: e.g., Tomasello, 2009.
188   clear signs of sympathy: Zahn-Waxler et al., 1979.
188   They initially require much prompting: Svetlova et al., 2010.
188   choices that avoid inequality: e.g., Fehr et al., 2008.
189   Thomas Hobbes observed: Pinker, 2011a.
190   what de Waal referred to as level 2: de Waal, 2006.
191   "In the last analysis, every kind: Einstein, 1950, p. 71.
191   Most groups prohibit: Haidt, 2007; Mikhail, 2007.
191   (Footnote 5) Game theory: Axelrod & Hamilton, 1981.
192   Ernst Fehr: e.g. Fehr & Gachter, 2002.
192   people are willing to punish: Fehr & Fischbacher, 2003; Henrich et al.,
          2006.
192   studies on hunter-gatherer societies: Hill et al., 2009.
192   experiments in economics: Fehr & Fischbacher, 2004.
192   We believe in a better world: e.g., Harre, 2011.
193   treat members of their own group differently: Hewstone et al., 2002.
193   rituals, ethnic signaling, and other: Hill et al., 2009.
194   Richard Shweder and colleagues: Shweder et al., 1987.
194   With the Enlightenment: Pinker, 2011a.
195   As man advances in civilisation: Darwin, 1871, pp. 122–123.
195   Universal Declaration of Human Rights: See http://www.un.org/en
          /documents/udhr/.
195   De Waal's third level: de Waal, 2006.
196   and later by Lawrence Kohlberg: Kohlberg, 1963.
196   imagine a situation in which you: Mikhail, 2007.
196   moral intuitions often precede: e.g., Haidt, 2007.
196   universal moral grammar: Mikhail, 2007; for a skeptical response, see
          Sterelny, 2010.
196   (Footnote 7) These draw in part: Trivers, 1971.
197   reasoning is marred by certain biases: Gilbert & Wilson, 2007.
197   motivated to choose future-directed actions: Suddendorf, 2011.
198   the marshmallow test: Mischel et al., 1989.
198   Differences in children's self-control: Casey et al., 2011.
198   (Footnote 8) Darwin wrote: Darwin, 1871, p.123.
200   in apparently self-deceptive ways: von Hippel & Trivers, 2011.
200   remember their own good behavior better: D'Argembeau & Van der
          Linden, 2008.
200   (Footnote 10) To call this self-deception: Pinker, 2011b.
201   The fact that man knows: Twain, 1906, p. vi.
201   (Footnote 11) A similar social explanation: Suddendorf, 2011.
202   Jane Goodall found: Goodall, 1986.
202   At 1710 Melissa, with: Goodall, 1986, p. 351.
203   Chimpanzee infanticide may happen: Murray et al., 2007.
204   comfort others who are suffering: de Waal, 1996; Goodall, 1986.
204   Researchers analyzed spontaneous: de Waal & Aureli, 1996.
204   Chimpanzees' physiological responses: Parr, 2001.

204   rodents are sensitive to pain of: Bartal et al., 2011; Rice & Gainer, 1962.
204   the gorilla was hand-reared: Silk, 2010.
205   stories of helping, such as Washoe: Fouts, 1997.
205   fundamentally good-natured: de Waal, 1996.
205   subsequently help each other: Yamamoto et al., 2009.
205   in one study chimpanzees: Melis et al., 2008.
205   Chimpanzees also help humans: Warneken & Tomasello, 2009.
205   Mothers rarely give: Ueno & Matsuzawa, 2004.
205   share because they are harassed: Gilby, 2006.
205   (Footnote 15) One study suggests that: Gomes & Boesch, 2009.
206   chimpanzees failed to help other: Silk et al., 2005.
206   study reported some prosocial choices: Horner et al., 2011.
206   marmosets, tamarins, and capuchin monkeys: e.g., Burkart et al., 2007.
206   Limits to sharing severely restricts: Melis et al., 2006b.
206   Interestingly, three-year-old: Hamann et al., 2011.
206   food and alarm calls: Cheney & Seyfarth, 1990; Hauser & Marler, 1993.
207   "a suite of interrelated: Bekoff & Pierce, 2009, p.7.
207   Recent work on rhesus macaques: Mahajan et al., 2011.
207   requires "shared intentionality,": Tomasello, 2009.
207   Shirley Strum: Strum, 2008; see also von Rohr et al., 2011.
208   Another prominent case: Brosnan & de Waal, 2003. De Waal (2006)
         discusses fairness in primates as part of level 1 morality.
208   work out of frustration: Roma et al., 2006; Wynne, 2004; but see van
         Wolkenten et al., 2007.
208   Other studies have found some: Brosnan et al., 2005; Range et al.,
         2009; but see Brauer et al., 2009.
208   primates breaking up fights: de Waal, 1996; Goodall, 1986.
208   "community concern": de Waal, 2006.
208   In one study rhesus monkeys: Hauser & Marler, 1993.
208   (Footnote 18) only humans blush: Darwin, 1873.
209   precursors of social norms: von Rohr et al., 2011.
209   A moral being is one: Darwin, 1871, p. 610.
209   level 3 clearly sets humans apart: Bekoff & Pierce, 2009; de Waal,
         2006.
209   normative self-government: Korsgaard, 2006.
210   lexigrams "good" and "bad": Lyn et al., 2008.
210   should be given legal personhood: Wise, 2000.
211   delay receiving a small reward: Dufour et al., 2007; Rosati et al., 2007.
211   In one study chimpanzees waited: Evans & Beran, 2007.
211   the Great Ape Project: Cavalieri & Singer, 1995.
212   in 1386 a court: Beirne, 1994.
212   needs and preferences into account: e.g., Stamp Dawkins, 2012.
212   Steven Pinker has documented: Pinker, 2011a.

Chapter 10: Mind the Gap

215   [Man] owes his success: Russell, 1954, p. 1.
216   rats can cognitively sweep ahead: Johnson & Redish, 2007.

217 **recorded during sleep and rest:** Wilson & McNaughton, 1994; Karlsson & Frank, 2009.

217 **the maze layout and its options:** Gupta et al., 2010.

217 **Great apes have a basic capacity to imagine:** Suddendorf & Whiten, 2001.

217 **not entirely unlike (adult) scientists:** Gopnik, 2012.

217 **They begin to deploy counterfactual reasoning:** Harris et al., 1996; Rafetseder et al., 2010.

218 **recursion is uniquely human:** Corballis, 2011.

218 **consciousness is a broadcasting system:** Baars, 2005.

218 **(Footnote 1) The theater metaphor does not:** see Dennett & Kinsbourne, 1992.

219 **improve the accuracy of their mental scenarios:** Gilbert, 2006.

219 **Humans have taken this sociality:** Frith & Frith, 2010.

219 **As Michael Tomasello and colleagues have:** Tomasello, 2009.

222 **sexual selection advantage:** Miller, 1998.

223 **on the evolution of our minds:** Nesse & Berridge, 1997.

224 **"enculturated," perform slightly better:** Tomasello et al., 1993b.

224 **No other animal passes:** Herschel, 1830, pp. 1–2.

224 **(Footnote 5) It is not entirely clear:** de Waal et al., 2008.

225 **Human brains grow more outside the womb:** DeSilva & Lesnik, 2006.

225 **Stephen Jay Gould has argued:** Gould, 1978.

225 **anthropologist Barry Bogin:** Bogin, 1999; Locke & Bogin, 2006.

225 **brain peaking in mass:** Cabana et al., 1993.

226 **Chimpanzee birth intervals:** Hill et al., 2009.

226 **cortical grey matter becomes thinner:** Huttenlocher, 1990.

226 **white matter increases:** Paus et al., 1999.

226 **Executive self-control gradually improves:** Luna et al., 2004.

227 **cooperative breeding:** Hill et al., 2009.

227 **(Footnote 7) some aspects of the brain:** e.g., Lebel et al., 2012.

228 **low survival rates:** Lancaster & Lancaster, 1983.

228 **While some great apes outlive:** Walker & Herndon, 2008.

228 **Grandmothers provide a range of support:** Hawkes, 2003; but see Hill et al., 2009.

228 **helps propagate genes:** Lahdenpera et al., 2004; Sear & Mace, 2008.

228 **surviving to age seventy and beyond:** Gurven & Kaplan, 2007.

228 **notions of wisdom:** Staudinger & Gluck, 2011.

229 **"the cognitive niche":** Pinker, 2010; Tooby & DeVore, 1987.

229 **humans frequently caused mass extinctions:** e.g., Holdaway & Jacomb, 2000; Turney et al., 2008.

229 **(Footnote 8) marmosets breed in small groups:** Tardif, 1997.

Chapter 11: The Real Middle Earth

232 **Australian Aboriginals have:** Flannery, 1994.

234 **Analyses of the Y chromosomes:** Thomson et al., 2000.

234 **Analyses of mitochondrial DNA:** Cann et al., 1987.

234 **The oldest anatomically modern:** McDougall et al., 2005.

234 **(Footnote 1) A recent study challenges:** Cruciani et al., 2011.

234   (Footnote 1) Genghis Khan: Zerjal et al., 2003.
235   Figure 11.1: Derevianko, 2012; Mellars, 2006; Soares et al., 2009; Stan-
           ford et al., 2013.
235   the birds rapidly went extinct: Holdaway & Jacomb, 2000.
235   the ecosystem collapsed: Diamond, 2005.
236   some interbreeding with the newcomers: Green et al., 2010; Reich et
           al., 2011.
236   *Australopithecus africanus*: Dart, 1925.
237   Some paleoanthropologists are more inclined: Robson & Wood, 2008.
237   A typical starting point: Suddendorf, 2004.
237   (Footnote 3) Even in biology there is: Groves, 2012b.
238   Figure 11.2: Groves, 2012a; Robson & Wood, 2008; Stanford et al., 2013.
239   initially split 7 million years ago: Patterson et al., 2006.
239   (Footnote 4) the aquatic ape theory: Morgan, 1982; for a critique, see
           Langdon, 2006.
240   more similar to that of the gorilla: Scally et al., 2012.
240   *Sahelanthropus tchadensis*: Brunet et al., 2002.
240   cranial capacity of: Unless otherwise specified hominin cranial capaci-
           ties throughout this chapter are based on Robson & Wood, 2008.
240   *Orronin tugenensis*: Senut et al., 2001.
240   *Ardipithecus kadabba*: Haile-Selassie, 2001.
240   *Ardipithecus ramidus*: White et al., 2009. For cranial capacity, see Suwa et
           al., 2009.
240   perhaps evolved independently: Kivell & Schmitt, 2009.
240   derived from bipedal clambering: Thorpe et al., 2007.
240   East Side Story: Coppens, 1994.
241   capable of power and precision grips: Young, 2003.
242   "man the hunted": Hart & Sussman, 2005.
242   famously called Lucy: Johanson, 2004.
243   the lunate sulcus: Dart, 1925.
243   fossilized footprints: Leakey & Hay, 1979.
243   (Footnote 7) In a personal history: Holloway, 2008.
243   (Footnote 8) 3.4-million-year-old foot: Haile-Selassie et al., 2012.
244   The pubic louse, DNA comparison: Reed et al., 2007.
244   louse that lives in clothes: Toups et al., 2011.
244   3.39-million-year-old bones: McPherron et al., 2010; but see
           Dominguez-Rodrigo et al., 2012.
244   oldest stone tools: Semaw, 2000; Dominguez-Rodrigo et al., 2005.
244   *Australopithecus garhi*: Asfaw et al., 1999.
244   *Australopithecus bahrelghazali*: Brunet et al., 1996.
245   *Kenyanthropus platyops*: Leakey et al., 2001.
245   *Australopithecus sediba*: Berger et al., 2010.
246   probably ate a variety of foods: Ungar & Sponheimer, 2011.
246   appeared around 2.4 million years ago: Stedman et al., 2004.
246   *Homo habilis*: Leakey et al., 1964.
247   1.95-million-year-old butchering site: Braun et al., 2010.
247   tools were of good throwing size: Cannell, 2002.
247   the hand of early *Homo*: Young, 2003.

248   *Homo rudolfensis*: McHenry & Coffing, 2000.
248   **really began with *Homo erectus*:** Wood & Collard, 1999.
248   **Their footprints are essentially:** Bennett et al., 2009.
248   **Nariokotome Boy:** Brown et al., 1985.
249   **more quickly than modern humans:** Graves et al., 2010.
249   **(Footnote 11) split into several species:** e.g., Groves, 2012a.
249   **(Footnote 12) Robson and Wood:** Robson & Wood, 2008.
250   **consumed more meat:** Stanford et al., 2013.
250   **efficient long-distance running:** Bramble & Lieberman, 2004.
250   **Reduced fur and whole-body:** Ruxton & Wilkinson, 2011.
250   **grandmother effect:** O'Connell et al., 1999.
250   **(Footnote 13) Overheating has also been:** Falk, 1990.
251   **cooking of food was the crucial step:** Wrangham, 2009.
251   **evidence for early fire use:** Goren-Inbar et al., 2004; Berna et al., 2012.
251   **(Footnote 15) evidence of heat treatment:** Brown et al., 2009.
252   **they are evident in Spain:** Carbonell et al., 2008.
252   **40,000 and 70,000 years ago:** Yokoyama et al., 2008.
252   **27,000 years ago:** Swisher et al., 1996.
252   **diversity of foods:** Ungar & Sponheimer, 2011.
252   **(Footnote 16) The role of refugia:** Stewart & Stringer, 2012.
253   **1.76 million years ago:** Lepre et al., 2011.
253   **tools became thinner and more:** Stout, 2011.
254   **aimed throwing to bring down prey:** Calvin, 1982.
254   **look like a Broca's area:** Wu et al., 2011.
254   **Robin Dunbar highlights the role:** Dunbar, 1992, 1996.
255   **skull from Dmanisi:** Lordkipanidze et al., 2005.
255   **(Footnote 20) possibly shared intentionality:** Shipton, 2010.
256   *Homo antecessor*: deCastro et al., 1997.
256   **cooperated to bring down big game:** Villa & Lenoir, 2009.
256   **four-hundred-thousand-year-old spears:** Thieme, 1997; see also
            Churchill & Rhodes, 2009.
256   **In 2010 evidence for stone blades:** Johnson & McBrearty, 2010.
256   **From around three hundred thousand years ago:** Ambrose, 2001;
            Lombard, 2012.
257   **early as five hundred thousand years ago:** Wilkins et al., 2012.
257   **over two hundred fossils:** Stanford et al., 2013; Trinkaus, 1995.
258   **pale skin and red hair:** Lalueza-Fox et al., 2007.
258   **up to western Siberia:** Krause et al., 2007b.
258   **including fish and dolphins:** Stringer et al., 2008.
258   **Neanderthals practiced cannibalism:** Defleur et al., 1999.
258   **draft sequence of their genome:** Green et al., 2010.
258   **DNA extracted from thirteen:** Dalen et al., 2012.
260   **they cared for the sick:** Dettwyler, 1991.
260   **Neanderthal burials:** Langley et al., 2008; but see Balter, 2012a.
260   **having made composite tools:** Langley et al., 2008.
260   **grown up more quickly:** Ramirez Rozzi & Bermudez De Castro, 2004.
260   **worn shells as jewelry:** Zilhao et al., 2010.
260   **Archaeologists are hotly debating:** Balter, 2012b.

260   **Philip Lieberman has long argued:** Lieberman, 1991.
260   **Neanderthal hyoid bone:** Arensburg et al., 1990.
260   **hyoid bone from an earlier archaic:** Martinez et al., 2008.
260   **hyoid bone of an *Australopithecus aferensis*:** Alemseged et al., 2006.
261   **version of a gene strongly implicated:** Krause et al., 2007a.
261   **gesturing preceded talking:** Corballis, 2003.
261   **modern human life history:** Smith et al., 2007.
261   **evidence for a fully modern human mind:** e.g., d'Errico & Stringer, 2011; Lombard, 2012.
261   **at Skuhl Cave:** Grun et al., 2005.
261   **82,000-year-old shell beads:** Bouzouggar et al., 2007.
261   **long-range projectile weapons:** Churchill & Rhodes, 2009.
261   **(Footnote 21) A mutation of the FOXP2 gene:** Enard et al., 2002.
262   **compound adhesives:** Wadley, 2010.
262   **40,000 years ago artifacts:** Balter, 2012c.
262   **prehistoric cave of Geissenklösterle:** Balter, 2012b.
263   **behavioral complexity increased:** Langley et al., 2008.
263   **In some instances the technologies:** d'Errico & Stringer, 2011.
263   **The archaeological record complicates:** Lombard, 2012.
263   **fossil teeth from over 700 individuals:** Caspari & Lee, 2004.
264   **appear to have interbred:** Reich et al., 2011.
264   ***Homo floresiensis*, tiny hominins:** Brown et al., 2004.
264   **debates about whether these specimens:** e.g., Brown, 2012; Oxnard et al., 2010.
264   **(Footnote 22) There are some problems:** O'Connell et al., 1999.
264   **(Footnote 23) In 2012, strange:** Curnoe et al., 2012.
265   **"at all times throughout the world:** Darwin, 1871, p. 132.
265   **conflict might select:** e.g., Bowles, 2009.
265   **(Footnote 25) Darwin wrote:** Darwin, 1871, p. 132.
265   **(Footnote 25) group selection continues to be:** e.g., see http://edge.org/conversation/the-false-allure-of-group-selection#edn9.
266   **oldest indication of violence:** Keeley, 1996; Thorpe, 2003.
266   **northern Iraq (Shanidar 3):** Churchill et al., 2009.
266   **The jaw of a 30,000-year-old:** Ramirez Rozzi et al., 2009.
266   **paradigm shift:** Gibbons, 2011.
266   **African data also suggest interbreeding:** Hammer et al., 2011.

Chapter 12: Quo Vadis?

269   **eight times the biomass:** Smil, 2002.
270   **the first script was:** Schmandt-Besserat, 1992.
270   **(Footnote 2) Those who continued:** Brody, 2000.
271   **historical novel *Creation*:** Vidal, 1981.
271   **(Footnote 3) Stanislas Dehaene:** Dehaene, 2009.
272   **"The library connects us:** Sagan, 1980, p. 282.
273   ***Philosophical Transactions of the Royal Society*:** See http://royalsociety publishing.org/journals/.
274   ***Systema Naturae*:** Linnaeus, 1758.

275 **to make models and predictions:** e.g., Borjeson et al., 2006.
275 **recognized as global problems:** Diamond, 2005; Singer, 2002.
278 **Darwinian man:** Gilbert & Sullivan, 2010.
278 **critical to complement the genetic data:** Varki et al., 1998.
278 **narrow down the search space:** Suddendorf & Butler, 2013.
280 **genetic changes in human populations:** Milot et al., 2011.
280 **brain sizes *decreased*:** McAuliffe, 2010.
280 **brain size and IQ:** McDaniel, 2005.
280 **"I do not know with what:** Johnson, 2005.
281 **Countless civilizations have collapsed:** Diamond, 2005.
281 **we have domesticated ourselves:** e.g., Hare et al., 2012.
281 **overall decline in violence:** Pinker, 2011a.
282 **Some evidence suggests that brain size:** Jantz, 2001.
283 **Plan it for the apes:** See the website of the Great Apes Survival Partnership (GRASP): http://www.un-grasp.org.

R E F E R E N C E S

Adams, D. (1979). *The hitchhiker's guide to the galaxy*. London: Pan.

Addis, D. R., et al. (2007). Remembering the past and imagining the future: Common and distinct neural substrates during event construction and elaboration. *Neuropsychologia, 45*, 1363–1377.

Addis, D. R., et al. (2008). Age-related changes in the episodic simulation of future events. *Psychological Science, 19*, 33–41.

Alemseged, Z., et al. (2006). A juvenile early hominin skeleton from Dikika, Ethiopia. *Nature, 443*, 296–301.

Alexander, R. D. (1989). Evolution of the human psyche. In P. Mellars & C. Stringer (Eds.). *The human revolution: behavioral and biological perspectives on the origins of modern humans* (pp. 455–513). Princeton, NJ: Princeton University Press.

Alloway, T. P., et al. (2006). Verbal and visuospatial short-term and working memory in children: Are they separable? *Child Development, 77*, 1698–1716.

Ambrose, S. H. (2001). Paleolithic technology and human evolution. *Science, 291*, 1748–1753.

Anderson, J. R., & Gallup, G. G. (2011). Which primates recognize themselves in mirrors. *PLOS Biology, 9*, 1–3.

Apperly, I. A., & Butterfill, S. A. (2009). Do humans have two systems to track beliefs and belief-like states? *Psychological Review, 116*, 953–970.

Arensburg, B., et al. (1990). A reappraisal of the anatomical basis for speech in middle paleolithic hominids. *American Journal of Physical Anthropology, 83*, 137–146.

Asfaw, B., et al. (1999). Australopithecus garhi: A new species of early hominid from Ethiopia. *Science, 284*, 629–635.

Astington, J. W., & Jenkins, J. M. (1999). A longitudinal study of the relation between language and theory-of-mind development. *Developmental Psychology, 35*, 1311–1320.

Axelrod, R., & Hamilton, W. D. (1981). The evolution of cooperation. *Science, 211*, 1390–1396.

Azevedo, F. A. C., et al. (2009). Equal numbers of neuronal and nonneuronal cells make the human brain an isometrically scaled-up primate brain. *Journal of Comparative Neurology, 513*, 532–541.

Baars, B. J. (2005). Global workspace theory of consciousness: Toward a cognitive neuroscience of human experience, *Progress in Brain Research* (Vol. 150, pp. 45–53).

Baddeley, A. (1992). Working memory. *Science, 255,* 556–559.

Baddeley, A. (2000). The episodic buffer: a new component of working memory? *Trends in Cognitive Sciences, 4,* 417–423.

Baillargeon, R. (1987). Object permanence in 3 1/2-month-old and 4 1/2-month-old infants. *Developmental Psychology, 23,* 655–664.

Balter, M. (2010). Did working memory spark creative culture? *Science, 328,* 160–163.

Balter, M. (2012a). Did Neandertals truly bury their dead? *Science, 337,* 1443–1444.

Balter, M. (2012b). Early dates for artistic Europeans. *Science, 336,* 1086–1087.

Balter, M. (2012c). Ice age tools hint at 40,000 years of bushman culture. *Science, 337,* 512.

Balter, M. (2012d). 'Killjoys' challenge claims of clever animals. *Science, 335,* 1036–1037.

Bar, M. (2011). *Predictions in the brain: Using our past to generate a future.* Oxford: Oxford University Press.

Bard, K. A. (1994). Evolutionary roots of intuitive parenting: Maternal competence in chimpanzees. *Early Development and Parenting, 3,* 19–28.

Bard, K. A. (2003). Are humans the only primates that cry? *Scientific American, June 16.*

Bard, K. A., et al. (2006). Self-awareness in human and chimpanzee infants: What is measured and what is meant by the mark and mirror test. *Infancy, 9,* 191–219.

Barkow, J. H., et al. (Eds.). (1992). *The adapted mind: Evolutionary psychology and the generation of culture.* Oxford: Oxford University Press.

Baron-Cohen, S. (1995). *Mindblindness: An essay on autism and theory of mind.* Cambridge, Mass.: Bradford/MIT Press.

Baron-Cohen, S. (2002). The extreme male brain theory of autism. *Trends in Cognitive Sciences, 6,* 248–254.

Baron-Cohen, S., et al. (1999). Recognition of faux pas by normally developing children and children with Asperger syndrome or high-functioning autism. *Journal of Autism and Developmental Disorders, 29,* 407–418.

Bartal, I. B.-A., et al. (2011). Empathy and pro-social behavior in rats. *Science, 334,* 1427–1430.

Bartlett, F. C. (1932). *Remembering: A study in experimental and social psychology.* Cambridge: Cambridge University Press.

Bateson, P. (1991). Assessment of pain in animals. *Animal Behaviour, 42,* 827–839.

Batki, A., et al. (2000). Is there an innate gaze module? Evidence from human neonates. *Infant Behavior and Development, 23,* 223–229.

Bauer, P. (2007). *Remembering the times of our lives: memory in infancy and beyond.* Mahwah, NJ: Laurence Erlbaum Associates.

Beirne, P. (1994). The law is an ass: Reading E. P. Evans' "The medieval prosecution and capital punishment of animals." *Society and Animals, 2,* 27–46.

Bekoff, M., & Pierce, J. (2009). *Wild justice: The moral lives of animals.* Chicago: University of Chicago Press.

Bennett, M. R., et al. (2009). Early hominin foot morphology based on 1.5-million-year-old footprints from Ileret, Kenya. *Science, 323,* 1197–1201.

Bentley-Condit, V. K., & Smith, E. O. (2010). Animal tool use: current definitions and an updated comprehensive catalog. *Behaviour, 147*, 185–221.

Berger, L. R., et al. (2010). *Australopithecus sediba*: A new species of Homo-like Australopith from South Africa. *Science, 328*, 195–204.

Berna, F., et al. (2012). Microstratigraphic evidence of in situ fire in the Acheulean strata of Wonderwerk Cave, Northern Cape province, South Africa. *Proceedings of the National Academy of Sciences of the United States of America, 109*, E1215–E1220.

Berwick, R. C., et al. (2013). Evolution, brain, and the nature of language. *Trends in Cognitive Sciences, 17*, 89–98.

Bird, C. D., & Emery, N. J. (2009). Rooks use stones to raise the water level to reach a floating worm. *Current Biology, 19*, 1410–1414.

Bischof, N. (1985). *Das Rätzel Ödipus [The Oedipus riddle]*. Munich: Piper.

Bischof-Köhler, D. (1985). Zur Phylogenese menschlicher Motivation [On the phylogeny of human motivation]. In L. H. Eckensberger & E. D. Lantermann (Eds.), *Emotion und Reflexivität* (pp. 3–47). Vienna: Urban & Schwarzenberg.

Blair, R. J. R. (2001). Neurocognitive models of aggression, the antisocial personality disorders, and psychopathy. *Journal of Neurology Neurosurgery and Psychiatry, 71*, 727–731.

Boesch, C. (1990). Tool use and tool making in wild chimpanzees. *Folia Primatologica, 54*, 86–99.

Boesch, C. (1991). Teaching among wild chimpanzees. *Animal Behaviour, 41*, 530–532.

Boesch, C. (1994). Chimpanzees-red colobus monkeys: A predator-prey system. *Animal Behaviour, 47*, 1135–1148.

Boesch, C., & Boesch, H. (1984). Mental map in wild chimpanzees: An analysis of hammer transports for nut cracking. *Primates, 25*, 160–170.

Bogin, B. (1999). Evolutionary perspective on human growth. *Annual Review of Anthropology, 28*, 109–153.

Boring, E. G. (1923). Intelligence as the test tests it. *New Republic, 35*, 35–37.

Borjeson, L., et al. (2006). Scenario types and techniques: Towards a user's guide. *Futures, 38*, 723–739.

Bouzouggar, A., et al. (2007). 82,000-year-old shell beads from North Africa and implications for the origins of modern human behavior. *Proceedings of the National Academy of Sciences of the United States of America, 104*, 9964–9969.

Bowles, S. (2009). Did warfare among ancestral hunter-gatherers affect the evolution of human social behaviors? *Science, 324*, 1293–1298.

Boyd, R., et al. (2011). The cultural niche: Why social learning is essential for human adaptation. *Proceedings of the National Academy of Sciences of the United States of America, 108*, 10918–10925.

Boysen, S. T., & Hallberg, K. I. (2000). Primate numerical competence: Contributions toward understanding nonhuman cognition. *Cognitive Science, 24*, 423–444.

Bramble, D. M., & Lieberman, D. E. (2004). Endurance running and the evolution of Homo. *Nature, 432*, 345–352.

Brauer, J., et al. (2009). Are apes inequity averse? New data on the token-exchange paradigm. *American Journal of Primatology, 71*, 175–181.

Braun, D. R., et al. (2010). Early hominin diet included diverse terrestrial and aquatic animals 1.95 Ma in East Turkana, Kenya. *Proceedings of the National Academy of Sciences of the United States of America, 107*, 10002–10007.

Breuer, T., et al. (2005). First observation of tool use in wild gorillas. *PLOS Biology, 3*, 2041–2043.

Brody, H. (2000). *The other side of Eden: Hunters, farmers and the shaping of the world*. London: Faber & Faber.

Brosnan, S. F., & de Waal, F. B. M. (2003). Monkeys reject unequal pay. *Nature, 425*, 297–299.

Brosnan, S. F., et al. (2005). Tolerance for inequity may increase with social closeness in chimpanzees. *Proceedings of the Royal Society B-Biological Sciences, 272*, 253–258.

Brown, F., et al. (1985). Early *Homo erectus* skeleton from West Lake Turkana, Kenya. *Nature, 316*, 788–792.

Brown, K. S., et al. (2009). Fire as an engineering tool of early modern humans. *Science, 325*, 859–862.

Brown, P. (2012). LB1 and LB6 *Homo floresiensis* are not modern human (Homo sapiens) cretins. *Journal of Human Evolution, 62*, 201–224.

Brown, P., et al. (2004). A new small-bodied hominin from the Late Pleistocene of Flores, Indonesia. *Nature, 431*, 1055–1061.

Browning, R. (1896). *The poetical works*. London: Smith, Elder & Co.

Brüne, M., & Brüne-Cohrs, U. (2006). Theory of mind-evolution, ontogeny, brain mechanisms and psychopathology. *Neuroscience and Biobehavioral Reviews, 30*, 437–455.

Brunet, M., et al. (1996). *Australopithecus bahrelghazali*, a new species of early hominid from Koro Toro region, Chad. *Comptes Rendus De L Academie Des Sciences Serie Ii Fascicule a—Sciences De La Terre Et Des Planetes, 322*, 907–913.

Brunet, M., et al. (2002). A new hominid from the Upper Miocene of Chad, Central Africa. *Nature, 418*, 145–151.

Burkart, J. M., et al. (2007). Other-regarding preferences in a non-human primate: Common marmosets provision food altruistically. *Proceedings of the National Academy of Sciences of the United States of America, 104*, 19762–19766.

Busby, J., & Suddendorf, T. (2005). Recalling yesterday and predicting tomorrow. *Cognitive Development, 20*, 362–372.

Busby Grant, J., & Suddendorf, T. (2009). Preschoolers begin to differentiate the times of events from throughout the lifespan. *European Journal of Developmental Psychology, 6*, 746–762.

Busby Grant, J., & Suddendorf, T. (2010). Young children's ability to distinguish past and future changes in physical and mental states. *British Journal of Developmental Psychology, 28*, 853–870.

Busby Grant, J., & Suddendorf, T. (2011). Production of temporal terms by 3-, 4-, and 5-year-old children. *Early Childhood Research Quarterly, 26*, 87–95.

Buss, D. M. (1999). *Evolutionary Psychology: The new science of the mind*. Boston: Allyn and Bacon.

Butler, D. L., et al. (2012). Mirror, mirror on the wall, how does my brain recognize my image at all? *PLOS One, 7*, e31452.

Byrne, R. W., & Russon, A. E. (1998). Learning by imitation: A hierarchical approach. *Behavioral and Brain Sciences, 21*, 667–721.

Byrne, R. W., & Tanner, J. (2006). Gestural imitation by a gorilla. *International Journal of Psychology and Psychological Therapy, 6*, 215–231.

Cabana, T., et al. (1993). Prenatal and postnatal growth and allometry of stature, head circumference, and brain weight in Quebec children. *American Journal of Human Biology, 5*, 93–99.

Call, J. (2001a). Chimpanzee social cognition. *Trends in Cognitive Sciences, 5*, 388–393.

Call, J. (2001b). Object permanence in orangutans (*Pongo pygmaeus*), chimpanzees (*Pan troglodytes*), and children (*Homo sapiens*). *Journal of Comparative Psychology, 115*, 159–171.

Call, J. (2004). Inferences about the location of food in the great apes (*Pan pansicus, Pan troglodytes, Gorilla gorilla*, and *Pongo pygmaeus*). *Journal of Comparative Psychology, 118*, 232–241.

Call, J. (2006). Inferences by exclusion in the great apes: The effect of age and species. *Animal Cognition, 9*, 393–403.

Call, J., & Tomasello, M. (1998). Distinguishing intentional from accidental actions in orangutans *(Pongo pygmaeus)*, chimpanzees *(Pan troglodytes)* and human children *(Homo sapiens)*. *Journal of Comparative Psychology, 112*, 192–206.

Call, J., et al. (1998). Chimpanzee gaze following in an object choice task. *Animal Cognition, 1*, 89–99.

Call, J., et al. (2004). 'Unwilling' versus 'unable': Chimpanzees' understanding of human intentional action. *Developmental Science, 7*, 488–498.

Calvin, W. H. (1982). Did throwing stones shape hominid brain evolution? *Ethology and Sociobiology, 3*, 115–124.

Cann, R. L., et al. (1987). Mitochondrial DNA and human evolution. *Nature, 325*, 31–36.

Cannell, A. (2002). Throwing behaviour and the mass distribution of geological hand samples, hand grenades and Olduvian manuports. *Journal of Archaeological Science, 29*, 335–339.

Carbonell, E., et al. (2008). The first hominin of Europe. *Nature, 452*, 465–469.

Carroll, L. (1871). *Through the looking glass*. London: Macmillan.

Casey, B. J., et al. (2011). Behavioral and neural correlates of delay of gratification 40 years later. *Proceedings of the National Academy of Sciences of the United States of America, 108*, 14998–15003.

Caspari, R., & Lee, S. H. (2004). Older age becomes common late in human evolution. *Proceedings of the National Academy of Sciences of the United States of America, 101*, 10895–10900.

Cavalieri, P., & Singer, P. (1995). *The great ape project: Equality beyond humanity*. New York: St. Martin's Griffin.

Chahl, J. S., et al. (2004). Landing strategies in honeybees and applications to uninhabited airborne vehicles. *The International Journal of Robotics Research, 23*, 101–110.

Chartrand, T. L., & Bargh, J. A. (1999). The Chameleon effect: The perception-behavior link and social interaction. *Journal of Personality and Social Psychology, 76*, 893–910.

Cheke, L. G., & Clayton, N. S. (2012). Eurasian jays (*Garrulus glandarius*) overcome their current desires to anticipate two distinct future needs and plan for them appropriately. *Biology Letters, 8*, 171–175.

Cheney, D. L., & Seyfarth, R. M. (1980). Vocal recognition in free-ranging vervet monkeys. *Animal Behaviour, 28*, 362–376.

Cheney, D. L., & Seyfarth, R. M. (1990). *How monkeys see the world*. Chicago: University of Chicago Press.

Churchill, S. E., & Rhodes, J. A. (2009). *The evolution of the human capacity for "killing at a distance."* Dordrecht: Springer.

Churchill, S. E., et al. (2009). Shanidar 3 Neandertal rib puncture wound and paleolithic weaponry. *Journal of Human Evolution, 57*, 163–178.

Ciochon, R. L. (1996). Dated co-occurrence of *Homo erectus* and Gigantopithecus from Tham Khuyen Cave, Vietnam. *Proceedings of the National Academy of Sciences of the United States of America, 93*, 3016–3020.

Clayton, N. S., & Dickinson, A. (1998). Episodic-like memory during cache recovery by scrub jays. *Nature, 395*, 272–278.

Clayton, N. S., et al. (2001). Elements of episodic-like memory in animals. *Philosophical Transactions: Royal Society of London, B, 356*, 1483–1491.

Clayton, N. S., et al. (2007). Social cognition by food-caching corvids: The western scrub-jay as a natural psychologist. *Philosophical Transactions of the Royal Society B-Biological Sciences, 362*, 507–522.

Clements, W. A., & Perner, J. (1994). Implicit understanding of belief. *Cognitive Development, 9*, 377–395.

Collier-Baker, E., & Suddendorf, T. (2006). Do chimpanzees (*Pan troglodytes*) and 2-year-old children (*Homo Sapiens*) understand double invisible displacement? *Journal of Comparative Psychology, 120*, 89–97.

Collier-Baker, E., et al. (2004). Do dogs (*Canis familiaris*) understand invisible displacement? *Journal of Comparative Psychology, 118*, 421–433.

Collier-Baker, E., et al. (2005). Do chimpanzees (*Pan troglodytes*) understand single invisible displacement? *Animal Cognition, 9*, 55–61.

Coppens, Y. (1994). East side story: The origin of humankind. *Scientific American, 270*, 62–69.

Corballis, M. C. (2003). *From hand to mouth: The origins of language*. New York: Princeton University Press.

Corballis, M. C. (2007a). Recursion, language, and starlings. *Cognitive Science, 31*, 697–704.

Corballis, M. C. (2007b). The uniqueness of human recursive thinking. *American Scientist, 95*, 242–250.

Corballis, M. C. (2011). *The recursive mind: The origins of human language, thought, and civilization*. Princeton, NJ: Princeton University Press.

Corballis, M. C., & Suddendorf, T. (2010). The evolution of concepts: A timely look. In D. Marshal et al. (Eds.), *The making of human concepts* (pp. 365–389). Oxford: Oxford University Press.

Corbey, R. (2005). *The metaphysics of apes*. Cambridge: Cambridge University Press.

Correia, S. P. C., et al. (2007). Western scrub-jays anticipate future needs independently of their current motivational state. *Current Biology, 17*, 856–861.

Cosmides, L., et al. (2005). Detecting cheaters. *Trends in Cognitive Sciences, 9*, 505–506.

Cowan, N. (2001). The magical number 4 in short-term memory: A reconsideration of mental storage capacity. *Behavioral and Brain Sciences, 24,* 87–114.

Crespi, B., & Badcock, C. (2008). Psychosis and autism as diametrical disorders of the social brain. *Behavioral and Brain Sciences, 31,* 241–261.

Cruciani, F., et al. (2011). A revised root for the human Y chromosomal phylogenetic tree: The origin of patrilineal diversity in Africa. *The American Journal of Human Genetics, 88,* 814–818.

Csibra, G., et al. (1999). Goal attribution without agency cues: The perception of 'pure reason' in infancy. *Cognition, 72,* 237–267.

Curnoe, D., et al. (2012). Human remains from the Pleistocene-Holocene transition of Southwest China suggest a complex evolutionary history for East Asians. *PLOS One, 7,* e31918.

Custance, D. M., et al. (1995). Can young chimpanzees imitate arbitrary actions? Hayes and Hayes (1952) revisited. *Behaviour, 132,* 839–858.

D'Argembeau, A., & Van der Linden, M. (2004). Phenomenal characteristics associated with projecting oneself back into the past and forward into the future: Influence of valence and temporal distance. *Consciousness and Cognition, 13,* 844–858.

D'Argembeau, A., & Van der Linden, M. (2008). Remembering pride and shame: Self-enhancement and the phenomenology of autobiographical memory. *Memory, 16,* 538–547.

D'Argembeau, A., et al. (2008). Remembering the past and imagining the future in schizophrenia. *Journal of Abnormal Psychology, 117,* 247–251.

d'Errico, F., & Stringer, C. B. (2011). Evolution, revolution or saltation scenario for the emergence of modern cultures? *Philosophical Transactions of the Royal Society B-Biological Sciences, 366,* 1060–1069.

Dalen, L., et al. (2012). Partial turnover in Neandertals: Continuity in the East and population replacement in the West. *Molecular Biology and Evolution, 29,* 1893–1897.

Daly, M., & Wilson, M. A. (1988). Evolutionary social psychology and family homicide. *Science, 242,* 519–524.

Dart, R. A. (1925). Australopithecus africanus: The man-ape of South Africa. *Nature, 115,* 195–199.

Darwin, C. (1859). *On the origin of species.* Cambridge, MA: Harvard University Press.

Darwin, C. (1871). *The descent of man, and selection in relation to sex* (2003 ed.). London: Gibson Square Books.

Darwin, C. (1873). *The expressions of the emotions in man and animal.* London: Murray.

Darwin, C. (1877). A biographical sketch of an infant. *Mind, 2,* 285–294.

Dawkins, R. (1976). *The selfish gene.* Oxford: Oxford University Press.

Dawkins, R. (2000). An open letter to Prince Charles. Retrieved from http://www .edge.org/3rd_culture/prince/prince_index.html

Deacon, T. (1997). *The symbolic species: The co-evolution of language and the brain.* New York: W.W. Norton.

Dean, L. G., et al. (2012). Identification of the social and cognitive processes underlying human cumulative culture. *Science, 335,* 1114–1118.

Deaner, R. O., et al. (2007). Overall brain size, and not encephalization quotient, best predicts cognitive ability across non-human primates. *Brain Behavior and Evolution, 70,* 115–124.

Deary, I., et al. (2010). The neuroscience of human intelligence differences. *Nature Reviews Neuroscience, 11*, 201–211.

deCastro, J. M. B., et al. (1997). A hominid from the lower Pleistocene of Atapuerca, Spain: Possible ancestor to Neanderthals and modern humans. *Science, 276*, 1392–1395.

Defleur, A., et al. (1999). Neanderthal cannibalism at Moula-Guercy, Ardeche, France. *Science, 286*, 128–131.

Dehaene, S. (2009). *Reading in the brain*. New York: Penguin Books.

Del Cul, A., et al. (2009). Causal role of prefrontal cortex in the threshold for access to consciousness. *Brain, 132*, 2531–2540.

DeLoache, J. S., & Burns, N. M. (1994). Early understanding of the representational function of pictures. *Cognition, 52*, 83–110.

Denault, L. K., & McFarlane, D. A. (1995). Reciprocal altruism between male vampire bats, Desmodus rotundus. *Animal Behaviour, 49*, 855–856.

Dennett, D. C. (1987). *The Intentional Stance*. Cambridge, Mass: Bradford Books, MIT Press.

Dennett, D. C. (1995). *Darwin's dangerous idea*. New York: Simon & Schuster.

Dennett, D. C., & Kinsbourne, M. (1992). Time and the observer: the where and when of consciousness in the brain. *Behavioral and Brain Sciences, 15*, 183–201.

Dere, E., et al. (2008). Animal episodic memory. In E. Dere et al. (Eds.), *Handbook of episodic memory* (pp. 155–184). Amsterdam: Elsevier.

Derevianko, A. P. (2012). *Recent discoveries in the Altai: Issues on the evolution of homo sapiens*. Novosibirsk: RAS Press.

de Saint-Exupéry, A. (1943). *The little prince* (I. Testot-Ferry, Trans.). London: Bibliophile Books.

DeSilva, J., & Lesnik, J. (2006). Chimpanzee neonatal brain size: Implications for brain growth in Homo erectus. *Journal of Human Evolution, 51*, 207–212.

Dettwyler, K. A. (1991). Can paleopathology provide evidence for compassion? *American Journal of Physical Anthropology, 84*, 375–384.

de Waal, F. B. M. (1982). *Chimpanzee politics*. London: Jonathan Cape.

de Waal, F. B. M. (1986). Deception in the natural communication of chimpanzees. In R. W. Mitchell & N. S. Thompson (Eds.), *Deception: Perspectives on human and non-human deceit* (pp. 221–244). Albany: State University of New York Press.

de Waal, F. B. M. (1989). Food sharing and reciprocal obligations among chimpanzees. *Journal of Human Evolution, 18*, 433–459.

de Waal, F. B. M. (1996). *Good natured*. Cambridge, MA: Harvard University Press.

de Waal, F. B. M. (2005). How animals do business. *Scientific American, 292*, 72–80.

de Waal, F. B. M. (2006). *Primates and philosophers: How morality evolved*. Princeton: Princeton University Press.

de Waal, F. B. M., & Aureli, F. (1996). Consolation, reconciliation, and a possible cognitive difference between macaque and chimpanzee. In A. E. Russon, K. A. Bard, & S. T. Parker (Eds.), *Reaching into thought: The minds of the great apes* (pp. 80–110). Cambridge: Cambridge University Press.

de Waal, F. B. M., et al. (2008). Comparing social skills of children and apes. *Science, 319*, 569.

Diamond, J. (1997). *Guns, germs, and steel: A short history of everybody for the last 13,000 years.* New York: Simon and Schuster.

Diamond, J. (2005). *Collapse: How societies choose to fail or succeed.* New York: Viking Press.

Diamond, J. (2010). The benefits of multilingualism. *Science, 330,* 332–333.

Dindo, M., et al. (2011). Observational learning in orangutan cultural transmission chains. *Biology Letters, 7,* 181–183.

Dominguez-Rodrigo, M., et al. (2012). Experimental study of cut marks made with rocks unmodified by human flaking and its bearing on claims of 3.4-million-year-old butchery evidence from Dikika, Ethiopia. *Journal of Archaeological Science, 39,* 205–214.

Dufour, V., & Sterck, E. H. M. (2008). Chimpanzees fail to plan in an exchange task but succeed in a tool-using procedure. *Behavioural Processes, 79,* 19–27.

Dufour, V., et al. (2007). Chimpanzee (*Pan troglodytes*) anticipation of food return: Coping with waiting time in an exchange task. *Journal of Comparative Psychology, 121,* 145–155.

Dunbar, R. I. M. (1992). Neocortex size as a constraint on group size in primates. *Journal of Human Evolution, 20,* 469–493.

Dunbar, R. I. M. (1996). *Grooming, gossip, and the evolution of language.* London: Faber.

Dunbar, R. I. M. (2007). Why are humans not just great apes? In C. Pasternak (Ed.), *What makes us human?* (pp. 37–48). Oxford: Oneworld.

Dunbar, R. I. M. (2010). The social role of touch in humans and primates: Behavioural function and neurobiological mechanisms. *Neuroscience and Biobehavioral Reviews, 34,* 260–268.

Dunn, M., et al. (2011). Evolved structure of language shows lineage-specific trends in word-order universals. *Nature, 473,* 79–82.

Eichenbaum, H., et al. (2005). Episodic recollection in animals: "If it walks like a duck and quacks like a duck. . . ." *Learning and Motivation, 36,* 190–207.

Einstein, A. (1950). Arms can bring no security. *Bulletin of the Atomic Scientist, 6,* 71.

Elston, G. N., et al. (2006). Specializations of the granular prefrontal cortex of primates: Implications for cognitive processing. *Anatomical Record Part A: Discoveries in Molecular Cellular and Evolutionary Biology, 288A,* 26–35.

Emery, N. J. (2000). The eyes have it: The neuroethology, function and evolution of social gaze. *Neuroscience and Biobehavioral Reviews, 24,* 581–604.

Emery, N. J., & Clayton, N. S. (2004). The mentality of crows: Convergent evolution of intelligence in corvids and apes. *Science, 306,* 1903–1907.

Enard, W., et al. (2002). Molecular evolution of FOXP2, a gene involved in speech and language. *Nature, 418,* 869–872.

Epstein, R., et al. (1981). "Self-awareness" in the pigeon. *Science, 212,* 695–696.

Evans, N., & Levinson, S. C. (2009). The myth of language universals: Language diversity and its importance for cognitive science. *Behavioral and Brain Sciences, 32,* 429–448.

Evans, T. A., & Beran, M. J. (2007). Chimpanzees use self-distraction to cope with impulsivity. *Biology Letters, 3,* 599–602.

Everett, D. L. (2005). Cultural constraints on grammar and cognition in Piraha: Another look at the design features of human language. *Current Anthropology, 46,* 621–646.

Fabre, J. H. (1915). *The hunting wasps*. New York: Dodd, Mead, and Company.

Falk, D. (1990). Brain evolution in Homo: The radiator theory. *Behavioral and Brain Sciences, 13*, 333–343.

Fedor, A., et al. (2008). Object permanence tests on gibbons (*Hylobatidae*). *Journal of Comparative Psychology, 122*, 403–417.

Fehr, E., & Fischbacher, U. (2003). The nature of human altruism. *Nature, 425*, 785–791.

Fehr, E., & Gachter, S. (2002). Altruistic punishment in humans. *Nature, 415*, 137–140.

Fehr, E., & Fischbacher, U. (2004). Social norms and human cooperation. *Trends in Cognitive Sciences, 8*, 185–190.

Fehr, E., et al. (2008). Egalitarianism in young children. *Nature, 454*, 1079–1083.

Feldman, R. (2012). Oxytocin and social affiliation in humans. *Hormones and Behavior, 61*, 380–391.

Feldman, R., et al. (2006). Microregulatory patterns of family interactions: Cultural pathways to toddlers' self-regulation. *Journal of Family Psychology, 20*, 614–623.

Ferrari, P. F., et al. (2006). Neonatal imitation in rhesus macaques. *PLOS Biology, 4*, 1501–1508.

Fiorito, G., & Scotto, P. (1992). Observational learning in *octopus vulgaris*. *Science, 256*, 545–547.

Fitch, W. T. (2000). The evolution of speech: A comparative review. *Trends in Cognitive Sciences, 4*, 258–265.

Fitch, W. T., & Hauser, M. D. (2004). Computational constraints on syntactic processing in a nonhuman primate. *Science, 303*, 377–380.

Fivush, T., et al. (2006). Elaborating on elaborations: Role of maternal reminiscing style in cognitive and socioemotional development. *Child Development, 77*, 1568–1588.

Flannery, T. (1994). *The future eaters: An ecological history of the Australian lands and people*. New York: Grove Press.

Flavell, J. H. (1963). *The developmental psychology of Jean Piaget*. New York: D. van Nostrand.

Flavell, J. H., et al. (1983). Development of the appearance-reality distinction. *Cognitive Psychology, 15*, 95–120.

Flemming, T. M., et al. (2008). What meaning means for same and different: Analogical reasoning in humans (*Homo sapiens*), chimpanzees (*Pan troglodytes*), and rhesus monkeys (*Macaca mulatta*). *Journal of Comparative Psychology, 122*, 176–185.

Flombaum, J. I., & Santos, L. R. (2005). Rhesus monkeys attribute perceptions to others. *Current Biology, 15*, 1–20.

Flynn, J. R. (2000). IQ gains, WISC subtests and fluid g: g theory and the relevance of Spearman's hypothesis to race. In G. R. Bock et al. (Eds.), *The nature of intelligence* (pp. 202–227). New York: Wiley.

Fossey, D. (1983). *Gorillas in the mist*. Boston: Houghton Mifflin.

Foster, D. J., & Wilson, M. A. (2006). Reverse replay of behavioural sequences in hippocampal place cells during the awake state. *Nature, 440*, 680–683.

Fouts, R. (1997). *Next of kin*. New York: William Morrow.

Friedman, W. J. (2005). Developmental and cognitive perspectives on humans' sense of the times of past and future events. *Learning and Motivation, 36,* 145–158.

Frith, U., & Frith, C. (2010). The social brain: allowing humans to boldly go where no other species has been. *Philosophical Transactions of the Royal Society B, 365,* 165–175.

Gagnon, S., & Doré, F. Y. (1994). Cross-sectional study of object permanence in domestic puppies (*Canis familiaris*). *Journal of Comparative Psychology, 108,* 220–232.

Galdikas, B. M. F. (1980). Living with the great orange apes. *National Geographic, 157,* 830–853.

Gallup, G. G. (1970). Chimpanzees: Self recognition. *Science, 167,* 86–87.

Gallup, G. G. (1998). Self-awareness and the evolution of social intelligence. *Behavioural Processes, 42,* 239–247.

Garcia, J., & Koelling, R. (1966). Relation of cue to consequence in avoidance learning. *Psychonomic Science, 4,* 123–124.

Garcia, J., et al. (1966). Learning with prolonged delay of reinforcement. *Psychonomic Science, 5,* 121–122.

Gardner, H. (1993). *Multiple intelligences: The theory in practice.* New York: Basic Books.

Gardner, R. A., & Gardner, B. T. (1969). Teaching sign language to a chimpanzee. *Science, 165,* 664–672.

Garland, E. C., et al. (2011). Dynamic horizontal cultural transmission of humpback whale song at the ocean basin scale. *Current Biology, 21,* 687–691.

Garrod, S., et al. (2007). Foundations of representation: Where might graphical symbol systems come from? *Cognitive Science, 31,* 961–987.

Geissmann, T. (2002). Taxonomy and evolution of gibbons. *Primatology and Anthropology, 11,* 28–31.

Gentner, T. Q., et al. (2006). Recursive syntactic pattern learning by songbirds. *Nature, 440,* 1204–1207.

Gergely, G., et al. (2002). Rational imitation in preverbal infants. *Nature, 415,* 755.

Gerrans, P. (2007). Mental time travel, somatic markers and "myopia for the future." *Synthese, 159,* 459–474.

Gibbons, A. (2008). The birth of childhood. *Science, 322,* 1040–1043.

Gibbons, A. (2011). African data bolster new view of modern human origins. *Science, 334,* 167.

Gilbert, D. T. (2006). *Stumbling on happiness.* New York: A. A. Knopf.

Gilbert, D. T., & Wilson, T. D. (2007). Prospection: Experiencing the future. *Science, 317,* 1351–1354.

Gilbert, W. S., & Sullivan, A. S. (2010). *The complete plays of Gilbert and Sullivan.* Rockville, MD: Wildside Press.

Gilby, I. C. (2006). Meat sharing among the Gombe chimpanzees: Harassment and reciprocal exchange. *Animal Behaviour, 71,* 953–963.

Gillan, D. J., et al. (1981). Reasoning in the chimpanzee. *Journal of Experimental Psychology-Animal Behavior Processes, 7,* 1–17.

Glickman, S., & Sroges, R. (1966). Curiosity in zoo animals. *Behaviour, 26,* 151–187.

Gomes, C. M., & Boesch, C. (2009). Wild chimpanzees exchange meat for sex on a long-term basis. *PLOS One, 4,* e5116.

Goodall, J. (1964). Tool using and aimed throwing in community of free living chimpanzees. *Nature, 201,* 1264–1266.

Goodall, J. (1986). *The chimpanzees of Gombe: Patterns of behavior.* Cambridge, MA: Harvard University Press.

Gopnik, A. (1993). How we know our minds: The illusion of first person knowledge of intentionality. *Behavioural and Brain Sciences, 16,* 1–14.

Gopnik, A. (2012). Scientific thinking in young children: Theoretical advances, empirical research, and policy implications. *Science, 337,* 1623–1627.

Gopnik, A., & Astington, J. W. (1988). Children's understanding of representational change and its relation to the understanding of false belief and the appearance-reality distinction. *Child Development, 59,* 26–37.

Gordon, A. C. L., & Olson, D. R. (1998). The relation between acquisition of a theory of mind and the capacity to hold in mind. *Journal of Experimental Child Psychology, 68,* 70–83.

Gordon, R. (1996). 'Radical' simulationism. In P. Carruthers & P. K. Smith (Eds.), *Theories of theories of mind* (pp. 11–21). Cambridge: Cambridge University Press.

Goren-Inbar, N., et al. (2004). Evidence of hominin control of fire at Gesher Benot Ya'aqov, Israel. *Science, 304,* 725–727.

Gottfredson, L. S. (1997). Why g matters: The complexity of everyday life. *Intelligence, 24,* 79–132.

Gould, S. J. (1978). *Ontogeny and phylogeny.* Boston, MA: Belknap.

Gould, S. J., & Eldredge, N. (1977). Punctuated equilibria: The tempo and mode of evolution reconsidered. *Paleobiology, 3,* 115–151.

Graves, R. R., et al. (2010). Just how strapping was KNM-WT 15000? *Journal of Human Evolution, 59,* 542–554.

Green, R. E., et al. (2010). A draft sequence of the Neandertal genome. *Science, 328,* 710–722.

Grice, H. P. (1989). *Studies in the way of words.* Cambridge, MA: Harvard University Press.

Groves, C. P. (1989). *A theory of human and primate evolution.* Oxford: Clarendon Press.

Groves, C. P. (2012a). Speciation in hominin evolution. In S. C. Reynolds & A. Gallagher (Eds.), *African genesis: Perspectives on hominin evolution.* (pp. 45–62). Cambridge: Cambridge University Press.

Groves, C. P. (2012b). Species concept in primates. *American Journal of Primatology, 74,* 687–691.

Grun, R., et al. (2005). U-series and ESR analyses of bones and teeth relating to the human burials from Skhul. *Journal of Human Evolution, 49,* 316–334.

Guinet, C., & Bouvier, J. (1995). Development of intentional stranding hunting techniques in killer whale (*Ornicus orca*) calves at Crozet Archipelago. *Canadian Journal of Zoology, 73,* 27–33.

Gupta, A. S., et al. (2010). Hippocampal replay is not a simple function of experience. *Neuron, 65,* 695–705.

Gurven, M., & Kaplan, H. (2007). Longevity among hunter-gatherers: A cross-cultural examination. *Population and Development Review, 33*, 321–365.

Haidt, J. (2007). The new synthesis in moral psychology. *Science, 316*, 998–1002.

Haile-Selassie, Y. (2001). Late Miocene hominids from the Middle Awash, Ethiopia. *Nature, 412*, 178–181.

Haile-Selassie, Y., et al. (2012). A new hominin foot from Ethiopia shows multiple Pliocene bipedal adaptations. *Nature, 483*, 565–569.

Halford, G. S., et al. (1998). Processing capacity defined by relational complexity: Implications for comparative, developmental and cognitive psychology. *Behavioral and Brain Sciences, 21*, 803–864.

Halford, G. S., et al. (2007). Separating cognitive capacity from knowledge: A new hypothesis. *Trends in Cognitive Sciences, 11*, 236–242.

Hamann, K., et al. (2011). Collaboration encourages equal sharing in children but not in chimpanzees. *Nature, 476*, 328–331.

Hamilton, W. D. (1964). The genetical evolution of social behaviour. *Journal of Theoretical Biology, 7*, 1–52.

Hamlin, J. K., et al. (2007). Social evaluation by preverbal infants. *Nature, 450*, 557–559.

Hammer, M. F., et al. (2011). Genetic evidence for archaic admixture in Africa. *Proceedings of the National Academy of Sciences of the United States of America, 108*, 15123–15128.

Hare, B., & Tomasello, M. (2004). Chimpanzees are more skillful in competitive than in cooperative cognitive tasks. *Animal Behaviour, 68*, 571–581.

Hare, B., et al. (2000). Chimpanzees know what conspecifics do and do not see. *Animal Behaviour, 59*, 771–785.

Hare, B., et al. (2001). Do chimpanzees know what conspecifics know? *Animal Behaviour, 61*, 139–151.

Hare, B., et al. (2006). Chimpanzees deceive a human competitor by hiding. *Cognition, 101*, 495–514.

Hare, B., et al. (2012). The self-domestication hypothesis: Evolution of bonobo psychology is due to selection against aggression. *Animal Behaviour, 83*, 573–585.

Harré, N. (2011). *Psychology for a better world.* Auckland: Department of Psychology.

Harman, O. S. (2010). *The price of altruism: George Price and the search for the origins of kindness.* New York: Norton.

Harris, P. L., et al. (1996). Children's use of counterfactual thinking in causal reasoning. *Cognition, 61*, 233–259.

Hart, D., & Sussman, R. W. (2005). *Man the hunted: Primates, predators, and human evolution.* Boulder: Westview Press.

Haun, D. B. M., & Call, J. (2008). Imitation recognition in great apes. *Current Biology, 18*, R288-R290.

Hauser, M. D. (1996). *The evolution of communication.* Cambridge, MA: MIT Press.

Hauser, M. D., & Marler, P. (1993). Food associated calls in rhesus macaques (*Macaca mulatta*). *Behavioral Ecology, 4*, 206–212.

Hauser, M. D., et al. (2002). The faculty of language: What is it, who has it, and how did it evolve? *Science, 298*, 1569–1579.

Hawkes, K. (2003). Grandmothers and the evolution of human longevity. *American Journal of Human Biology, 15*, 380–400.

Hayes, C. (1951). *The ape in our house*. New York: Harper.

Hayes, K. J., & Hayes, C. (1952). Imitation in a home-raised chimpanzee. *Journal of Comparative and Physiological Psychology, 45*, 450–459.

Hazlitt, W. (1805). *Essay on the principles of human action and some remarks on the systems of Hartley and Helvetius*. London: J. Johnson.

Heinrich, B. (1995). An experimental investigation of insight in common ravens (*Corvus corax*). *Auk, 112*, 994–1003.

Henrich, J., et al. (2006). Costly punishment across human societies. *Science, 312*, 1767–1770.

Herculano-Houzel, S. (2009). The human brain in numbers: A linearly scaled-up primate brain. *Frontiers in Human Neuroscience, 3*, 1–11.

Herman, L. (2002). Vocal, social and self-imitation by bottlenosed dolphins. In Dautenhahn, K., & Nehaniv, C. (Eds.), *Imitation in Animals and Artifacts*. (pp. 63–108). Cambridge, MA: MIT Press.

Herman, L. M., et al. (1993). Representational and conceptual skills of dolphins. In H. L. Roitblat, L. M. Herman, & P. E. Nachtigall (Eds.), *Language and communication: Comparative perspectives* (pp. 403–442). Hillsdale, NJ: Erlbaum.

Herrmann, E., et al. (2007). Humans have evolved specialized skills of social cognition: The cultural intelligence hypothesis. *Science, 317*, 1360–1366.

Herschel, J. (1830). *A preliminary discourse on the study of natural philosophy*. London: Longman et al.

Hewstone, M., et al. (2002). Intergroup bias. *Annual Review of Psychology, 53*, 575–604.

Heyes, C. M. (1994). Reflections on self-recognition in primates. *Animal Behaviour, 47*, 909–919.

Heyes, C. M. (1998). Theory of mind in nonhuman primates. *Behavioral and Brain Sciences, 21*, 101–134.

Hill, A., et al. (2011). Inferential reasoning by exclusion in great apes, lesser apes, and spider monkeys. *Journal of Comparative Psychology, 125*, 91–103.

Hill, K., et al. (2009). The emergence of human uniqueness: Characters underlying behavioral modernity. *Evolutionary Anthropology, 18*, 187–200.

Hoffer, E. (1973). Reflections on the human condition. New York: Harper & Row.

Hofreiter, M., et al. (2010). Vertebrate DNA in fecal samples from bonobos and gorillas: Evidence for meat consumption or artefact? *PLOS One, 5*, e9419.

Holdaway, R. N., & Jacomb, C. (2000). Rapid extinction of the moas (Aves: dinorinthiformes): Model, test, and implications. *Science, 287*, 2250–2254.

Holden, C. (2005). Time's up on time travel. *Science, 308*, 1110.

Holloway, R. L. (2008). The human brain evolving: A personal retrospective. *Annual Review of Anthropology, 37*, 1–19.

Holmes, R. (2008). *The age of wonder: How the romantic generation discovered the beauty and terror of science*. London: Harper Press.

Holzhaider, J. C., et al. (2010). The development of pandanus tool manufacture in wild New Caledonian crows. *Behaviour, 147*, 553–586.

Hoppitt, W. J. E., et al. (2008). Lessons from animal teaching. *Trends in Ecology & Evolution, 23*, 486–493.

Horner, V., & Whiten, A. (2005). Causal knowledge and imitation/emulation switching in chimpanzees (*Pan troglodytes*) and children (*Homo sapiens*). *Animal Cognition, 8*, 164–181.

Horner, V., et al. (2010). Prestige affects cultural learning in chimpanzees. *PLOS One, 5, e10625.*

Horner, V., et al. (2011). Spontaneous prosocial choice by chimpanzees. *Proceedings of the National Academy of Sciences of the United States of America, 108,* 13847–13851.

Hudson, J. A. (2006). The development of future time concepts through mother-child conversation. *Merrill-Palmer Quarterly, 52,* 70–95.

Huffman, M. A. (1997). Current evidence for self-medication in primates: A multidisciplinary perspective. In *Yearbook of Physical Anthropology* (Vol. 40, pp. 1–30). Wilmington, DE: Wiley-Liss.

Humphrey, N. (1976). The social function of intellect. In P. P. G. Bateson & R. A. Hinde (Eds.), *Growing points in ethology* (pp. 303–313). Cambridge: Cambridge University Press.

Hunt, G., & Gray, R. (2003). Diversification and cumulative evolution in New Caledonian crow tool manufacture. *Proceedings of the Royal Society B-Biological Sciences, 270,* 867–874.

Huttenlocher, P. R. (1990). Morphometric study of human cerebral cortex development. *Neuropsychologia, 28,* 517–527.

Huxley, A. (1956). *Adonis and the alphabet.* London: Chatto & Windus.

Hyatt, C. W. (1998). Responses of gibbons (*Hylobates lar*) to their mirror images. *American Journal of Primatology, 45,* 307–311.

Inoue, S., & Matsuzawa, T. (2007). Working memory of numerals in chimpanzees. *Current Biology, 17,* R1004-R1005.

Isanski, B., & West, C. (2010). The body of knowledge: Understanding embodied cognition. *Observer, 23,* 13–18.

Jackendoff, R., & Pinker, S. (2005). The nature of the language faculty and its implications for evolution of language. *Cognition, 97,* 211–225.

James, W. (1890). *The principles of psychology.* London: Macmillan.

Jantz, R. L. (2001). Cranial change in Americans: 1850–1975. *Journal of Forensic Sciences, 46,* 784–787.

Jerison, H. J. (1973). *The evolution of the brain and intelligence.* New York: Academic Press.

Johanson, D. C. (2004). Lucy, thirty years later: An expanded view of *Australopithecus afarensis. Journal of Anthropological Research, 60,* 465–486.

Johnson, A., & Redish, A. D. (2007). Neural ensembles in CA3 transiently encode paths forward of the animal at a decision point. *Journal of Neuroscience, 27,* 12176–12189.

Johnson, C. R., & McBrearty, S. (2010). 500,000 year old blades from the Kapthurin Formation, Kenya. *Journal of Human Evolution, 58,* 193–200.

Johnson, M. A. (April 18, 2005). The culture of Einstein: Achievements in science gave him a platform to address the world. *NBC News.* Retrieved from http://www.msnbc.msn.com/id/7406337/#.UD2tYUQuKHk

Jones, B. W., & Nishiguchi, M. K. (2004). Counterillumination in the Hawaiian bobtail squid, Euprymna scolopes Berry (Mollusca: Cephalopoda). *Marine Biology, 144,* 1151–1155.

Kaertner, J., et al. (2012). The development of mirror self-recognition in different sociocultural contexts. *Monographs of the Society for Research in Child Development, 77,* 1–101.

Kafka, F. (2009). A report for an academy. (I. Johnston, Trans.). Retrieved from http://records.viu.ca/~johnstoi/kafka/reportforacademy.htm

Kaminski, J., et al. (2004). Word learning in a domestic dog: Evidence for "fast mapping." *Science, 5677,* 1682–1683.

Kaminski, J., et al. (2008). Chimpanzees know what others know, but not what they believe. *Cognition, 109,* 224–234.

Kana, R. K., et al. (2011). A systems level analysis of the mirror neuron hypothesis and imitation impairments in autism spectrum disorders. *Neuroscience and Biobehavioral Reviews, 35,* 894–902.

Karlsson, M. P., & Frank, L. M. (2009). Awake replay of remote experiences in the hippocampus. *Nature Neuroscience, 12,* 913–918.

Kawai, M. (1965). Newly-acquired pre-cultural behaviour of the natural troop of Japanese monkeys on Koshima Islets. *Primates, 6,* 1–30.

Kawai, N., & Matsuzawa, T. (2000). Numerical memory span in a chimpanzee. *Nature, 403,* 39–40.

Keeley, L. H. (1996). *War before civilization.* London: Oxford University Press.

Kivell, T. L., & Schmitt, D. (2009). Independent evolution of knuckle-walking in African apes shows that humans did not evolve from a knuckle-walking ancestor. *Proceedings of the National Academy of Sciences of the United States of America, 106,* 14241–14246.

Klein, S. B., et al. (2002). Memory and temporal experience: The effects of episodic memory loss on an amnesic patient's ability to remember the past and imagine the future. *Social Cognition, 20,* 353–379.

Kohlberg, L. (1963). Development of children's orientation toward a moral order. *Vita Humana, 6,* 11–33.

Köhler, W. (1917/1925). *The mentality of apes.* London: Routledge & Kegan Paul.

Korsgaard, C. (2006). Morality and the distinctiveness of human action. In F. B. M. de Waal (Ed.), *Primates and philosophers: How morality evolved* (pp. 98–119). Princeton: Princeton University Press.

Krachun, C., et al. (2009a). Can chimpanzees (*Pan troglodytes*) discriminate appearance from reality? *Cognition, 112,* 435–450.

Krachun, C., et al. (2009b). A competitive nonverbal false belief task for children and apes. *Developmental Science, 12,* 521–535.

Krause, J., et al. (2007a). The derived FOXP2 variant of modern humans was shared with Neandertals. *Current Biology, 17,* 1908–1912.

Krause, J., et al. (2007b). Neanderthals in central Asia and Siberia. *Nature, 449,* 902–904.

Krause, J., et al. (2010). The complete mitochondrial DNA genome of an unknown hominin from southern Siberia. *Nature, 464,* 894–897.

Kuhlmeier, V. A., & Boysen, S. T. (2002). Chimpanzees (*Pan troglodytes*) recognize spatial and object correspondences between a scale model and its referent. *Psychological Science, 13,* 60–63.

Kundera, M. (1992). *Immortality.* (P. Kussi, Trans.). New York: HarperCollins.

Lahdenpera, M., et al. (2004). Fitness benefits of prolonged post-reproductive lifespan in women. *Nature, 428,* 178–181.

Lalueza-Fox, C., et al. (2007). A melanocortin 1 receptor allele suggests varying pigmentation among Neanderthals. *Science, 318,* 1453–1455.

Lancaster, J. B., & Lancaster, C. S. (1983). Parental investment: The hominid adaptation. In D. J. Ortner (Ed.), *How humans adapt: A biocultural odyssey.* Washington: Smithsonian Institution.

Langdon, J. H. (2006). Has an aquatic diet been necessary for hominin brain evolution and functional development? *British Journal of Nutrition, 96,* 7–17.

Langley, M. C., et al. (2008). Behavioural complexity in Eurasian Neanderthal populations: A chronological examination of the archaeological evidence. *Cambridge Archaeological Journal, 18,* 289–307.

Larick, R., & Ciochon, R. L. (1996). The African emergence and early Asian dispersals of the genus *Homo. American Scientist, 84,* 538–551.

Lea, S. E. G. (2001). Anticipation and memory as criteria for special welfare consideration. *Animal Welfare, 10,* S195–208.

Leakey, L. S. B., et al. (1964). A new species of the genus Homo from Olduvai gorge. *Nature, 202,* 7–9.

Leakey, M. D., & Hay, R. L. (1979). Pliocene footprints in the Laetolil beds at Laetoli, Northern Tanzania. *Nature, 278,* 317–323.

Leakey, M. G., et al. (2001). New hominin genus from eastern Africa shows diverse middle Pliocene lineages. *Nature, 410,* 433–440.

Leaver, L. A., et al. (2007). Audience effects on food caching in grey squirrels (*Sciurus carolinensis*): Evidence for pilferage avoidance strategies. *Animal Cognition, 10,* 23–27.

Lebel, C., et al. (2012). Diffusion tensor imaging of white matter tract evolution over the lifespan. *NeuroImage, 60,* 340–352.

Lepre, C. J., et al. (2011). An earlier origin for the Acheulian. *Nature, 477,* 82–85.

Leslie, A. M. (1987). Pretense and representation in infancy: The origins of "theory of mind." *Psychological Review, 94,* 412–426.

Lethmate, J., & Dücker, G. (1973). Untersuchungen zum Selbsterkennen im Spiegel bei Orang-Utans und einigen anderen Affenarten [Investigations into self-recognition in orangutans and some other apes]. *Zeitschrift für Tierpsychologie, 33,* 248–269.

Levinson, S. C., & Gray, R. D. (2012). Tools from evolutionary biology shed new light on the diversification of languages. *Trends in Cognitive Sciences, 16,* 167–173.

Lewis, M., & Ramsay, D. (2004). Development of self-recognition, personal pronoun use, and pretend play during the 2nd year. *Child Development, 75,* 1821–1831.

Lewis, M., et al. (1989). Self development and self-conscious emotions. *Child Development, 60,* 146–156.

Lieberman, P. (1991). *Uniquely human: The evolution of speech, thought and selfless behavior.* Cambridge, MA: Harvard University Press.

Lindsay, W. L. (1880). *Mind in the lower animals, in health and disease.* New York: Appleton and Co.

Linnaeus, C. (1758). *Systema naturae* (10th edition). Stockholm: Laurentii Sylvii.

Liszkowski, U., et al. (2004). Twelve-month-olds point to share attention and interest. *Developmental Science, 7,* 297–307.

Liszkowski, U., et al. (2009). Prelinguistic infants, but not chimpanzees, communicate about absent entities. *Psychological Science, 20,* 654–660.

Locke, J. L., & Bogin, B. (2006). Language and life history: A new perspective on the development and evolution of human language. *Behavioral and Brain Sciences, 29,* 259–280.

Loftus, E. F. (1992). When a lie becomes memory's truth: Memory distortion after exposure to misinformation. *Current Directions in Psychological Science, 1,* 121–123.

Lombard, M. (2012). Thinking through the Middle Stone Age of sub-Saharan Africa. *Quaternary International, 270,* 140–155.

Lordkipanidze, D., et al. (2005). The earliest toothless hominin skull. *Nature, 434,* 717–718.

Luna, B., et al. (2004). Maturation of cognitive processes from late childhood to adulthood. *Child Development, 75,* 1357–1372.

Lyn, H., et al. (2008). Precursors of morality in the use of the symbols "good" and "bad" in two bonobos (*Pan paniscus*) and a chimpanzee (*Pan troglodytes*). *Language & Communication, 28,* 213–224.

Lyn, H., et al. (2011). Nonhuman primates do declare! A comparison of declarative symbol and gesture use in two children, two bonobos, and a chimpanzee. *Language & Communication, 31,* 63–74.

Madsen, E. A., et al. (2007). Kinship and altruism: A cross-cultural experimental study. *British Journal of Psychology, 98,* 339–359.

Mahajan, N., et al. (2011). The evolution of intergroup bias: Perceptions and attitudes in rhesus macaques. *Journal of Personality and Social Psychology, 100,* 387–405.

Martinez, I., et al. (2008). Human hyoid bones from the middle Pleistocene site of the Sima de los Huesos (Sierra de Atapuerca, Spain). *Journal of Human Evolution, 54,* 118–124.

Mäthger, L. M., et al. (2009). Do cephalopods communicate using polarized light reflections from their skin? *Journal of Experimental Biology, 212,* 2133–2140.

Matsuzawa, T. (2009). Symbolic representation of number in chimpanzees. *Current Opinion in Neurobiology, 19,* 92–98.

Maynard, A. E. (2002). Cultural teaching: The development of teaching skills in Maya sibling interactions. *Child Development, 73,* 969–982.

McAuliffe, K. (2010, September). The incredible shrinking brain. *Discover Magazine, 31,* 54–59.

McDaniel, M. A. (2005). Big-brained people are smarter: A meta-analysis of the relationship between in vivo brain volume and intelligence. *Intelligence, 33,* 337–346.

McDougall, I., et al. (2005). Stratigraphic placement and age of modern humans from Kibish, Ethiopia. *Nature, 433,* 733–736.

McGuigan, F., & Salmon, K. (2004). The time to talk: The influence of the timing of adult-child talk on children's event memory. *Child Development, 75,* 669–686.

McHenry, H. M., & Coffing, K. (2000). Australopithecus to Homo: Transformations in body and mind. *Annual Review of Anthropology, 29,* 125–146.

McIlwain, D. (2003). Bypassing empathy: A Machiavellian theory of mind and sneaky power. In B. Repacholi & V. P. Slaughter (Eds.), *Individual differences in theory of mind* (pp. 39–66). New York: Psychology Press.

McPherron, S. P., et al. (2010). Evidence for stone-tool-assisted consumption of animal tissues before 3.39 million years ago at Dikika, Ethiopia. *Nature, 466*, 857–860.

Mealey, L. (1995). The sociobiology of sociopathy: An integrated evolutionary model. *Behavioral and Brain Sciences, 18*, 523–541.

Melis, A. P., et al. (2006a). Chimpanzees recruit the best collaborators. *Science, 311*, 1297–1300.

Melis, A. P., et al. (2006b). Engineering cooperation in chimpanzees: Tolerance constraints on cooperation. *Animal Behaviour, 72*, 275–286.

Melis, A. P., et al. (2008). Do chimpanzees reciprocate received favours? *Animal Behaviour, 76*, 951–962.

Mellars, P. (2006). Going east: New genetic and archaeological perspectives on the modern human colonization of Eurasia. *Science, 313*, 796–800.

Meltzoff, A. N. (1988). Infant imitation and memory: Nine-month-olds and immediate and deferred tests. *Child Development, 59*, 217–225.

Meltzoff, A. N. (1995). Understanding the intentions of others: Re-enactment of intended acts by 18-month-old children. *Developmental Psychology, 31*, 838–850.

Meltzoff, A. N., & Moore, M. K. (1977). Imitation of facial and manual gestures by human neonates. *Science, 198*, 75–78.

Mendes, N., & Huber, L. (2004). Object-permanence in common marmosets (*Callithrix jacchus*). *Journal of Comparative Psychology, 118*, 103–112.

Mendes, N., et al. (2007). Raising the level: Orangutans use water as a tool. *Biology Letters, 3*, 453–455.

Menzel, E. (2005). Progress in the study of chimpanzee recall and episodic memory. In H. S. Terrace & J. Metcalfe (Eds.), *The missing link in cognition* (pp. 188–224). Oxford: Oxford University Press.

Mercader, J., et al. (2007). 4,300-year-old chimpanzee sites and the origins of percussive stone technology. *Proceedings of the National Academy of Sciences of the United States of America, 104*, 3043–3048.

Mesoudi, A., et al. (2006). Towards a unified science of cultural evolution. *Behavioral and Brain Sciences, 29*, 329-347.

Mikhail, J. (2007). Universal moral grammar: Theory, evidence and the future. *Trends in Cognitive Sciences, 11*, 143–152.

Miles, H. L., et al. (1996). Simon says: The development of imitation in an enculturated orangutan. In A. E. Russon, S.T. Parker, & K. A. Bard (Eds.), *Reaching into thought: The minds of the great apes* (pp. 278–299). Cambridge: Cambridge University Press.

Miles, L. (1994). The cognitive foundations for references in a single orangutan. In S. T. Parker & K. R. Gibson (Eds.), *'Language' and intelligence in monkeys and apes* (pp. 511–539). Cambridge: Cambridge University Press.

Miller, G. A. (2003). The cognitive revolution. *Trends in Cognitive Sciences, 7*, 141–144.

Miller, G. F. (1998). How mate choice shaped human nature: A review of sexual selection and human evolution. *Handbook of Evolutionary Psychology*, 87–129.

Milot, E., et al. (2011). Evidence for evolution in response to natural selection in a contemporary human population. *Proceedings of the National Academy of Sciences of the United States of America, 108*, 17040–17045.

Mischel, W., et al. (1989). Delay of gratification in children. *Science, 244,* 933–938.

Mitchell, A., et al. (2009). Adaptive prediction of environmental changes by microorganisms. *Nature, 460,* 220–224.

Mitchell, R. W., & Anderson, J. R. (1993). Discrimination-learning of scratching, but failure to obtain imitation and self-recognition in a long-tailed macaque. *Primates, 34,* 301–309.

Moore, C. (2006). *The development of common sense psychology.* Mahwah, NJ: Lawrence Erlbaum Associates.

Moore, C. (2013). Homology through development of triadic interaction and language. *Developmental Psychobiology, 55,* 59–66.

Morete, M. E., et al. (2003). A novel behavior observed in humpback whales on wintering grounds at Abrolhos Bank (Brazil). *Marine Mammal Science, 19,* 694–707.

Morgan, E. (1982). *The aquatic ape: A theory of human evolution.* London: Souvenir.

Mulcahy, N. J., & Call, J. (2006a). Apes save tools for future use. *Science, 312,* 1038–1040.

Mulcahy, N. J., & Call, J. (2006b). How great apes perform on a modified trap-tube task. *Animal Cognition, 9,* 193–199.

Mulcahy, N. J., & Call, J. (2009). The performance of bonobos (*Pan paniscus*), chimpanzees (*Pan troglodytes*), and orangutans (*Pongo pygmaeus*) in two versions of an object-choice task. *Journal of Comparative Psychology, 123,* 304–309.

Mulcahy, N. J., & Suddendorf, T. (2011). An obedient orangutan (*Pongo abelii*) performs perfectly in peripheral object-choice tasks but fails the standard centrally presented versions. *Journal of Comparative Psychology, 125,* 112–115.

Mulcahy, N. J., et al. (2005). Gorillas (*Gorilla gorilla*) and orangutans (*Pongo pygmaeus*) encode relevant problem features in a tool-using task. *Journal of Comparative Psychology, 119,* 23–32.

Mulcahy, N. J., et al. (2013). Orangutans (*Pongo pygmaeus* and *Pongo abelii*) understand connectivity in the skewered grape tool task. *Journal of Comparative Psychology, 127,* 109–113.

Murray, C. M., et al. (2007). New case of intragroup infanticide in the chimpanzees of Gombe National Park. *International Journal of Primatology, 28,* 23–37.

Myowa-Yamakoshi, M., et al. (2004). Imitation in neonatal chimpanzees (*Pan troglodytes*). *Developmental Science, 7,* 437–442.

Naqshbandi, M., & Roberts, W. A. (2006). Anticipation of future events in squirrel monkeys (*Saimiri sciureus*) and rats (*Rattus norvegicus*): Tests of the Bischof-Kohler hypothesis. *Journal of Comparative Psychology, 120,* 345–357.

Neisser, U. (1997). The roots of self-knowledge. In J. G. Snodgrass & R. L. Thompson (Eds.), *The Self Across Psychology* (Vol. 818, pp. 19–33). New York: New York Academy of Sciences.

Neisser, U., et al. (1996). Intelligence: Knowns and unknowns. *American Psychologist, 51,* 77–101.

Nelson, K. D., & Fivush, R. (2004). The emergence of autobiographical memory: A social cultural developmental theory. *Psychological Review, 111,* 486–511.

Nesse, R. M., & Berridge, K. C. (1997). Psychoactive drug use in evolutionary perspective. *Science, 278,* 63–66.

Nielsen, M. (2006). Copying actions and copying outcomes: Social learning through the second year. *Developmental Psychology, 42,* 555–565.

Nielsen, M., & Dissanayake, C. (2004). Pretend play, mirror self-recognition and imitation: A longitudinal investigation through the second year. *Infant Behavior and Development, 27,* 342–365.

Nielsen, M., & Tomaselli, K. (2010). Overimitation in Kalahari bushman children and the origins of human cultural cognition. *Psychological Science, 21,* 729–736.

Nielsen, M., & Widjojo, E. (2011). Failure to find over-imitation in captive orangutans (*Pongo pygmaeus*): Implications for our understanding of cross-generational information transfer. In J Håkansson (Ed.), *Developmental Psychology* (pp.153–167). New York: Nova Science Publishers.

Nielsen, M., et al. (2005). Imitation recognition in a captive chimpanzee (*Pan troglodytes*). *Animal Cognition, 8,* 31–36.

Nielsen, M., et al. (2006). Mirror self-recognition beyond the face. *Child Development, 77,* 176–185.

Nimchinsky, E. A., et al. (1999). A neuronal morphologic type unique to humans and great apes. *Proceedings of the National Academy of Sciences of the United States of America, 96,* 5268–5273.

Noack, R. A. (2012). Solving the "human problem": The frontal feedback model. *Consciousness and Cognition, 21,* 1043–1067.

Noad, M. J., et al. (2000). Cultural revolution in whale songs. *Nature, 408,* 537.

O'Connell, J. F., et al. (1999). Grandmothering and the evolution of Homo erectus. *Journal of Human Evolution, 36,* 461–485.

O'Neill, D. K., et al. (1992). Young children's understanding of the role that sensory experiences play in knowledge acquisition. *Child Development, 63,* 474–490.

Oberauer, K., et al. (2005). Working memory and intelligence: Their correlation and their relation. *Psychological Bulletin, 131,* 61–65.

Oberauer, K., et al. (2008). Which working memory functions predict intelligence? *Intelligence, 36,* 641–652.

Okuda, J., et al. (2003). Thinking of the future and past: The roles of the frontal pole and the medial temporal lobes. *NeuroImage, 19,* 1369–1380.

Onishi, K. H., & Baillargeon, R. (2005). Do 15-month-old infants understand false beliefs? *Science, 308,* 255–258.

Ostrom, E. (2009). Beyond markets and states: Polycentric governance of complex economic systems. *Nobel Prize Lectures.* Retrieved from http://www.nobel prize.org/nobel_prizes/economics/laureates/2009/ostrom-lecture.html

Osvath, M. (2009). Spontaneous planning for future stone throwing by a male chimpanzee. *Current Biology, 19,* R190–R191.

Osvath, M., & Osvath, H. (2008). Chimpanzee (*Pan troglodytes*) and orangutan (*Pongo abelii*) forethought: Self-control and pre-experience in the face of future tool use. *Animal Cognition, 11,* 661–674.

Oxnard, C., et al. (2010). Post-cranial skeletons of hypothyroid cretins show a similar anatomical mosaic as Homo floresiensis. *PLOS One, 5,* e13018.

Parker, C. E. (1974a). The antecedents of man the manipulator. *Journal of Human Evolution, 3,* 493–500.

Parker, C. E. (1974b). Behavioral diversity in ten species of nonhuman primates. *Journal of Comparative and Physiological Psychology, 5*, 930–937.

Parr, L. A. (2001). Cognitive and physiological markers of emotional awareness in chimpanzees (*Pan troglodytes*). *Animal Cognition, 4*, 223–229.

Patterson, F. (1991). Self-awareness in the gorilla Koko. *Gorilla, 14*, 2–5.

Patterson, F., & Linden, E. (1981). *The education of Koko.* New York: Holt, Rinehart, & Winston.

Patterson, N., et al. (2006). Genetic evidence for complex speciation of humans and chimpanzees. *Nature, 441*, 1103–1108.

Paukner, A., et al. (2009). Capuchin monkeys display affiliation toward humans who imitate them. *Science, 325*, 880–883.

Paus, T., et al. (1999). Structural maturation of neural pathways in children and adolescents: In vivo study. *Science, 283*, 1908–1911.

Penn, D. C., & Povinelli, D. J. (2007). On the lack of evidence that non-human animals possess anything remotely resembling a 'theory of mind.' *Philosophical Transactions of the Royal Society B-Biological Sciences, 362*, 731–744.

Penn, D. C., et al. (2008). Darwin's mistake: Explaining the discontinuity between human and nonhuman minds. *Behavioral and Brain Sciences, 31*, 109–178.

Pepperberg, I. M. (1987). Acquisition of the same different concept by an African grey parrot. *Animal Learning & Behavior, 15*, 423–432.

Perner, J. (1991). *Understanding the representational mind.* Cambridge, MA: MIT Press.

Peterson, C. C., & Siegal, M. (2000). Insights into theory of mind from deafness and autism. *Mind and Language, 15*, 123–145.

Peterson, C. C., et al. (2000). Factors influencing the development of a theory of mind in blind children. *British Journal of Developmental Psychology, 18*, 431–447.

Pinker, S. (1994). *The language instinct.* London: Penguin.

Pinker, S. (1997). *How the mind works.* London: Penguin.

Pinker, S. (2010). The cognitive niche: Coevolution of intelligence, sociality, and language. *Proceedings of the National Academy of Sciences of the United States of America, 107*, 8993–8999.

Pinker, S. (2011a). *The better angels of our nature: Why violence has declined.* New York: Penguin.

Pinker, S. (2011b). Representations and decision rules in the theory of self-deception. *Behavioral and Brain Sciences, 34*, 35–37.

Plotnik, J. M., et al. (2006). Self-recognition in an Asian elephant. *Proceedings of the National Academy of Sciences of the United States of America, 103*, 17053–17057.

Posada, S., & Colell, M. (2007). Another gorilla recognizes himself in a mirror. *American Journal of Primatology, 69*, 576–583.

Povinelli, D. J. (2000). *Folk physics for apes.* Oxford: Oxford University Press.

Povinelli, D. J., & Eddy, T. J. (1996). What young chimpanzees know about seeing. *Monographs of the Society for Research in Child Development, 61*, 1–198.

Povinelli, D. J., et al. (1990). Inferences about guessing and knowing by chimpanzees (*Pan troglodytes*). *Journal of Comparative Psychology, 104*, 203–210.

Povinelli, D. J., et al. (1992). Comprehension of role reversal in chimpanzees: Evidence of empathy? *Animal Behaviour, 43*, 633–640.

Povinelli, D. J., et al. (1996). Self-recognition in young children using delayed versus live feedback: Evidence for a developmental asynchrony. *Child Development, 67,* 1540–1554.

Povinelli, D. J., et al. (1997). Exploitation of pointing as a referential gesture in young children, but not adolescent chimpanzees. *Cognitive Development, 12,* 327–365.

Povinelli, D. J., et al. (2000). Toward a science of other minds: Escaping the argument by analogy. *Cognitive Science, 24,* 509–542.

Premack, D. (2007). Human and animal cognition: Continuity and discontinuity. *Proceedings of the National Academy of Sciences of the United States of America, 104,* 13861–13867.

Premack, D. (2012). Why humans are unique: Three theories. *Perspectives on Psychological Science, 5,* 22–32.

Premack, D., & Premack, A. (1983). *The mind of an ape.* New York: Norton.

Premack, D., & Woodruff, G. (1978). Does the chimpanzee have a theory of mind? *Behavioral and Brain Sciences, 1,* 515–526.

Preuss, T. M. (2000). What's human about the human brain? In M. S. Gazzaniga (Ed.), *The new cognitive neurosciences* (pp. 1219–1234). Cambridge, MA: MIT Press.

Preuss, T. M., et al. (1999). Distinctive compartmental organization of human primary visual cortex. *Proceedings of the National Academy of Sciences of the United States of America, 96,* 11601–11606.

Priel, B., & Deschonen, S. (1986). Self-recognition: A study of a population without mirrors. *Journal of Experimental Child Psychology, 41,* 237–250.

Prior, H., et al. (2008). Mirror-induced behavior in the magpie (*Pica pica*): Evidence of self-recognition. *PLOS Biology, 6,* 1642–1650.

Proffitt, D. (2006). Embodied perception and the economy of action. *Perspectives on Psychological Science, 1,* 110–122.

Pruetz, J. D., & Bertolani, P. (2007). Savanna chimpanzees, *Pan troglodytes verus,* hunt with tools. *Current Biology, 17,* 412–417.

Radick, G. (2007). *The simian tongue.* Chicago: University of Chicago Press.

Rafetseder, E.R., et al. (2010). Counterfactual reasoning: Developing a sense of "nearest possible world." *Child Development, 81,* 376–389.

Ramirez Rozzi, F. V., & Bermudez De Castro, J. M. (2004). Surprisingly rapid growth in Neanderthals. *Nature, 428,* 936–939.

Ramirez Rozzi, F. V., et al. (2009). Cutmarked human remains bearing Neandertal features and modern human remains associated with the Aurignacian at Les Rois. *Journal of Anthropological Sciences = Rivista di antropologia : JASS / Istituto italiano di antropologia, 87,* 153–185.

Range, F., et al. (2009). The absence of reward induces inequity aversion in dogs. *Proceedings of the National Academy of Sciences of the United States of America, 106,* 340–345.

Ranlet, P. (2000). The British, the Indians, and smallpox: What actually happened at Fort Pitt in 1763? *Pennsylvania History, 67,* 427–441.

Read, D. W. (2008). Working memory: A cognitive limit to non-human primate recursive thinking prior to hominid evolution. *Evolutionary Psychology, 6,* 676–714.

Reader, S. M., & Laland, K. N. (Eds.). (2003). *Animal innovation*. Oxford: Oxford University Press.

Reed, D. L., et al. (2007). Pair of lice lost or parasites regained: The evolutionary history of anthropoid primate lice. *BMC Biology, 5:7*.

Reich, D., et al. (2011). Denisova admixture and the first modern human dispersals into Southeast Asia and Oceania. *American Journal of Human Genetics, 89*, 516–528.

Reiss, D., & Marino, L. (2001). Mirror self-recognition in the bottlenose dolphin: A case of cognitive convergence. *Proceedings of the National Academy of Sciences of the United States of America, 98*, 5937–5942.

Rendell, L., & Whitehead, H. (2001). Culture in whales and dolphins. *Behavioral and Brain Sciences, 24*, 309–324.

Rice, G. E., & Gainer, P. (1962). "Altruism" in the albino rat. *Journal of Comparative and Physiological Psychology, 55*, 123–125.

Ridley, M. (1997). *The origins of virtue: Human instincts and the evolution of cooperation*. New York: Viking.

Rizzolatti, G., et al. (1996). Premotor cortex and the recognition of motor actions. *Cognitive Brain Research, 3*, 131–141.

Roberts, W. A. (2002). Are animals stuck in time? *Psychological Bulletin, 128*, 473–489.

Robson, S. L., & Wood, B. (2008). Hominin life history: Reconstruction and evolution. *Journal of Anatomy, 212*, 394–425.

Roediger, H. L., & McDermott, K. B. (2011). Remember when? *Science, 333*, 47–48.

Roma, P. G., et al. (2006). Capuchin monkeys, inequity aversion, and the frustration effect. *Journal of Comparative Psychology, 120*, 67–73.

Roma, P. G., et al. (2007). Mark tests for self-recognition in Capuchin monkeys (*Cebus paella*) trained to touch marks. *American Journal of Primatology, 69*, 989–1000.

Rosati, A. G., et al. (2007). The evolutionary origins of human patience: Temporal preferences in chimpanzees, bonobos, and human adults. *Current Biology, 17*, 1663–1668.

Roth, G., & Dicke, U. (2005). Evolution of the brain and intelligence. *Trends in Cognitive Sciences, 9*, 250–257.

Ruffman, T., et al. (1998). Older (but not younger) siblings facilitate false belief understanding. *Developmental Psychology, 34*, 161–174.

Russell, B. (1954). *Human society in ethics and politics*. New York: Allan and Unwin.

Russell, B. (2009). The basic writings of Bertrand Russell. New York: Routledge.

Russon, A. E., & Galdikas, B. M. (1993). Imitation in free-ranging rehabilitant orangutans (*Pongo pygmaeus*). *Journal of Comparative Psychology, 107*, 147–161.

Ruxton, G. D., & Wilkinson, D. M. (2011). Avoidance of overheating and selection for both hair loss and bipedality in hominins. *Proceedings of the National Academy of Sciences of the United States of America, 108*, 20965–20969.

Sagan, C. (1980). *Cosmos*. New York: Random House.

Salovey, P., & Mayer, J. D. (1990). Emotional intelligence. *Imagination, Cognition and Personality, 9*, 185–211.

Savage-Rumbaugh, E. S. (1986). *Ape language*. New York: Columbia University Press.

Savage-Rumbaugh, E. S., et al. (1980). Do apes use language? *American Scientist, 68*, 49–61.

Savage-Rumbaugh, E. S., et al. (1993). Language comprehension in ape and child. *Monographs of the Society for Research in Child Development, 58*, 1–222.

Scally, A., et al. (2012). Insights into hominid evolution from the gorilla genome sequence. *Nature, 483*, 169–175.

Scarf, D., et al. (2012). Social evaluation or simple association? *PLOS One, 7*, e42698.

Schacter, D. L. (1999). The seven sins of memory: Insights from psychology and cognitive neuroscience. *American Psychologist, 54*, 182–203.

Schacter, D. L., et al. (2007). Remembering the past to imagine the future: The prospective brain. *Nature Reviews Neuroscience, 8*, 657–661.

Schmandt-Besserat, D. (1992). *Before writing*. Austin: University of Texas Press.

Schusterman, R. J., & Gisiner, R. (1988). Artificial language comprehension in dolphins and sea lions: Essential cognitive skills. *Psychological Record, 38*, 311–348.

Schwarz, E. (1929). The occurrence of the chimpanzee south of the Congo River. *Revue de Zoologie et de Botanique Africaines, 16*, 425–426.

Sear, R., & Mace, R. (2008). Who keeps children alive? A review of the effects of kin on child survival. *Evolution and Human Behavior, 29*, 1–18.

Seed, A. M., et al. (2009). Chimpanzees solve the trap problem when the confound of tool-use is removed. *Journal of Experimental Psychology—Animal Behavior Processes, 35*, 23–34.

Semaw, S. (2000). The world's oldest stone artefacts from Gona, Ethiopia: Their implications for understanding stone technology and patterns of human evolution between 2.6–1.5 million years ago. *Journal of Archaeological Science, 27*, 1197–1214.

Senut, B., et al. (2001). First hominid from the Miocene (Lukeino Formation, Kenya). *Comptes Rendus de l'Academie des Sciences Serie Ii Fascicule A-Sciences de la Terre et des Planetes, 332*, 137–144.

Seyfarth, R. M., & Cheney, D. L. (2012). The evolutionary origins of friendship. *Annual Review of Psychology, 63*, 153–177.

Shahaeian, A., et al. (2011). Culture and the sequence of steps in theory of mind development. *Developmental Psychology, 47*, 1239–1247.

Shermer, M. (1997). *Why people believe weird things*. New York: W. H. Freeman.

Shettleworth, S. J. (2010). Clever animals and killjoy explanations in comparative psychology. *Trends in Cognitive Sciences, 14*, 477–481.

Shipton, C. (2010). Imitation and shared intentionality in the Acheulean. *Cambridge Archaeological Journal, 20*, 197–210.

Shweder, R. A., et al. (1987). Cultural and moral development. In J. Kagan & S. Lamb (Eds.), *The emergence of morality in young children* (pp. 1–83). Chicago: University of Chicago Press.

Silberberg, A., & Kearns, D. (2009). Memory for the order of briefly presented numerals in humans as a function of practice. *Animal Cognition, 12*, 405–407.

Silk, J. B. (2010). Fellow feeling. *American Scientist, 98*, 158–160.

Silk, J. B., et al. (2005). Chimpanzees are indifferent to the welfare of unrelated group members. *Nature, 437*, 1357–1359.

Singer, P. (2002). *One world: The ethics of globalization*. New Haven, CT: Yale University Press.

Skinner, B. F. (1957). *Verbal behavior*. New York: Appleton-Century-Crofts.

Sleator, R. D. (2010). The human superorganism: Of microbes and men. *Medical Hypotheses, 74*, 214–215.

Slobodchikoff, C. N., et al. (2009). Prairie dog alarm calls encode labels about predator colors. *Animal Cognition, 12*, 435–439.

Smil, V. (2002). *The earth's biosphere*. Cambridge, MA: MIT Press.

Smith, J. D., et al. (1995). The uncertain response in the bottlenosed dolphin (*Tusiops truncatus*). *Journal of Experimental Psychology-General, 124*, 391–408.

Smith, J. D., et al. (2012). The highs and lows of theoretical interpretation in animal-metacognition research. *Philosophical Transactions of the Royal Society B-Biological Sciences, 367*, 1297–1309.

Smith, J. N., et al. (2008). Songs of male humpback whales, *Megaptera novaeangliae*, are involved in intersexual interactions. *Animal Behaviour, 76*, 467–477.

Smith, T. M., et al. (2007). Earliest evidence of modern human life history in North African early *Homo sapiens*. *Proceedings of the National Academy of Sciences of the United States of America, 104*, 6128–6133.

Soares, P., et al. (2009). Correcting for purifying selection: An improved human mitochondrial molecular clock. *American Journal of Human Genetics, 84*, 740–759.

Stamp Dawkins, M. (2012). What do animals want? *The Edge*. Retrieved from http://edge.org/conversation/what-do-animals-want

Stanford, C., et al. (2013). *Biological Anthropology (3rd ed.)*. Boston: Pearson.

Staudinger, U. M., & Gluck, J. (2011). Psychological wisdom research: Commonalities and differences in a growing field. *Annual Review of Psychology, 62*, 215–241.

Stedman, H. H., et al. (2004). Myosin gene mutation correlates with anatomical changes in the human lineage. *Nature, 428*, 415–418.

Sterelny, K. (2003). *Thought in a hostile world: The evolution of human cognition*. Malden, MA: Blackwell.

Sterelny, K. (2010). Moral nativism: A sceptical response. *Mind and Language, 25*, 279–297.

Sternberg, R. J. (1999). Successful intelligence: Finding a balance. *Trends in Cognitive Sciences, 3*, 436–442.

Stevens, J. R., & Hauser, M. D. (2004). Why be nice? Psychological constraints on the evolution of cooperation. *Trends in Cognitive Sciences, 8*, 60–65.

Stewart, J. R., & Stringer, C. B. (2012). Human evolution out of Africa: The role of refugia and climate change. *Science, 335*, 1317–1321.

Stone, R. (2011). Last-ditch effort to save embattled ape. *Science, 331*, 390.

Stout, D. (2011). Stone toolmaking and the evolution of human culture and cognition. *Philosophical Transactions of the Royal Society B-Biological Sciences, 366*, 1050–1059.

Stringer, C. B., et al. (2008). Neanderthal exploitation of marine mammals in Gibraltar. *Proceedings of the National Academy of Sciences of the United States of America, 105*, 14319–14324.

Strum, S. C. (2008). Perspectives on de Waal's *Primates and philosophers: How morality evolved. Current Anthropology, 49*, 701–702.

Suddendorf, T. (1999a). Children's understanding of the relation between delayed video representation and current reality: A test for self-awareness? *Journal of Experimental Child Psychology, 72*, 157–176.

Suddendorf, T. (1999b). The rise of the metamind. In M. C. Corballis & S. E. G. Lea (Eds.), *The descent of mind: Psychological perspectives on hominid evolution* (pp. 218–260). London: Oxford University Press.

Suddendorf, T. (2003). Early representational insight: Twenty-four-month-olds can use a photo to find an object in the world. *Child Development, 74*, 896–904.

Suddendorf, T. (2004). How primatology can inform us about the evolution of the human mind. *Australian Psychologist, 39*, 180–187.

Suddendorf, T. (2006). Foresight and evolution of the human mind. *Science, 312*, 1006–1007.

Suddendorf, T. (2008). Explaining human cognitive autapomorphies. *Behavioral and Brain Sciences, 31*, 147–148.

Suddendorf, T. (2010a). Episodic memory versus episodic foresight: Similarities and differences. *Wiley Interdisciplinary Reviews Cognitive Science, 1*, 99–107.

Suddendorf, T. (2010b). Linking yesterday and tomorrow: Preschoolers' ability to report temporally displaced events. *British Journal of Developmental Psychology, 28*, 491–498.

Suddendorf, T. (2011). Evolution, lies and foresight biases. *Behavioral and Brain Sciences, 34*, 38–39.

Suddendorf, T., & Busby, J. (2003). Mental time travel in animals? *Trends in Cognitive Sciences, 7*, 391–396.

Suddendorf, T., & Busby, J. (2005). Making decisions with the future in mind: Developmental and comparative identification of mental time travel. *Learning and Motivation, 36*, 110–125.

Suddendorf, T., & Butler, D. L. (2013). The nature of visual self-recognition. *Trends in Cognitive Sciences, 17*, 121–127.

Suddendorf, T., & Collier-Baker, E. (2009). The evolution of primate visual self-recognition: evidence of absence in lesser apes. *Proceedings of the Royal Society B-Biological Sciences, 276*, 1671–1677.

Suddendorf, T., & Corballis, M. C. (1997). Mental time travel and the evolution of the human mind. *Genetic Social and General Psychology Monographs, 123*, 133–167.

Suddendorf, T., & Corballis, M. C. (2007). The evolution of foresight: What is mental time travel and is it unique to humans? *Behavioral and Brain Sciences, 30*, 299–313+335–351.

Suddendorf, T., & Corballis, M. C. (2008a). Episodic memory and mental time travel. In E. Dere et al. (Eds.), *Handbook of Episodic Memory* (Vol. 18, pp. 31–42) Amsterdam: Elsevier.

Suddendorf, T., & Corballis, M. C. (2008b). New evidence for animal foresight? *Animal Behaviour, 75*, e1–e3.

Suddendorf, T., & Corballis, M. C. (2010). Behavioural evidence for mental time travel in nonhuman animals. *Behavioural Brain Research, 215*, 292–298.

Suddendorf, T., & Dong, A. (2013). On the evolution of imagination and design. In M. Taylor (Ed.), *Oxford handbook of the development of imagination* (pp. 453–467). Oxford: Oxford University Press.

Suddendorf, T., & Fletcher-Flinn, C. M. (1999). Children's divergent thinking improves when they understand false beliefs. *Creativity Research Journal, 12,* 115–128.

Suddendorf, T., & Moore, C. (2011). Introduction to the special issue: The development of episodic foresight. *Cognitive Development, 26,* 295–298.

Suddendorf, T., & Whiten, A. (2001). Mental evolution and development: Evidence for secondary representation in children, great apes and other animals. *Psychological Bulletin, 127,* 629–650.

Suddendorf, T., & Whiten, A. (2003). Reinterpreting the mentality of apes. In K. Sterelny & J. Fitness (Eds.), *From mating to mentality: Evaluating evolutionary psychology* (pp. 173–196). New York: Psychology Press.

Suddendorf, T., et al. (2007). Visual self-recognition in mirrors and live videos: Evidence for a developmental asynchrony. *Cognitive Development, 22,* 185–196.

Suddendorf, T., et al. (2009a). How great is great ape foresight? *Animal Cognition, 12,* 751–754.

Suddendorf, T., et al. (2009b). Mental time travel and the shaping of the human mind. *Philosophical Transactions of the Royal Society B-Biological Sciences, 364,* 1317–1324.

Suddendorf, T., et al. (2011). Children's capacity to remember a novel problem and to secure its future solution. *Developmental Science, 14,* 26–33.

Suddendorf, T., et al. (2012). If I could talk to the animals. *Metascience, 21,* 253–267.

Suddendorf, T., et al. (2013). Is newborn imitation developmentally homologous to later social-cognitive skills? *Developmental Psychobiology, 55,* 52–58.

Surbeck, M., & Hohmann, G. (2008). Primate hunting by bonobos at LuiKotale, Salonga National Park. *Current Biology, 18,* R906–R907.

Suwa, G., et al. (2009). The *Ardipithecus ramidus* skull and its implications for hominid origins. *Science, 326,* 68e1–68e7.

Svetlova, M., et al. (2010). Toddlers' prosocial behavior: From instrumental to empathic to altruistic helping. *Child Development, 81,* 1814–1827.

Swartz, K. B., et al. (1999). Comparative aspects of mirror self-recognition in great apes. In S. T. Parker, R.W. Mitchell, & M.L. Boccia (Eds.), *The mentalities of gorillas and orangutans* (pp. 283–294). Cambridge: Cambridge University Press.

Swisher, C. C., et al. (1996). Latest *Homo erectus* of Java: Potential contemporaneity with Homo sapiens in southeast Asia. *Science, 274,* 1870–1874.

Szagun, G. (1978). On the frequency of use of tenses in English and German children's spontaneous speech. *Child Development, 49,* 898–901.

Tardif, S. D. (1997). The bioenergetics of parental behavior and the evolution of alloparental care in marmosets and tamarins: In N. G. Solomon & J. A. French (Eds.), *Cooperative breeding in mammals* (pp. 11–33). Cambridge: Cambridge University Press.

Taylor, A. H., et al. (2007). Spontaneous metatool use by New Caledonian crows. *Current Biology, 17,* 1504–1507.

Taylor, A. H., et al. (2009). Do New Caledonian crows solve physical problems through causal reasoning? *Proceedings of the Royal Society B-Biological Sciences, 276,* 247–254.

Taylor, A. H., et al. (2010). An investigation into the cognition behind spontaneous string pulling in New Caledonian crows. *PLOS One, 5,* e9345.

Taylor, A. H., et al. (2011). New Caledonian Crows learn the functional properties of novel tool types. *PLOS One, 6,* e26887.

Taylor, A. H., et al. (2012). New Caledonian crows reason about hidden causal agents. *Proceedings of the National Academy of Sciences of the United States of America, 109,* 16389–16391.

Taylor, M., et al. (1994). Children's understanding of knowledge acquisition: The tendency for children to report that they have always known what they have just learned. *Child Development, 65,* 1581–1604.

Taylor, T. (2010). *The artificial ape: How technology changed the course of human evolution.* New York: Palgrave Macmillan.

Tedeschi, J. T. (1981). *Impression management theory and social psychological research.* New York: Academic Press.

Tennie, C., et al. (2004). Imitation versus emulation in great apes. *Folia Primatologica, 75,* 728.

Terrace, H. S. (1979). *Nim.* New York: Knopf.

Thieme, H. (1997). Lower Palaeolithic hunting spears from Germany. *Nature, 385,* 807–810.

Thompson-Cannino, J., et al. (2009). *Picking cotton: Our memoir of injustice and redemption.* New York: St. Martin's Press.

Thomson, R., et al. (2000). Recent common ancestry of human Y chromosomes: Evidence from DNA sequence data. *Proceedings of the National Academy of Sciences of the United States of America, 97,* 7360–7365.

Thorpe, I. J. N. (2003). Anthropology, archaeology, and the origin of warfare. *World Archaeology, 35,* 145–165.

Thorpe, S. K. S., et al. (2007). Origin of human bipedalism as an adaptation for locomotion on flexible branches. *Science, 316,* 1328–1331.

Tinbergen, N. (1963). On aims and methods of ethology. *Zeitschrift für Tierpsychologie, 20,* 410–433.

Tolman, E. C. (1948). Cognitive maps in rats and men. *Psychological Review, 55,* 189–208.

Tomasello, M. (1999). The human adaptation for culture. *Annual Review of Anthropology, 28,* 509–529.

Tomasello, M. (2009). *Why we cooperate.* Cambridge, MA: MIT Press.

Tomasello, M., & Call, J. (1997). *Primate cognition.* New York: Oxford University Press.

Tomasello, M., & Carpenter, M. (2005). The emergence of social cognition in three young chimpanzees. *Monographs of the Society for Research in Child Development, 70,* 1–132.

Tomasello, M., et al. (1993a). Cultural learning. *Behavioral and Brain Sciences, 16,* 495–552.

Tomasello, M., et al. (1993b). Imitative learning of actions on objects by children, chimpanzees, and enculturated chimpanzees. *Child Development, 64,* 1688–1705.

Tomasello, M., et al. (1999). Chimpanzees, *Pan troglodytes,* follow gaze direction geometrically. *Animal Behaviour, 58,* 769–777.

Tomasello, M., et al. (2005). Understanding and sharing intentions: The origins of cultural cognition. *Behavioral and Brain Sciences, 28,* 675–691.

Tooby, J., & DeVore, I. (1987). The reconstruction of hominid behavioral evolution through strategic modelling. In W. Kinzey (Ed.), *The evolution of human behavior: Primate models* (pp. 183–238). Albany: State University of New York Press.

Toups, M. A., et al. (2011). Origin of clothing lice indicates early clothing use by anatomically modern humans in Africa. *Molecular Biology and Evolution, 28,* 29–32.

Trinkaus, E. (1995). Neanderthal mortality patterns. *Journal of Archaeological Science, 22,* 121–142.

Trivers, R. L. (1971). Evolution of reciprocal altruism. *Quarterly Review of Biology, 46,* 35–57.

Tulving, E. (1985). Memory and consciousness. *Canadian Psychology, 26,* 1–12.

Tulving, E. (2005). Episodic memory and autonoesis: Uniquely human? In H. S. Terrace & J. Metcalfe (Eds.), *The missing link in cognition* (pp. 3–56). Oxford: Oxford University Press.

Turney, C. S. M., et al. (2008). Late-surviving megafauna in Tasmania, Australia, implicate human involvement in their extinction. *Proceedings of the National Academy of Sciences of the United States of America, 105,* 12150–12153.

Tversky, A., & Kahneman, D. (1974). Judgment under uncertainty: Heuristics and biases. *Science, 185,* 1124–1131.

Twain, M. (1906). What is man? Retrieved from: http://www.gutenberg.org/ebooks/70

Ueno, A., & Matsuzawa, T. (2004). Food transfer between chimpanzee mothers and their infants. *Primates, 45,* 231–239.

Ujhelyi, M., et al. (2000). Observations on the behavior of gibbons (*Hylobates leucogenys, H. gabriellae,* and *H. lar*) in the presence of mirrors. *Journal of Comparative Psychology, 114,* 253–262.

Ungar, P. S., & Sponheimer, M. (2011). The diets of early hominins. *Science, 334,* 190–193.

Utami, S. S., et al. (2002). Male bimaturism and reproductive success in Sumatran orang-utans. *Behavioral Ecology, 13,* 643–652.

van Baaren, R. B., et al. (2004). Mimicry and prosocial behavior. *Psychological Science, 15,* 71–74.

van der Vaart, E., et al. (2012). Corvid re-caching without 'theory of mind': A model. *PLOS One, 7,* e32904.

van Schaik, C. P., et al. (2003). Orangutan cultures and the evolution of material culture. *Science, 299,* 102–105.

van Wolkenten, M., et al. (2007). Inequity responses of monkeys modified by effort. *Proceedings of the National Academy of Sciences of the United States of America, 104,* 18854–18859.

Varki, A., et al. (1998). Great ape phenome project? *Science, 282,* 239–240.

Vidal, G. (1981). *Creation.* New York: Doubleday.

Villa, P., & Lenoir, M. (2009). Hunting and Hunting Weapons of the Lower and Middle Paleolithic of Europe. In J. J. Hublin & M. P. Richard (Eds.), *Evolution of hominin diets: Integrating approaches to the study of palaeolithic subsistence* (pp. 59–85). Dordrecht: Springer.

Visalberghi, E., & Limongelli, L. (1994). Lack of comprehension of cause-effect relations in tool-using capuchin monkeys (*Cebus apella*). *Journal of Comparative Psychology, 108,* 15–22.

von Hippel, W., & Trivers, R. (2011). The evolution and psychology of self-deception. *Behavioral and Brain Sciences, 34*, 1–56.

von Rohr, C. R., et al. (2011). Evolutionary precursors of social norms in chimpanzees: A new approach. *Biology & Philosophy, 26*, 1–30.

Wadley, L. (2010). Compound-adhesive manufacture as a behavioral proxy for complex cognition in the Middle Stone Age. *Current Anthropology, 51*, S111-S119.

Walker, M. L., & Herndon, J. G. (2008). Menopause in nonhuman primates? *Biology of Reproduction, 79*, 398–406.

Warneken, F., & Tomasello, M. (2009). Varieties of altruism in children and chimpanzees. *Trends in Cognitive Sciences, 13*, 397–402.

Warneken, F., et al. (2006). Cooperative activities in young children and chimpanzees. *Child Development, 77*, 640–663.

Wearing, D. (2005). *Forever today: A memoir of love and amnesia.* New York: Doubleday.

Wellman, H. M., & Liu, D. (2004). Scaling of theory-of-mind tasks. *Child Development, 75*, 523–541.

Wellman, H. M., et al. (2001). Meta-analysis of theory-of-mind development: The truth about false belief. *Child Development, 72*, 655–684.

White, T. D., et al. (2009). *Ardipithecus ramidus* and the paleobiology of early hominids. *Science, 326*, 75–86.

Whitehead, A. N. (1956). *Dialogues of Alfred North Whitehead, as recorded by Lucien Price.* New York: New American Library.

Whiten, A. (2005). The second inheritance system of chimpanzees and humans. *Nature, 437*, 52–55.

Whiten, A., & Byrne, R. W. (1988). Tactical deception in primates. *Behavioral and Brain Sciences, 11*, 233–273.

Whiten, A., & Ham, R. (1992). On the nature and evolution of imitation in the animal kingdom: Reappraisal of a century of research. In P. J. B. Slater, J. S. Rosenblatt, C. Beer, & M. Milinski (Eds.), *Advances in the study of behavior* (pp. 239–283). San Diego: Academic Press.

Whiten, A., & McGrew, W. C. (2001). Is this the first portrayal of tool use by a chimp? *Nature, 409*, 12.

Whiten, A., & Mesoudi, A. (2008). Establishing an experimental science of culture: animal social diffusion experiments. *Philosophical Transactions of the Royal Society B-Biological Sciences, 363*, 3477–3488.

Whiten, A., & Suddendorf, T. (2007). Great ape cognition and the evolutionary roots of human imagination. In I. Roth (Ed.), *Imaginative minds* (pp. 31–60). Oxford: Oxford University Press.

Whiten, A., et al. (1996). Imitative learning of artificial fruit processing in children (*Homo sapiens*) and chimpanzees (*Pan troglodytes*). *Journal of Comparative Psychology, 110*, 3–14.

Whiten, A., et al. (1999). Cultures in chimpanzees. *Nature, 399*, 682–685.

Whiten, A., et al. (2005). Conformity to cultural norms of tool use in chimpanzees. *Nature, 437*, 737–740.

Wilcox, S., & Jackson, R. (2002). Jumping spider tricksters: Deceit, predation, and cooperation. In M. Bekoff et al. (Eds.), *The cognitive animal: Empirical and theoretical perspectives on animal cognition* (pp. 27–45). Cambridge, MA: MIT Press.

Wildman, D. E., et al. (2003). Implications of natural selection in shaping 99.4% nonsynchronous DNA identity between humans and chimpanzees: Enlarging genus Homo. *Proceedings of the National Academy of Sciences of the United States of America, 100*, 7181–7188.

Wilkins, J., et al. (2012). Evidence for early hafted hunting technology. *Science, 338*, 942–946.

Williams, J. M. G., et al. (1996). The specificity of autobiographical memory and imageability of the future. *Memory and Cognition, 24*, 116–125.

Williams, J. H. G., et al. (2001). Imitation, mirror neurons and autism. *Neuroscience and Biobehavioral Reviews, 25*, 287–295.

Wilson, M. A., & McNaughton, B. L. (1994). Reactivation of hippocampal ensemble memories during sleep. *Science, 265*, 676–679.

Wimmer, H., & Perner, J. (1983). Beliefs about beliefs: Representation and constraining function of wrong beliefs in young children's understanding of deception. *Cognition, 13*, 103–128.

Wise, S. M. (2000). *Rattling the cage: Toward legal rights for animals.* Cambridge, MA: Perseus Books.

Wood, B., & Collard, M. (1999). Anthropology: The human genus. *Science, 284*, 65–71.

Wrangham, R. (2009). *Catching fire: How cooking made us human.* New York: Basic Books.

Wrangham, R., & Peterson, D. (1996). *Demonic males.* London: Bloomsbury.

Wu, X., et al. (2011). A new brain endocast of Homo erectus from Hulu Cave, Nanjing, China. *American Journal of Physical Anthropology, 145*, 452–460.

Wynne, C. D. L. (2001). *Animal cognition.* New York: Palgrave.

Wynne, C. D. L. (2004). Fair refusal by capuchin monkeys. *Nature, 428*, 140.

Yamamoto, S., et al. (2009). Chimpanzees help each other upon request. *PLOS One, 4*, e7416.

Yocom, A. M., & Boysen, S. T. (2011). Comprehension of functional support by enculturated chimpanzees *Pan troglodytes. Current Zoology, 57*, 429–440.

Yokoyama, Y., et al. (2008). Gamma-ray spectrometric dating of late *Homo erecrus* skulls from Ngandong and Sambungmacan, Central Java, Indonesia. *Journal of Human Evolution, 55*, 274–277.

Young, R. W. (2003). Evolution of the human hand: The role of throwing and clubbing. *Journal of Anatomy, 202*, 165–174.

Zahn-Waxler, C., et al. (1979). Child rearing and children's prosocial imitations towards victims of distress. *Child Development, 50*, 319–330.

Zerjal, T., et al. (2003). The genetic legacy of the Mongols. *American Journal of Human Genetics, 72*, 717–721.

Zilhao, J., et al. (2010). Symbolic use of marine shells and mineral pigments by Iberian Neandertals. *Proceedings of the National Academy of Sciences of the United States of America, 107*, 1023–1028.

Zimbardo, P. G., & Boyd, J. N. (1999). Putting time in perspective: A valid, reliable individual-differences metric. *Journal of Personality and Social Psychology, 77*, 1271–1288.

Zimmer, C. (2003). How the mind reads other minds. *Science, 300*, 1079–1080.

I N D E X

Absence of evidence, vs. evidence of
    absence, 57, 60, 82, 208
Accounting, invention of, 270, 271
Acheulean stone tool industry,
    252–254, 253(fig), 254n17, 256,
    259(fig)
Adolescence, 226–227, 227n7
Aesthetics, 143–144, 277n6
Afar Triangle, Ethiopia, 242, 244
Africa, early humans in, 233–234,
    235(fig), 236, 240–248, 249n,
    253, 256, 261–262, 266
Age ratio, of human populations,
    263–264, 264n22
Aggression. *See also* Violence; War
    cooperative, 12–13, 30–31,
    202–203, 208. *See also* Hunting
    behavior
    and eye contact, 124
    human, 12, 189–190, 194, 265n26,
    281
Agriculture, advent of, 103, 270, 281
Alcohol, drinking, 223
Altruism, 160n3, 162, 163, 172,
    192n6, 198
    reciprocal, 161, 163, 164, 172, 187,
    188, 191–192, 191n5, 194, 198
Alzheimer's dementia, 69n6
Amnesia, 91, 91n1, 94
    infantile, 100
Analogical reasoning, 68–70, 136,
    149–150
Anthropogenesis, 239, 239n4
Anthropomorphism, 41–42
Apes, great. *See* Bonobos;
    Chimpanzees; Gorillas; Great
    apes; Orangutans

Apes, small. *See* Gibbons
Aquatic ape theory, 239n4
Ardipithecus, 240–242
Aristotle, 30, 139, 272
Art, of prehistoric humans, 262–263,
    263(fig)
Artificial fruit tasks, 178–179, 180(fig).
    See also Puzzle box
Artificial mutation, 282
Artificial selection, 281, 281–282
Asia, 17, 20, 21, 22, 159, 234n1, 235,
    236, 249n11, 254n17, 256, 264,
    267
Associative learning, 42, 43n2, 44n4,
    51, 73, 73n10, 74–75, 83, 93, 104,
    105, 107, 108, 109n6, 110, 110n6,
    146–147, 148, 147–148, 151,
    152, 152n12, 176–177
    vs. insightful reasoning, 152
Australia, 7n4, 58, 134, 139, 164,
    166n9, 166, 173, 175, 177, 181,
    218, 235, 285, 286
    languages of, 69n6, 75
    Rockhampton Zoo 29, 30(fig), 47,
    286
Australian aboriginals, 45, 69n6, 75,
    166, 166n9, 232–233
Australophithecines, 242–246, 248, 263
Australopithecus afarensis, 238(fig),
    242, 242(fig), 243, 243n7, 244,
    260
Australopithecus africanus, 236,
    238(fig), 243, 249n12
Australopithecus anamensis, 238(fig),
    243
Australopithecus bahrelghazali,
    238(fig), 244

341